Multi-Scale Phenomena in Complex Fluids

Modeling, Analysis and Numerical Simulation

Series in Contemporary Applied Mathematics   CAM

Honorary Editor: Chao-Hao Gu (*Fudan University*)
Editors: P. G. Ciarlet (*City University of Hong Kong*),
Ta-Tsien Li (*Fudan University*)

---

1. Mathematical Finance —— Theory and Practice
   (Eds. Yong Jiongmin, Rama Cont)
2. New Advances in Computational Fluid Dynamics
   —— Theory, Methods and Applications
   (Eds. F. Dubois, Wu Huamo)
3. Actuarial Science —— Theory and Practice
   (Eds. Hanji Shang, Alain Tosseti)
4. Mathematical Problems in Environmental Science and Engineering
   (Eds. Alexandre Ern, Liu Weiping)
5. Ginzburg-Landau Vortices
   (Eds. Haïm Brezis, Ta-Tsien Li)
6. Frontiers and Prospects of Contemporary Applied Mathematics
   (Eds. Ta-Tsien Li, Pingwen Zhang)
7. Mathematical Methods for Surface and Subsurface Hydrosystems
   (Eds. Deguan Wang, Christian Duquennoi, Alexandre Ern)
8. Some Topics in Industrial and Applied Mathematics
   (Eds. Rolf Jeltsch, Ta-Tsien Li, Ian H. Sloan)
9. Differential Geometry: Theory and Applications
   (Eds. Philippe G. Ciarlet, Ta-Tsien Li)
10. Industrial and Applied Mathematics in China
    (Eds. Ta-Tsien Li, Pingwen Zhang)
11. Modeling and Dynamics of Infectious Diseases
    (Eds. Zhien Ma, Yicang Zhou, Jianhong Wu)
12. Multi-scale Phenomena in Complex Fluids: Modeling, Analysis
    and Numerical Simulation
    (Eds. Tomas Y. Hou, Chun Liu, Jianguo Liu)
13. Nonlinear Conservation Laws, Fluid Systems and Related Topics
    (Eds. Gui-Qiang Chen, Ta-Tsien Li, Chun Liu)

Series in Contemporary Applied Mathematics CAM 12

Multi-Scale Phenomena in Complex Fluids

Modeling, Analysis and Numerical Simulation

editors

Thomas Y Hou
California Institute of Technology, USA

Chun Liu
Penn State University, USA

Jian-Guo Liu
University of Maryland, USA

Higher Education Press

World Scientific

NEW JERSEY · LONDON · SINGAPORE · BEIJING · SHANGHAI · HONG KONG · TAIPEI · CHENNAI

Thomas Y Hou

Applied Mathematics 217-50

California Institute of Technology

Pasadena, CA 91125

USA

Chun Liu

Department of Mathematics

The Penn State University

University Park, PA 16802

USA

Jian-Guo Liu

Department of Mathematics

University of Maryland

College Park, MD 20742-4015

USA

Editorial Assistant: Zhou Chun-Lian

图书在版编目 (CIP) 数据

复杂流体中的多尺度问题: 建模、分析与数值模拟=
Multi-Scale Phenomena in Complex Fluids: Modeling,
Analysis and Numerical Simulation: 英文 / 侯一钊,
柳春, 刘建国主编. — 北京: 高等教育出版社, 2009.2
(现代应用数学丛书)
ISBN 978–7–04–017358–1

I.复··· II.①侯···②柳···③刘··· III.①流体力学–数值
模拟–研究–英文 IV. O35

中国版本图书馆 CIP 数据核字 (2009) 第 006998 号

Copyright © 2009 by

Higher Education Press

4 Dewai Dajie, Beijing 100120, P. R. China, and

World Scientific Publishing Co Pte Ltd

5 Toh Tuch Link, Singapore 596224

ISBN 978–7–04–017358–1

Printed in P. R. China

Preface

This volume is a collection of lecture notes generated from the first two series of mini-courses of "Shanghai Summer School on Analysis and Numerics in Modern Sciences" held during the summers of 2004 and 2006 at Fudan University, Shanghai, China. The summer school programs attracted more than 130 participants each year, including graduate students, postdoctors, and junior faculty members from more than 30 universities in China and USA.

The purpose of the summer school is to promote the interaction and collaboration of researchers with expertise in scientific modeling, mathematical analysis and numerical simulations. The focus of the year 2004's program was on the study of the multi-scale phenomena in complex fluids. The focus of the year 2006's program was on multi-scale analysis in nonlinear partial differential equations and their applications.

The summer school hosted several mini-courses each year. During the summer of 2004, the instructors are Weizhu Bao (National University of Singapore), Thomas Hou (California Institute of Technology, USA), Chun Liu (Penn State University, USA), Jianguo Liu (University of Maryland, USA), Tiehu Qin (Fudan University, PRC) and Qi Wang (Florida State University, USA). During the summer of 2006, the mini-courses were taught by Zhaojun Bai and Albert Fannjiang (University of California at Davis, USA), Thomas Hou, Wenbin Chen and Feng Qiu (Fudan University, PRC), Chun Liu, and Xiaoming Wang (Florida State University, USA). There are also short lectures given by many distinguished visitors from around the world.

There are five chapters in this volume, covering a wide range of topics in both analysis and numerical simulation methods, as well as their applications.

Chapter 1, by Zhaojun Bai, Wenbin Chen, Richard Scalettar and Ichitaro Yamazaki, is on the numerical methods for quantum Monte Carlo simulations of the Hubbard Models.

Chapter 2, by Albert Fannjiang, is on the wave propagation and imaging in random media.

Chapter 3, by Thomas Hou, is on multi-scale computations for flow and transport in porous media.

Chapter 4, by Chun Liu, is on the energetic variational approaches of elastic complex fluids.

Chapter 5, by Qi Wang, is on the kinetic theories of complex fluids.

We would like to express our gratitude to all the authors for their contributions to this volume, all the instructors for their contributions to the Shanghai Summer Schools in 2004 and 2006 and, in particular, thanks also go to all the participants in the Summer School programs. We want to thank Ms. Chunlian Zhou for her assistance, without which will be impossible for the success of the Summer School. The editors are grateful to Fudan University, the Mathematical Center of Ministry of Education of China, the National Natural Science Foundation of China (NSFC) and the Institut Sino-Francais de Mathematiques Appliquees (ISFMA) for their help and support. Finally, the editor wish to thank Tianfu Zhao (Senior Editor, Higher Education Press) for his patience and professional assistance.

Tomas Y. Hou, Chun Liu and Jianguo Liu

Editors

April 2008

Contents

Numerical Methods for Quantum Monte Carlo Simulations of the Hubbard Model*

Zhaojun Bai

Department of Computer Science and
Department of Mathematics
University of California
Davis, CA 95616, USA
E-mail: bai@cs.ucdavis.edu

Wenbin Chen

School of Mathematical Sciences
Fudan University
Shanghai 200433, China
E-mail: wbchen@fudan.edu.cn

Richard Scalettar

Department of Physics
University of California
Davis, CA 95616, USA
E-mail: scalettar@physics.ucdavis.edu

Ichitaro Yamazaki

Department of Computer Science
University of California
Davis, CA 95616, USA
E-mail: yamazaki@cs.ucdavis.edu

Abstract

One of the core problems in materials science is how the interactions between electrons in a solid give rise to properties like magnetism, superconductivity, and metal-insulator transitions. Our

*This work was partially supported by the National Science Foundation under
Grant 0313390, and Department of Energy, Office of Science, SciDAC grant DE-
FC02 06ER25793. Wenbin Chen was also supported in part by the China Basic
Research Program under the grant 2005CB321701.

ability to solve this central question in quantum statistical mechanics numerically is presently limited to systems of a few hundred electrons. While simulations at this scale have taught us a considerable amount about certain classes of materials, they have very significant limitations, especially for recently discovered materials which have mesoscopic magnetic and charge order.

In this paper, we begin with an introduction to the Hubbard model and quantum Monte Carlo simulations. The Hubbard model is a simple and effective model that has successfully captured many of the qualitative features of materials, such as transition metal monoxides, and high temperature superconductors. Because of its voluminous contents, we are not be able to cover all topics in detail; instead we focus on explaining basic ideas, concepts and methodology of quantum Monte Carlo simulation and leave various part for further study. Parts of this paper are our recent work on numerical linear algebra methods for quantum Monte Carlo simulations.

1   Hubbard model and QMC simulations

The Hubbard model is a fundamental model to study one of the core problems in materials science: How do the interactions between electrons in a solid give rise to properties like magnetism, superconductivity, and metal-insulator transitions? In this lecture, we introduce the Hubbard model and outline quantum Monte Carlo (QMC) simulations to study many-electron systems. Subsequent lectures will describe computational kernels of the QMC simulations.

1.1   Hubbard model

The two-dimensional Hubbard model [8, 9] we shall study is defined by the Hamiltonian:

$$\mathcal{H} = \mathcal{H}_K + \mathcal{H}_\mu + \mathcal{H}_V, \tag{1.1}$$

where $\mathcal{H}_K$, $\mathcal{H}_\mu$ and $\mathcal{H}_V$ stand for kinetic energy, chemical energy and potential energy, respectively, and are defined as

$$\mathcal{H}_K = -t \sum_{\langle i,j \rangle, \sigma} (c_{i\sigma}^\dagger c_{j\sigma} + c_{j\sigma}^\dagger c_{i\sigma}),$$

$$\mathcal{H}_\mu = -\mu \sum_i (n_{i\uparrow} + n_{i\downarrow})$$

$$\mathcal{H}_V = U \sum_i \left(n_{i\uparrow} - \frac{1}{2} \right) \left(n_{i\downarrow} - \frac{1}{2} \right)$$

and

- i and j label the spatial sites of the lattice. $\langle i, j \rangle$ represents a pair of nearest-neighbor sites in the lattice and indicates that the electrons only hopping to nearest neighboring sites,

- the operators $c_{i\sigma}^\dagger$ and $c_{i\sigma}$ are the fermion creation and annihilation operators for electrons located on the ith lattice site with z component of spin-up ($\sigma = \uparrow$) or spin-down ($\sigma = \downarrow$), respectively,

- the operators $n_{i\sigma} = c_{i\sigma}^\dagger c_{i\sigma}$ are the number operators which count the number of electrons of spin σ on site i,

- t is the hopping parameter for the kinetic energy of the electrons, and is determined by the overlap of atomic wave functions on neighboring sites,

- U is the repulsive Coulomb interaction between electrons on the same lattice site. The term $U n_{i\uparrow} n_{i\downarrow}$ represents an energy cost U for the site i has two electrons and describes a local repulsion between electrons,

- μ is the chemical potential parameter which controls the electron numbers (or density).

Note that we consider the case of a half-filled band. Hence the Hamiltonian is explicitly written in particle-hole symmetric form.

The expected value of a physical observable $\mathcal{O}$ of interest, such as density-density correlation, spin-spin correlation or magnetic susceptibility, is given by

$$\langle \mathcal{O} \rangle = \mathrm{Tr}(\mathcal{OP}), \tag{1.2}$$

where $\mathcal{P}$ is a distribution operator defined as

$$\mathcal{P} = \frac{1}{\mathcal{Z}} e^{-\beta \mathcal{H}}, \tag{1.3}$$

and $\mathcal{Z}$ is the partition function defined as

$$\mathcal{Z} = \mathrm{Tr}(e^{-\beta \mathcal{H}}), \tag{1.4}$$

and β is proportional to the inverse of the product of the Boltzmann's constant k_B and the temperature T:

$$\beta = \frac{1}{k_B T}.$$

β is referred to as an *inverse temperature*.

"Tr" is a trace over the Hilbert space describing all the possible occupation states of the lattice:

$$\mathrm{Tr}(e^{-\beta \mathcal{H}}) = \sum_i \langle \psi_i | e^{-\beta \mathcal{H}} | \psi_i \rangle,$$

where $\{|\psi_i\rangle\}$ is an orthonormal basis of the Hilbert space. Note that the trace does not depend on the choice of the basis. A convenient choice of the basis is the so-called "occupation number basis (local basis)" as described below.

In a classical problem where $\mathcal{H} = E$ is the energy, a real variable, then $\exp(-\beta E)/Z$ is the probability, where $Z = \int e^{-\beta E}$. In quantum mechanics, as we shall see, we will need to recast the operator $\exp(-\beta \mathcal{H})$ into a real number. The "path integral representation" of the problem to do this was introduced by Feynman and Hibbs [3].

Remark 1.1. According to Pauli exclusion principle of electrons, there are four possible states at every site:

$$\begin{array}{ll} |\cdot\rangle & \text{no particle,} \\ |\uparrow\rangle & \text{one spin up particle,} \\ |\downarrow\rangle & \text{one spin down particle,} \\ |\uparrow\downarrow\rangle & \text{two particles with different spin directions.} \end{array}$$

Therefore the dimension of the Hilbert space is 4^N, where N is the number of sites.

The actions of the spin creation operators $c_\sigma^\dagger$ on the four states are

	$\lvert\cdot\rangle$	$\lvert\uparrow\rangle$	$\lvert\downarrow\rangle$	$\lvert\uparrow\downarrow\rangle$
$c_\uparrow^\dagger$	$\lvert\uparrow\rangle$	0	$\lvert\uparrow\downarrow\rangle$	0
$c_\downarrow^\dagger$	$\lvert\downarrow\rangle$	$\lvert\uparrow\downarrow\rangle$	0	0

The actions of the spin annihilation operators c_σ are

	$\lvert\cdot\rangle$	$\lvert\uparrow\rangle$	$\lvert\downarrow\rangle$	$\lvert\uparrow\downarrow\rangle$
$c_\uparrow$	0	$\lvert\cdot\rangle$	0	$\lvert\downarrow\rangle$
$c_\downarrow$	0	0	$\lvert\cdot\rangle$	$\lvert\uparrow\rangle$

Remark 1.2. The states $|\cdot\rangle$ and $|\uparrow\rangle$ are the eigen-states of the number operator $n_\uparrow = c_\uparrow^\dagger c_\uparrow$:

$$n_\uparrow|\cdot\rangle = 0|\cdot\rangle = 0, \quad n_\uparrow|\uparrow\rangle = |\uparrow\rangle.$$

When the operator $n_\uparrow$ takes the actions on the states $|\downarrow\rangle$ and $|\uparrow\downarrow\rangle$, we have

$$n_\uparrow|\downarrow\rangle = 0, \quad n_\uparrow|\uparrow\downarrow\rangle = |\uparrow\downarrow\rangle.$$

The states $|\cdot\rangle$ and $|\downarrow\rangle$ are the eigen-states of the number operator $n_\downarrow = c_\downarrow^\dagger c_\downarrow$:

$$n_\downarrow|\cdot\rangle = 0|\cdot\rangle = 0, \quad n_\downarrow|\downarrow\rangle = |\downarrow\rangle.$$

When the operator $n_\downarrow$ on the state $|\uparrow\rangle$ and $|\uparrow\downarrow\rangle$, we have

$$n_\downarrow|\uparrow\rangle = 0, \quad n_\downarrow|\uparrow\downarrow\rangle = |\uparrow\downarrow\rangle.$$

The operator $U(n_\uparrow - \frac{1}{2})(n_\downarrow - \frac{1}{2})$ describes the potential energy of two electrons with different spin directions at the same site:

$$U(n_\uparrow - \tfrac{1}{2})(n_\downarrow - \tfrac{1}{2}) \ : |\cdot\rangle = +\tfrac{U}{4}|\cdot\rangle, \quad |\uparrow\rangle = -\tfrac{U}{4}|\uparrow\rangle,$$
$$|\downarrow\rangle = -\tfrac{U}{4}|\downarrow\rangle, |\uparrow\downarrow\rangle = +\tfrac{U}{4}|\uparrow\downarrow\rangle.$$

These eigenenergies immediately illustrate a key aspect of the physics of the Hubbard model: The single occupied states $|\uparrow\rangle$ and $|\downarrow\rangle$ are lower in energy by U (and hence more likely to occur). These states are the ones which have nonzero magnetic moment $m^2 = (n_\uparrow - n_\downarrow)^2$. One therefore says that the Hubbard interaction U favors the presence of magnetic moments. As we shall see, a further question (when t is nonzero) is whether these moments will order in special patterns from site to site.

Remark 1.3. The creation operators $c_{i\sigma}^\dagger$ and the annihilation operators $c_{i\sigma}$ anticommute:

$$\{c_{j\sigma}, c_{\ell\sigma'}^\dagger\} = \delta_{j\ell}\delta_{\sigma\sigma'},$$
$$\{c_{j\sigma}^\dagger, c_{\ell\sigma'}^\dagger\} = 0,$$
$$\{c_{j\sigma}, c_{\ell\sigma'}\} = 0,$$

where the anticommutator of two operators a and b is defined by $ab + ba$, i.e., $\{a, b\} = ab + ba$, and $\delta_{j\ell} = 1$ if $j = \ell$, and otherwise, $\delta_{j\ell} = 0$.

If we choose $\ell = j$ and $\sigma = \sigma'$ in the second anticommutation relation, we conclude that $(c_{j\sigma}^\dagger)^2 = 0$. That is, one cannot create two electrons on the same site with the same spin (Pauli exclusion principle). Thus the anticommutation relations imply the Pauli principle. If the site or spin indices are different, the anticommutation relations tell us that exchanging the order of the creation (or destruction) of two electrons introduces a minus sign. In this way the anticommutation relations also guarantee that the wave function of the particles being described is *antisymmetric*, another attribute of electrons (fermions). Bosonic particles (which have *symmetric* wave functions) are described by creation and destruction operators which commute.

Remark 1.4. When the spin direction σ and the site i are omitted, a quantization to describe the states is

$$|0\rangle : \text{no particle},$$
$$|1\rangle : \text{one particle}.$$

The actions of the creation and destruction operators on the states are

$$c \ : |0\rangle \rightarrow 0, \quad |1\rangle \rightarrow |0\rangle,$$
$$c^\dagger : |0\rangle \rightarrow |1\rangle, |1\rangle \rightarrow 0. \tag{1.5}$$

Subsequently, the eigen-states of the number operator $n = c^\dagger c$ are

$$n \ : \ |0\rangle = 0, \quad |1\rangle = |1\rangle.$$

In addition, the operator $c_i^\dagger c_{i+1}$ describes the kinetic energy of the electrons on nearest neighbor sites:

$$c_i^\dagger c_{i+1} : |00\rangle \to 0, |01\rangle \to |10\rangle,$$
$$|10\rangle \to 0, |11\rangle \to c_i^\dagger |10\rangle \to 0.$$

Therefore, if there is one particle on the $(i+1)$th site, and no particle on the ith site, the operator $c_i^\dagger c_{i+1}$ annihilates the particle on the $(i+1)$th site and creates one particle on the ith site. We say that the electron hops from site $i+1$ to site i after the action of the operator $c_i^\dagger c_{i+1}$.

1.1.1 Hubbard model with no hopping

Let us consider a special case of the Hubbard model, namely, there is only one site and no hopping, $t = 0$. Then the Hamiltonian $\mathcal{H}$ is

$$\mathcal{H} = U \left(n_\uparrow - \frac{1}{2} \right) \left(n_\downarrow - \frac{1}{2} \right) - \mu(n_\uparrow + n_\downarrow).$$

It can be verified that the orthonormal eigen-states ψ_i of the operator n_σ are the eigen-states of the Hamiltonian $\mathcal{H}$:

$$\mathcal{H} : |\cdot\rangle = \tfrac{U}{4}|\cdot\rangle, \qquad\qquad |\uparrow\rangle = \left(\tfrac{U}{4} - (\mu + \tfrac{U}{2}) \right) |\uparrow\rangle,$$
$$|\downarrow\rangle = \left(\tfrac{U}{4} - (\mu + \tfrac{U}{2}) \right) |\downarrow\rangle, \quad |\uparrow\downarrow\rangle = \left(\tfrac{U}{4} - 2\mu \right) |\uparrow\downarrow\rangle.$$

The Hamiltonian $\mathcal{H}$ is diagonalized under the basis $\{\psi_i\}$:

$$\mathcal{H} \longrightarrow \left(\langle \psi_i | \mathcal{H} | \psi_j \rangle \right) = \begin{bmatrix} \frac{U}{4} & & & \\ & \frac{U}{4} - (\mu + \frac{U}{2}) & & \\ & & \frac{U}{4} - (\mu + \frac{U}{2}) & \\ & & & \frac{U}{4} - 2\mu \end{bmatrix}.$$

Consequently, the operator $e^{-\beta\mathcal{H}}$ is diagonalized:

$$e^{-\beta\mathcal{H}} \longrightarrow e^{-\frac{U\beta}{4}} \, \mathrm{diag} \left(1, e^{\beta(U/2+\mu)}, e^{\beta(U/2+\mu)}, e^{2\mu\beta} \right).$$

The partition function $\mathcal{Z}$ becomes

$$\mathcal{Z} = \mathrm{Tr}(e^{-\beta\mathcal{H}}) = \sum_i \langle \psi_i | e^{-\beta\mathcal{H}} | \psi_i \rangle \longrightarrow Z = e^{-\frac{U\beta}{4}} \left(1 + 2e^{\left(\frac{U}{2}+\mu\right)\beta} + e^{2\mu\beta} \right).$$

The operators $\mathcal{H}e^{-\beta\mathcal{H}}$, $n_\uparrow e^{-\beta\mathcal{H}}$, $n_\downarrow e^{-\beta\mathcal{H}}$ and $n_\uparrow n_\downarrow e^{-\beta\mathcal{H}}$ required for calculating physical observables $\mathcal{O}$ of interest become

$$\mathcal{H}e^{-\beta\mathcal{H}} \longrightarrow e^{-\frac{U\beta}{4}} \text{diag}\left(\frac{U}{4}, \left(-\mu - \frac{U}{4}\right)e^{\beta(U/2+\mu)},\right.$$

$$\left.\left(-\mu - \frac{U}{4}\right)e^{\beta(U/2+\mu)}, \left(\frac{U}{4} - 2\mu\right)e^{2\mu\beta}\right),$$

$$n_\uparrow e^{-\beta\mathcal{H}} \longrightarrow e^{-\frac{U\beta}{4}} \text{diag}\left(0, e^{\beta(U/2+\mu)}, 0, e^{2\mu\beta}\right),$$

$$n_\downarrow e^{-\beta\mathcal{H}} \longrightarrow e^{-\frac{U\beta}{4}} \text{diag}\left(0, 0, e^{\beta(U/2+\mu)}, e^{2\mu\beta}\right),$$

$$n_\uparrow n_\downarrow e^{-\beta\mathcal{H}} \longrightarrow e^{-\frac{U\beta}{4}} \text{diag}\left(0, 0, 0, e^{2\mu\beta}\right).$$

The traces of these operators are

$$\text{Tr}(\mathcal{H}e^{-\beta\mathcal{H}}) = e^{-\frac{U\beta}{4}}\left(\frac{U}{4} + 2\left(-\mu - \frac{U}{4}\right)e^{\beta(U/2+\mu)}\right.$$

$$\left. + \left(\frac{U}{4} - 2\mu\right)e^{2\mu\beta}\right),$$

$$\text{Tr}((n_\uparrow + n_\downarrow)e^{-\beta\mathcal{H}}) = e^{-\frac{U\beta}{4}}\left(2e^{\beta(U/2+\mu)} + 2e^{2\mu\beta}\right),$$

$$\text{Tr}(n_\uparrow n_\downarrow e^{-\beta\mathcal{H}}) = e^{-\frac{U\beta}{4}}e^{2\mu\beta}.$$

By definition (1.2), the following physical observables $\mathcal{O}$ can be computed exactly:

1. The one-site density $\rho = \langle n_\uparrow \rangle + \langle n_\downarrow \rangle$ to measure the average occupation of each site:

$$\rho = \langle n_\uparrow \rangle + \langle n_\downarrow \rangle = \frac{\text{Tr}\left((n_\uparrow + n_\downarrow)e^{-\beta\mathcal{H}}\right)}{\text{Tr}(\mathcal{Z})}$$

$$= \frac{2e^{\left(\frac{U}{2}+\mu\right)\beta} + 2e^{2\mu\beta}}{1 + 2e^{\left(\frac{U}{2}+\mu\right)\beta} + e^{2\mu\beta}}.$$

When there is no chemical potential, i.e., $\mu = 0$, $\rho = 1$ for any U and β, it is referred to as "half-filling" because the density is one-half the maximal possible value.

2. The one-site energy $E = \langle \mathcal{H} \rangle$:

$$E = \langle \mathcal{H} \rangle = \frac{\text{Tr}(\mathcal{H}e^{-\beta\mathcal{H}})}{\text{Tr}(\mathcal{Z})}$$

$$= \frac{U}{4} - \frac{(2\mu + U)e^{\left(\frac{U}{2}+\mu\right)\beta} + 2\mu e^{2\mu\beta}}{1 + 2e^{\left(\frac{U}{2}+\mu\right)\beta} + e^{2\mu\beta}}.$$

When there is no chemical potential, i.e., $\mu = 0$,

$$E = \frac{U}{4} - \frac{U}{2(1 + e^{-\frac{U\beta}{2}})}.$$

Figure 1.1 shows the plot of E versus U and β.

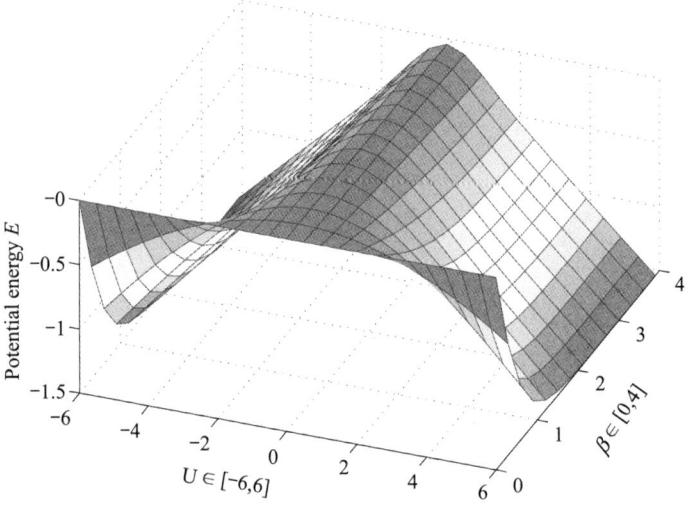

Figure 1.1. Potential energy E for $t = 0, \mu = 0$.

3. The double occupancy $\langle n_\uparrow n_\downarrow \rangle$ is

$$\langle n_\uparrow n_\downarrow \rangle = \frac{\text{Tr}(n_\uparrow n_\downarrow e^{-\beta \mathcal{H}})}{\text{Tr}(\mathcal{Z})} = \frac{e^{2\mu\beta}}{1 + 2e^{(\frac{U}{2}+\mu)\beta} + e^{2\mu\beta}}.$$

When there is no chemical potential, i.e., $\mu = 0$,

$$\langle n_\uparrow n_\downarrow \rangle = \frac{1}{2(1 + e^{\frac{U}{2}\beta})}.$$

Note that as U or β increases, the double occupancy goes to zero.

1.1.2 Hubbard model without interaction

When there is no interaction, $U = 0$, the spin-up and spin-down spaces are independent. $\mathcal{H}$ breaks into the spin-up ($\uparrow$) and spin-down ($\downarrow$) terms. We can consider each spin space separately. Therefore, by omitting the spin, the Hamiltonian $\mathcal{H}$ becomes

$$\mathcal{H} = -t \sum_{\langle i,j \rangle} (c_i^\dagger c_j + c_j^\dagger c_i) - \mu \sum_i n_i.$$

It can be recast as a bilinear form:

$$\mathcal{H} = \boldsymbol{c}^\dagger(-tK - \mu I)\,\boldsymbol{c},$$

where

$$\boldsymbol{c} = \begin{bmatrix} c_1 \\ c_2 \\ \vdots \\ c_N \end{bmatrix} \quad \text{and} \quad \boldsymbol{c}^\dagger = [c_1^\dagger, \ c_2^\dagger, \ \cdots, \ c_N^\dagger],$$

and I is the identity matrix, $K = (k_{ij})$ is a matrix to describe the hopping lattice geometry $\langle i, j \rangle$:

$$k_{ij} = \begin{cases} 1, \text{ if } i \text{ and } j \text{ are nearest neighbors,} \\ 0, \text{ otherwise.} \end{cases}$$

For instance, for a one-dimensional (1D) lattice of N_x sites, K is an $N_x \times N_x$ matrix given by

$$K = K_x = \begin{bmatrix} 0 & 1 & & & 1 \\ 1 & 0 & 1 & & \\ & \ddots & \ddots & \ddots & \\ & & 1 & 0 & 1 \\ 1 & & & 1 & 0 \end{bmatrix}.$$

The $(1, N_x)$ and $(N_x, 1)$ elements of K incorporate the so-called "periodic boundary conditions (PBCs)" in which sites 1 and N_x are connected by t. The use of PBC reduces finite size effects. For example, the energy on a finite lattice of length N with open boundary conditions (OBCs) differs from the value in the thermodynamic limit ($N \to \infty$) by a correction of order $1/N$ while with PBCs, the correction is order $1/N^2$.[①] The use of PBCs also makes the system translationally invariant. The density of electrons per site, and other similar quantities, will not depend on the site in question. With OBCs quantities will vary with the distance from the edges of the lattice.

For a two-dimensional (2D) rectangle lattice of $N_x \times N_y$ sites, K is an $N_x N_y \times N_x N_y$ matrix given by

$$K = K_{xy} = I_y \otimes K_x + K_y \otimes I_x,$$

where I_x and I_y are identity matrices with dimensions N_x and N_y, respectively; $\otimes$ is the matrix Kronecker product.

[①] A simple analogy is this: Consider numerical integration of $f(x)$ on an interval $a \leqslant x \leqslant b$. The only difference between the rectangle and trapezoidal rules is in their treatment of the boundary point contributions $f(a)$ and $f(b)$, yet the integration error changes from linear in the mesh size to quadratic.

The matrix K of 1D or 2D lattice has the exact eigen-decomposition

$$K = F^T \Lambda F, \quad F^T F = I,$$

where $\Lambda = \text{diag}(\lambda_k)$ is a diagonal eigenvalue matrix, see Lemma 2.1. Let

$$\tilde{c} = Fc \quad \text{and} \quad \tilde{c}^\dagger = (F\vec{c})^\dagger.$$

Then the Hamiltonian $\mathcal{H}$ is diagonalized:

$$\mathcal{H} = \tilde{c}^\dagger(-t\Lambda - \mu I)\tilde{c} = \sum_k \epsilon_k \tilde{n}_k,$$

where $\epsilon_k \equiv -t\lambda_k - \mu$, and $\tilde{n}_k = \tilde{c}_k^\dagger \tilde{c}_k$.

It can be shown that the operators $\tilde{c}_k$ obey the same anticommutation relations as the original operators c_i. Hence they too appropriately describe electrons. Indeed, the original operators create and destroy particles on particular spatial sites i while the new ones create and destroy with particular momenta k. Either set is appropriate to use, however, the interaction term in the Hubbard model is fairly complex when written in momentum space.

Lemma 1.1. *If the operator $\mathcal{H}$ is in a quadratic form of fermion operators*

$$\mathcal{H} = c^\dagger H c,$$

where H is an $N \times N$ Hermitian matrix, then

$$\text{Tr}(e^{-\beta \mathcal{H}}) = \prod_{i=1}^{N}(1 + e^{-\beta \lambda_{k_i}}),$$

where λ_{k_i} are the eigenvalues of H.

Proof. First let us assume that $H = \text{diag}(\lambda_{k_1}, \lambda_{k_2}, \cdots, \lambda_{k_N})$. Then the Hamiltonian $\mathcal{H}$ is

$$\mathcal{H} = c^\dagger H c = c^\dagger \text{diag}(\lambda_{k_1}, \lambda_{k_2}, \cdots, \lambda_{k_N})c = \sum_{i=1}^{N} \lambda_{k_i} n_{k_i}.$$

The lemma can be proved by induction. When $N = 1$, for the two eigen-states $|0\rangle$ and $|1\rangle$ of the number operator n_{k_1}, we have

$$\text{Tr}(e^{-\beta \mathcal{H}}) = \langle 0|e^{-\beta \lambda_{k_1} n_{k_1}}|0\rangle + \langle 1|e^{-\beta \lambda_{k_1} n_{k_1}}|1\rangle = e^0 + e^{-\beta \lambda_{k_1}}.$$

Assume that for $N - 1$, we have

$$\text{Tr}(e^{-\beta \sum_{i=1}^{N-1} \lambda_{k_i} n_{k_i}}) = \prod_{i-1}^{N-1}(1 + e^{-\beta \lambda_{k_i}}). \tag{1.6}$$

Then, by the definition of the trace, we have

$$\mathrm{Tr}(e^{-\beta \sum_{i=1}^{N} \lambda_{k_i} n_{k_i}})$$

$$= \sum_{k_1,\cdots,k_N} \langle \psi_1^{k_1} \cdots \psi_N^{k_N} | e^{-\beta \sum_{i=1}^{N} \lambda_{k_i} n_{k_i}} | \psi_1^{k_1} \cdots \psi_N^{k_N} \rangle \qquad (1.7)$$

$$= \sum_{k_1,\cdots,k_{N-1}} \left\{ \langle \psi_1^{k_1} \cdots \psi_{N-1}^{k_{N-1}} 0 | e^{-\beta \sum_{i=1}^{N-1} \lambda_{k_i} n_{k_i}} e^{-\beta \lambda_{k_N} n_{k_N}} | \psi_1^{k_1} \cdots \psi_{N-1}^{k_{N-1}} 0 \rangle \right.$$

$$\left. + \langle \psi_1^{k_1} \cdots \psi_{N-1}^{k_{N-1}} 1 | e^{-\beta \sum_{i=1}^{N-1} \lambda_{k_i} n_{k_i}} e^{-\beta \lambda_{k_N} n_{k_N}} | \psi_1^{k_1} \cdots \psi_{N-1}^{k_{N-1}} 1 \rangle \right\}$$

$$= (1 + e^{-\beta \lambda_{k_N}}) \sum_{k_1,\cdots,k_{N-1}} \langle \psi_1^{k_1} \cdots \psi_{N-1}^{k_{N-1}} | e^{-\beta \sum_{i=1}^{N-1} \lambda_{k_i} n_{k_i}} | \psi_1^{k_1} \cdots \psi_{N-1}^{k_{N-1}} \rangle$$

$$= (1 + e^{-\beta \lambda_{k_N}}) \prod_{i=1}^{N-1} (1 + e^{-\beta \lambda_{k_i}}) = \prod_{i=1}^{N} (1 + e^{-\beta \lambda_{k_i}}). \qquad (1.8)$$

For a general Hermitian matrix H, there exists a unitary matrix Q such that

$$Q^* H Q = \Lambda = \mathrm{diag}(\lambda_{k_1}, \lambda_{k_2}, \cdots, \lambda_{k_N}).$$

Let $\tilde{c} = Qc$ and $\tilde{n}_i = \tilde{c}_i^\dagger \tilde{c}_i$. Then we have

$$\mathcal{H} = c^\dagger H c = \tilde{c}^\dagger \Lambda \tilde{c} = \sum_{i=1}^{N} \lambda_{k_i} \tilde{n}_{k_i}.$$

Since the trace is independent of the basis functions and by equation (1.8), we have

$$\mathrm{Tr}(e^{-\beta \mathcal{H}}) = \mathrm{Tr}\left(\prod_{i=1}^{N} e^{-\beta \lambda_{k_i} \tilde{n}_{k_i}} \right) = \prod_{i=1}^{N} (1 + e^{-\beta \lambda_{k_i}}).$$

The lemma is proved.   □

By Lemma 1.1, the partition function $\mathcal{Z}$ is given by

$$Z = \prod_{k} (1 + e^{-\beta \epsilon_k}).$$

Subsequently, we have the exact expressions for the following physical observables $\mathcal{O}$:

1. The density ρ, the average occupation of each site:

$$\rho = \langle n \rangle = \langle \tilde{n} \rangle = \frac{1}{N} \sum_{k=1}^{N} \langle \tilde{n}_k \rangle = \frac{1}{N} \sum_{k} \frac{1}{1 + e^{\beta \epsilon_k}}.$$

2. The energy E:

$$E = \langle \mathcal{H} \rangle = \frac{1}{N} \sum_k \frac{\epsilon_k}{e^{\beta \epsilon_k} + 1}. \tag{1.9}$$

For the sake of completeness, let us write down the expression for the Green's function, which plays a key role for computing the physical observables:

$$G_{\ell,j} = \langle c_\ell c_j^\dagger \rangle = \frac{1}{N} \sum_k e^{ik \cdot (j - \ell)} (1 - f_k), \tag{1.10}$$

where $f_k = 1/[1 + e^{\beta(\epsilon_k - \mu)}]$. Notice that G is a function of the difference $j - \ell$. This is a consequence of the fact that the Hamiltonian is translationally invariant, that is, with PBCs, there is no special site which is singled out as the origin of the lattice. All sites are equivalent.

At $T = 0$ ($\beta = \infty$), the contours in the right side of Figure 1.2 separate the k values where the states are occupied $f_k = 1$ (inside the contour) from those where the states are empty $f_k = 0$ (outside the contour). The contour is often referred to as the "Fermi surface". There is actually a lot of interesting physics which follows from the geometry of the contour plot of ϵ_k. For example, one notes that the vector (π, π) connects large regions of the contour in the case when $\rho = 1$ and the contour is the rotated square connecting the points $(\pi, 0), (0, \pi), (-\pi, 0)$, and $(0, -\pi)$. One refers to this phenomenon as "nesting" and to (π, π) as the "nesting wave vector". Because the Fermi surface describes the location where the occupation changes from 0 to 1, the electrons are most active there. If there is a wave vector k which connects big expanses of these active regions, special order is likely to occur with that wave vector. Thus the contour plot of ϵ_k is one way of understanding the tendency of the Hubbard model to have *antiferromagnetic* order (magnetic order at $k = (\pi, \pi)$ for half-filling).

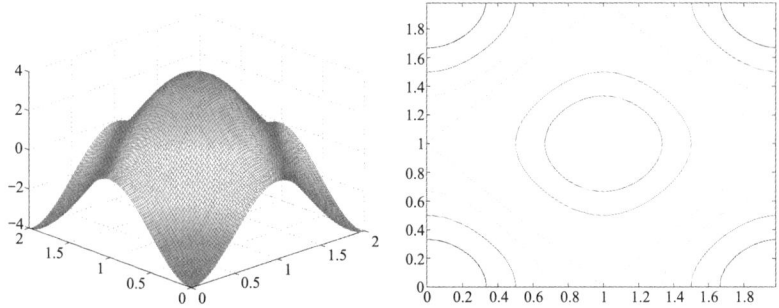

Figure 1.2. ϵ_k for $U = 0$ and $\mu = 0$ (left). Contour plot of ϵ_k (right).

1.2 Determinant QMC

In this section, we first introduce a computable approximation of the distribution operator $\mathcal{P}$ defined in (1.3) by using a discrete Hubbard-Stratonovich transformation, and then a so-called determinant QMC (DQMC) algorithm to generate samples that follow the distribution. For simplicity, we assume that the chemical potential $\mu = 0$ which corresponds to the important half-filled-band case (and there is no the "sign problem"). It turns out that many of the most interesting phenomena of the Hubbard model, like magnetic ordering and insulating-metal transition, occur at half-filling.

1.2.1 Computable approximation of distribution operator $\mathcal{P}$

The gist of a computable approximation of the distribution operator $\mathcal{P}$ defined in (1.3) is on the approximation of the partition function $\mathcal{Z}$. Since the operators $\mathcal{H}_K$ and $\mathcal{H}_V$ do not commute, we apply the Trotter-Suzuki decomposition to approximate $\mathcal{Z}$. Specifically, by dividing the imaginary-time interval $[0, \beta]$ into L equal subintervals of the width $\Delta\tau = \frac{\beta}{L}$, then $\mathcal{Z}$ can be written as

$$\mathcal{Z} = \text{Tr}\left(e^{-\beta\mathcal{H}}\right)$$

$$= \text{Tr}\left(\prod_{\ell=1}^{L} e^{-\Delta\tau\mathcal{H}}\right)$$

$$= \text{Tr}\left(\prod_{\ell=1}^{L} e^{-\Delta\tau\mathcal{H}_K} e^{-\Delta\tau\mathcal{H}_V}\right) + O(\Delta\tau^2). \qquad (1.11)$$

The kinetic energy term $e^{-\Delta\tau\mathcal{H}_K}$ is quadratic in the fermion operators and the spin-up and spin-down operators are independent. Therefore, it can be written as

$$e^{-\Delta\tau\mathcal{H}_K} = e^{-\Delta\tau\mathcal{H}_{K_+}} e^{-\Delta\tau\mathcal{H}_{K_-}},$$

where the operators $\mathcal{H}_{K_+}$ and $\mathcal{H}_{K_-}$ correspond to kinetic energy with spin-up and spin-down respectively, and are of the forms

$$\mathcal{H}_{K_\sigma} = -t\, c_\sigma^\dagger K c_\sigma.$$

On the other hand, the potential energy term $e^{-\Delta\tau\mathcal{H}_V}$ is quartic in the fermion operators. It is necessary to recast it in a quadratic form to use something like Lemma 1.1. To do so, first, noting that the number operators $n_{i\sigma}$ are independent on different sites, we have

$$e^{-\Delta\tau\mathcal{H}_V} = e^{-U\Delta\tau \sum_{i=1}^{N}(n_{i+}-\frac{1}{2})(n_{i-}-\frac{1}{2})}$$

$$= \prod_{i=1}^{N} e^{-U\Delta\tau(n_{i+}-\frac{1}{2})(n_{i-}-\frac{1}{2})}.$$

To treat the term $e^{-U\Delta\tau(n_{i+}-\frac{1}{2})(n_{i-}-\frac{1}{2})}$, we use the following discrete Hubbard-Stratonovich transformation. It replaces the interaction quartic term $(n_{i+}-\frac{1}{2})(n_{i-}-\frac{1}{2})$ by the quadratic one $(n_{i+}-n_{i-})$.

Lemma 1.2 (Discrete Hubbard-Stratonovich transformation [2,5]). *If $U > 0$, then*

$$e^{-U\Delta\tau(n_{i+}-\frac{1}{2})(n_{i-}-\frac{1}{2})} = C_1 \sum_{h_i=\pm 1} e^{\nu h_i(n_{i+}-n_{i-})}, \qquad (1.12)$$

where the constant $C_1 = \frac{1}{2}e^{-\frac{U\Delta\tau}{4}}$ and the scalar ν is defined by $\cosh\nu = e^{\frac{U\Delta\tau}{2}}$.

Proof. First, the following table lists the results of actions of the operators $(n_{i+}-\frac{1}{2})(n_{i-}-\frac{1}{2})$ and $(n_{i+}-n_{i-})$ on the four possible eigen-states $|\cdot\rangle, |\uparrow\rangle, |\downarrow\rangle$ and $|\uparrow\downarrow\rangle$:

ψ	$(n_{i+}-\frac{1}{2})(n_{i-}-\frac{1}{2})$	$n_{i+}-n_{i-}$
$|\cdot\rangle$	$\frac{1}{4}|\cdot\rangle$	$0\,|\cdot\rangle$
$|\uparrow\rangle$	$-\frac{1}{4}|\uparrow\rangle$	$|\uparrow\rangle$
$|\downarrow\rangle$	$-\frac{1}{4}|\downarrow\rangle$	$|\downarrow\rangle$
$|\uparrow\downarrow\rangle$	$\frac{1}{4}|\uparrow\downarrow\rangle$	$0\,|\uparrow\downarrow\rangle$

For the operator of the left-hand side of (1.12):

$$e^{-U\Delta\tau(n_{i+}-\frac{1}{2})(n_{i-}-\frac{1}{2})}\psi = e^{-\frac{U\Delta\tau}{4}}\psi \quad \text{if} \quad \psi = |\cdot\rangle \text{ or } |\uparrow\downarrow\rangle,$$

and

$$e^{-U\Delta\tau(n_{i+}-\frac{1}{2})(n_{i-}-\frac{1}{2})}\psi = e^{\frac{U\Delta\tau}{4}}\psi \quad \text{if} \quad \psi = |\uparrow\rangle \text{ or } |\downarrow\rangle.$$

On the other hand, for the operator of the left-hand side of (1.12):

$$C_1 \sum_{h_i=\pm 1} e^{\nu h_i(n_{i+}-n_{i-})}\psi = e^{-\frac{U\Delta\tau}{4}}\psi \quad \text{if} \quad \psi = |\cdot\rangle \text{ or } |\uparrow\downarrow\rangle,$$

and

$$C_1 \sum_{h_i=\pm 1} e^{\nu h_i(n_{i+}-n_{i-})}\psi = \frac{1}{2}e^{-\frac{U\Delta\tau}{4}}(e^\nu + e^{-\nu})\psi \quad \text{if} \quad \psi = |\uparrow\rangle \text{ or } |\downarrow\rangle.$$

Since

$$\cosh\nu = \frac{e^\nu + e^{-\nu}}{2} = e^{\frac{U\Delta\tau}{2}},$$

the discrete Hubbard-Stratonovich transformation (1.12) holds. $\qquad\square$

Remark 1.5. Note that in the previous proof, U is required to be positive, otherwise there is no real number ν such that $\cosh \nu = e^{\frac{U \Delta \tau}{2}}$. When $U < 0$, the Hubbard model is called the attractive Hubbard model. A similar discrete Hubbard-Stratonovich transformation exists [6,7]. Other transformations for treating the quartic term can also be founded in [13,17].

Let us continue to reformulate the term $e^{-\Delta \tau \mathcal{H}_V}$. By the discrete Hubbard-Stratonovich transformation (1.12), we have

$$e^{-\Delta \tau \mathcal{H}_V} = \prod_{i=1}^{N} \left(C_1 \sum_{h_i = \pm 1} e^{\nu h_i (n_{i+} - n_{i-})} \right). \tag{1.13}$$

$\{h_i\}$ are referred to as auxiliary variables. The collection of all these variables is called the Hubbard-Stratonovich field or configurations.

For the sake of simplicity, let us consider the case $N = 2$ of the expression (1.13):

$$e^{-\Delta \tau \mathcal{H}_V} = (C_1)^2 \left(\sum_{h_i = \pm 1} e^{\nu h_i (n_{1+} - n_{1-})} \right) \left(\sum_{h_i = \pm 1} e^{\nu h_i (n_{2+} - n_{2-})} \right)$$

$$= (C_1)^2 \sum_{h_i = \pm 1} e^{\sum_{i=1}^{2} \nu h_i (n_{i+} - n_{i-})}$$

$$\equiv (C_1)^2 \mathrm{Tr}_h e^{\sum_{i=1}^{2} \nu h_i (n_{i+} - n_{i-})},$$

where the new notation Tr_h represents the trace for $h_i = \pm 1$.

In general, we have

$$e^{-\Delta \tau \mathcal{H}_V} = (C_1)^N \mathrm{Tr}_h e^{\sum_{i=1}^{N} \nu h_i (n_{i+} - n_{i-})}$$

$$= (C_1)^N \mathrm{Tr}_h \left(e^{\sum_{i=1}^{N} \nu h_i n_{i+}} e^{-\sum_{i=1}^{N} \nu h_i n_{i-}} \right)$$

$$\equiv (C_1)^N \mathrm{Tr}_h (e^{\mathcal{H}_{V_+}} e^{\mathcal{H}_{V_-}}), \tag{1.14}$$

where $\mathcal{H}_{V_+}$ and $\mathcal{H}_{V_-}$ correspond to spin-up and spin-down, respectively, and are of the forms

$$\mathcal{H}_{V_\sigma} = \sum_{i=1}^{N} \nu h_i n_{i\sigma} = \sigma \nu \, \mathbf{c}_\sigma^\dagger V(h) \mathbf{c}_\sigma,$$

and $V(h)$ is a diagonal matrix $V(h) = \mathrm{diag}(h_1, h_2, \cdots, h_N)$.

Taking into account the partition of the inverse temperature β into L imaginary time slices, $L = \beta / \Delta \tau$, the Hubbard-Stratonovich variables h_i are changed to have two subindices $h_{\ell,i}$, where i is for the spatial site and ℓ is for the imaginary time slice. Correspondingly, the index ℓ is also introduced for the diagonal matrix V and operators $\mathcal{H}_{V_\sigma}$:

$$h_i \longrightarrow h_{\ell,i}, \qquad V \longrightarrow V_\ell, \qquad \mathcal{H}_{V_\sigma} \longrightarrow \mathcal{H}_{V_\sigma}^\ell.$$

Subsequently, by applying the Trotter-Suzuki approximation (1.11) and expression (1.14) and interchanging the traces, the partition function $\mathcal{Z}$ can be approximated by

$$\mathcal{Z} = (C_1)^{NL} \mathrm{Tr}_h \mathrm{Tr} \left(\prod_{\ell=1}^L e^{-\Delta\tau\mathcal{H}_{K_+}} e^{\mathcal{H}_{V_+}^\ell} \right) \left(\prod_{\ell=1}^L e^{-\Delta\tau\mathcal{H}_{K_-}} e^{\mathcal{H}_{V_-}^\ell} \right), \quad (1.15)$$

where for $\sigma = \pm$,

$$\mathcal{H}_{K_\sigma} = -t\, \mathbf{c}_\sigma^\dagger K \mathbf{c}_\sigma,$$

$$\mathcal{H}_{V_\sigma}^\ell = \sigma \sum_{i=1}^N \nu h_{\ell,i} n_{i+} = \sigma\nu \mathbf{c}_\sigma^\dagger V_\ell(h_\ell) \mathbf{c}_\sigma$$

and $V_\ell(h_\ell)$ is a diagonal matrix

$$V_\ell(h_\ell) = \mathrm{diag}(\, h_{\ell,1}, h_{\ell,2}, \cdots, h_{\ell,N}\,).$$

At this point, all operators $\mathcal{H}_{K_+}$, $\mathcal{H}_{K_-}$, $\mathcal{H}_{V+}^\ell$ and $\mathcal{H}_{V-}^\ell$ are quadratic in the fermion operators. We can apply the following lemma presented in [2,5][1]:

Lemma 1.3. *If operators $\mathcal{H}_\ell$ are in the quadratic forms of the fermion operators*

$$\mathcal{H}_\ell = \sum_{i,j} c_i^\dagger (H_\ell)_{ij} c_j,$$

where H_ℓ are matrices of real numbers, then

$$\mathrm{Tr}(e^{-\mathcal{H}_1} e^{-\mathcal{H}_2} \cdots e^{-\mathcal{H}_L}) = \det(I + e^{-H_L} e^{-H_{L-1}} \cdots e^{-H_1}). \quad (1.16)$$

Note that while "Tr" is over the quantum mechanical Hilbert space whose dimension is 4^N, since by Pauli exclusion principle, there are 4 possible states in every lattice: no electron, one electron with spin-up, one electron with spin-down and two electron with different spin. The "det" is the determinant of a matrix.

By using identity (1.16), the partition function $\mathcal{Z}$ described in (1.15) is turned into the following computable form

$$Z_h = (C_1)^{NL} \mathrm{Tr}_h \det[M_+(h)] \det[M_-(h)], \quad (1.17)$$

---

[1] We are unable to provide a rigorous proof for this important identity. The special case $L = 1$ is Lemma 1.1.

where for $\sigma = \pm$ and $h = (h_1, h_2, \cdots, h_L)$, the fermion matrices

$$M_\sigma(h) = I + B_{L,\sigma}(h_L)B_{L-1,\sigma}(h_{L-1}) \cdots B_{1,\sigma}(h_1), \qquad (1.18)$$

and matrices $B_{\ell,\sigma}(h_\ell)$ are associated with the operators $e^{-\Delta_\tau \mathcal{H}_K} e^{-\Delta_\tau \mathcal{H}_{V_\sigma}^\ell}$, and are defined as

$$B_{\ell,\sigma}(h_\ell) = e^{t\Delta_\tau K} e^{\sigma\nu V_\ell(h_\ell)}.$$

By expression (1.17), we have a computable approximation of the distribution operator $\mathcal{P}$ defined in (1.3):

$$P(h) = \frac{\eta_d}{Z_h} \det[M_+(h)] \det[M_-(h)], \qquad (1.19)$$

where $\eta_d = (C_1)^{NL}$ is a normalizing constant.

Remark 1.6. When $U = 0$, $\nu = 0$ and $M_\sigma(h)$ is a constant matrix and does not depend on the configuration h. The Trotter-Suzuki approximation is exact. The Hubbard Hamiltonian is computed exactly after a single evaluation of the matrix $M_\sigma(h)$.

Remark 1.7. It is a rather amazing thing that a quantum problem can be re-written as a classical one. The price for this is that the classical problem is in one higher dimension than the original quantum one: the degrees of freedom in the quantum problem c_i had a single spatial index i while the Hubbard-Stratonovich variables which replace them have an additional imaginary time index ℓ. This mapping is by no means restricted to the Hubbard Hamiltonian, but is generally true for all quantum mechanics problems.

1.2.2   Algorithm

Our computational task now becomes a classical Monte Carlo problem: sample Hubbard-Stratonovich variables (configurations) h that follow the probability distribution function $P(h)$ defined in (1.19). Recall that the dimension of the configuration sample space $\{h\}$ is 2^{NL}. For an efficient Monte Carlo procedure there are two essential questions:

1. How to move to a new configuration h' from the current h? A simple strategy is to flip only at one selected site (ℓ, i),

$$h'_{\ell,i} = -h_{\ell,i},$$

   and leave the rest components of h unchanged.

2. How to ensure that the accepted sample configuration h follows the desired distribution $P(h)$? This is answered by the Metropolis-Hasting algorithm, for example, see [10, p.111].

Combining the answers of these two questions, we have the following so-called determinant QMC (DQMC) algorithm, first presented in [2].

DQMC
- Initialize $h = (h_{\ell,i}) = (\pm 1)$
- MC loop (total steps = warm up + measurement)
  1. set $(\ell, i) = (1, 1)$
  2. (ℓ, i)–loop:
     (a) propose a new configuration h' by flipping at the site (ℓ, i): $h'_{\ell,i} = -h_{\ell,i}$
     (b) compute the Metropolis ratio
     $$r_{\ell,i} = \frac{\det[M_+(h')] \det[M_-(h')]}{\det[M_+(h)] \det[M_-(h)]}$$
     (c) Metropolis acceptance-rejection:
     $$h = \begin{cases} h', & \text{if } r \leqslant \min\{1, r_{\ell,i}\}, \\ h, & \text{otherwise}, \end{cases}$$
     where r is a random number and $r \sim \text{Uniform}[0, 1]$
     (d) go to the next site (ℓ, i), where
        - if $i < N$, then $\ell := \ell$, $i := i + 1$,
        - if $i = N$ and $\ell < L$, then $\ell := \ell + 1, i = 1$,
        - if $i = N$ and $\ell = L$, then $\ell := 1, i = 1$
  3. after the warm up steps, perform physical measurements, see section 1.2.3.

Note that the one-site update at the inner loop leads to a simple rank-one updating of the matrix $M_\sigma(h)$. Based on this observation, one can efficiently compute the Metropolis ratio $r_{\ell,i}$, see Appendix A for detail.

1.2.3   Physical measurements

How is the physics extracted from QMC simulations? Two of the most elementary physical observables of interest are the *density* and *kinetic energy*, which can be obtained from the single-particle Green's function,

$$\begin{aligned} G_{ij}^\sigma &= \langle c_{i\sigma} c_{j\sigma}^\dagger \rangle \\ &= \left(M_\sigma^{-1}(h) \right)_{ij} \\ &= \left([I + B_{L,\sigma}(h_L) B_{L-1,\sigma}(h_{L-1}) \cdots B_{1,\sigma}(h_1)]^{-1} \right)_{ij}. \end{aligned}$$

The density of electrons of spin σ on site i is

$$\rho_{i,\sigma} = \langle n_{i,\sigma} \rangle = \langle c_{i,\sigma}^\dagger c_{i,\sigma} \rangle = 1 - \langle c_{i,\sigma} c_{i,\sigma}^\dagger \rangle = 1 - G_{ii}^\sigma,$$

where the third identity arises from the use of the anticommutation relations in interchanging the order of creation and destruction operators.

The Hubbard Hamiltonian is translationally invariant, so one expects $\rho_{i,\sigma}$ to be independent of spatial site i. Likewise, up and down spin species are equivalent. Thus to reduce the statistical errors, one usually averages all the values and have the code report back

$$\rho = \frac{1}{2N} \sum_\sigma \sum_{i=1}^N \rho_{i,\sigma}.$$

We will not emphasize this point further, but such averaging is useful for most observables.[①] As is true in any classical or quantum Monte Carlo, these expectation values are also averaged over the sequence of Hubbard-Stratonovich configurations generated in the simulation.

The kinetic energy is obtained from the Green's function for pairs of sites i, j which are near neighbors,

$$\langle \mathcal{H}_K \rangle = -t \left\langle \sum_{\langle i,j \rangle, \sigma} (c_{i\sigma}^\dagger c_{j\sigma} + c_{j\sigma}^\dagger c_{i\sigma}) \right\rangle$$

$$= +t \sum_{\langle i,j \rangle, \sigma} (G_{ij}^\sigma + G_{ji}^\sigma).$$

An extra minus sign arose from interchanging the fermion operator order so that the creation operator was at the right.

Extended physical measurements. Interesting types of magnetic, charge, and superconducting order, and associated phase transitions, are determined by looking at correlation functions of the form:

$$c(j) = \langle \mathcal{O}_{i+j} \mathcal{O}_i^\dagger \rangle - \langle \mathcal{O}_{i+j} \rangle \langle \mathcal{O}_i^\dagger \rangle \tag{1.20}$$

where, for example,

- for spin order in z direction (magnetism):

$$\mathcal{O}_i = n_{i,\uparrow} - n_{i,\downarrow}, \quad \mathcal{O}_i^\dagger = n_{i,\uparrow} - n_{i,\downarrow};$$

[①]There are some situations where translation invariance is broken, for example if randomness is included in the Hamiltonian.

- for spin order in x/y direction (magnetism):

$$\mathcal{O}_i = c_{i,\downarrow}^\dagger c_{i,\uparrow}, \quad \mathcal{O}_i^\dagger = c_{i,\uparrow}^\dagger c_{i,\downarrow};$$

- for charge order:

$$\mathcal{O}_i = n_{i,\uparrow} + n_{i,\downarrow}, \quad \mathcal{O}_i^\dagger = n_{i,\uparrow} + n_{i,\downarrow};$$

- for pair order (superconductivity):

$$\mathcal{O}_i = c_{i,\downarrow} c_{i,\uparrow}, \quad \mathcal{O}_i^\dagger = c_{i,\uparrow}^\dagger c_{i,\downarrow}^\dagger.$$

In words, what such correlation functions probe is the relationship between spin, charge, pairing on an initial site i with that on a site $i + j$ separated by a distance j. It is plausible that at high temperatures where there is a lot of random thermal vibration, the values of $\mathcal{O}_i^\dagger$ and $\mathcal{O}_{i+j}$ will not 'know about' each other for large j. In such a case, the expectation value $\langle \mathcal{O}_{i+j} \mathcal{O}_i^\dagger \rangle$ factorizes to $\langle \mathcal{O}_{i+j} \rangle \langle \mathcal{O}_i^\dagger \rangle$ and $c(j)$ vanishes. The more precise statement is that at high temperatures $c(j)$ decays exponentially, $c(j) \propto e^{-l/\xi}$. The quantity ξ is called the "correlation length". On the other hand, at low temperatures $c(j) \propto \alpha^2$, a nonzero value, as $j \to \infty$. The quantity α is called the 'order parameter'.[1]

As one can well imagine from this description, one needs to be very careful in analyzing data on finite lattices if the $j \to \infty$ behavior is what is crucial to determining the physics. The techniques of 'finite size scaling' provide methods for accomplishing this.

How are these correlation functions actually evaluated? As commented above in describing measurements of the density and kinetic energy, expectation values of two fermion operators are simply expressed in terms of the Green's function. The general rule for expectation values of more than two fermion creation and destruction operators is that they reduce to products of expectation values of pairs of creation and destruction operators, the famous "Wick's Theorem" of many body physics. For example, for spin order in the x/y direction,

$$\langle c(j) \rangle = \langle c_{i+j,\downarrow}^\dagger c_{i+j,\uparrow} c_{i,\uparrow}^\dagger c_{i,\downarrow} \rangle = G_{i+j,i}^\uparrow G_{i,i+j}^\downarrow.$$

Similarly, for superconducting order,

$$\langle c(j) \rangle = \langle c_{i+j,\downarrow} c_{i+j,\uparrow} c_{i,\uparrow}^\dagger c_{i,\downarrow}^\dagger \rangle = G_{i+j,i}^\uparrow G_{i+j,i}^\downarrow.$$

---

[1]It is interesting to note what happens right at the critical point T_c separating the high temperature disordered phase from the low temperature ordered one. In what are termed 'second order' phase transitions, the correlation length diverges $\xi \propto 1/(T - T_c)^\nu$. Right at T_c the correlation function decays as a power law, $c(j) \propto 1/j^\eta$, a behavior intermediate between its high temperature exponential decay and its low temperature nonzero value.

We conclude with two comments. First, it is useful to look at correlation functions where the operators $\mathcal{O}_{i+j}$ and $\mathcal{O}_i^\dagger$ are separated in imaginary time as well as in space. We will come to this point when we discuss measurements in the hybrid QMC algorithm. Second, one often considers the Fourier transform of the real space correlation function $S(q) = \sum_j e^{iqj} c(j)$. This quantity is often referred to as the 'structure factor', and is important because it is in fact the quantity measured by experimentalists. For example, the scattering rate of a neutron off the electron spins in a crystal is proportional to $S(q)$ where q is the change in momentum of the neutron and the $c(l)$ under consideration is the spin correlation function.

1.3 Hybrid QMC

The procedure summarized in section 1.2.2 is the one used in most DQMC codes today. Many interesting physical results have been obtained with it. However, it has a crucial limitation. At the heart of the procedure is the need to compute the ratio of determinants of matrices which have a dimension N, the spatial size of the system. Thus the algorithm scales as N^3. In practice, this means simulations are limited to a few hundred sites. In order to circumvent this bottleneck and develop an algorithm which potentially scales linearly with N, we reformulate our problem as the following:

1. Replace the discrete Hubbard-Stratonovich field by a continuous one.

2. Express the determinant of the dense $N \times N$ matrices $M_\sigma(h)$ as Gaussian integrals over NL-dimensional space to lead an $NL \times NL$ sparse matrix calculations.

We shall now describe these steps in detail.

1.3.1 Computable approximation of distribution operator $\mathcal{P}$

Instead of using discrete Hubbard-Stratonovich transformation as described in section 1.2.1, one can use a continuous Hubbard-Stratonovich transformation to derive a computable approximation of the distribution operator $\mathcal{P}$. First, recall the identity[①]:

$$e^{\frac{1}{2}a^2} = \frac{1}{\sqrt{2\pi}} \int_{-\infty}^{\infty} e^{-\frac{1}{2}z^2 - za} dz \qquad (1.21)$$

for any scalar $a > 0$.

Lemma 1.4 (Continuous Hubbard-Stratonovich transformation). *For $U > 0$, we have*

$$e^{-U\Delta\tau(n_{i+} - \frac{1}{2})(n_{i-} - \frac{1}{2})} = C_2 \int_{-\infty}^{\infty} e^{-\Delta\tau[x^2 + (2U)^{\frac{1}{2}}x(n_{i+} - n_{i-})]} dx, \qquad (1.22)$$

where $C_2 = (\Delta\tau e^{-\frac{U\Delta\tau}{2}}/\pi)^{\frac{1}{2}}$.

Proof. It is easy to verify that

$$\left(n_{i+} - \frac{1}{2}\right)\left(n_{i-} - \frac{1}{2}\right) = -\frac{1}{2}(n_{i+} - n_{i-})^2 + \frac{1}{4}.$$

Note that $(n_{i+} - n_{i-})^2$ and $(n_{i+} - n_{i-})$ can be diagonalized based on the eigen-states of the operators $n_{i\sigma}$. Then identity (1.21) holds if we replace the scalar α by the operator $(n_{i+} - n_{i-})$:

$$e^{\frac{U\Delta\tau}{2}(n_{i+} - n_{i-})^2} = \frac{1}{\sqrt{2\pi}} \int_{-\infty}^{\infty} e^{-\frac{1}{2}x^2 - (U\Delta\tau)^{\frac{1}{2}}(n_{i+} - n_{i-})x} dx.$$

Let $x' = \frac{x}{\sqrt{2\Delta\tau}}$. Then we have

$$e^{\frac{U\Delta\tau}{2}(n_{i+} - n_{i-})^2} = \frac{\sqrt{\Delta\tau}}{\sqrt{\pi}} \int_{-\infty}^{\infty} e^{-\Delta\tau(x^2 + (2U)^{\frac{1}{2}}(n_{i+} - n_{i-})x)} dx.$$

Combining the above equations, we obtain identity (1.22). □

[①]Note that the identity can be easily verified by using the following well-known identity

$$\int_{-\infty}^{\infty} e^{-z^2} dz = \sqrt{\pi}.$$

In fact, we have

$$\frac{1}{\sqrt{2\pi}} \int_{-\infty}^{\infty} e^{-\frac{1}{2}z^2 - za} dz = \frac{1}{\sqrt{2\pi}} \int_{-\infty}^{\infty} e^{-\frac{1}{2}(z^2 + 2za + a^2 - a^2)} dz$$

$$= e^{\frac{1}{2}a^2} \frac{1}{\sqrt{2\pi}} \int_{-\infty}^{\infty} e^{-\frac{1}{2}(z+a)^2} dz$$

$$= e^{\frac{1}{2}a^2}.$$

Returning to the approximation of the partition function $\mathcal{Z}$ by the Trotter-Suzuki decomposition (1.11), by the continuous Hubbard-Stratonovich identity (1.22), we have

$$
\begin{aligned}
e^{-\Delta\tau\mathcal{H}_V^\ell} &= (C_2)^N \int_{-\infty}^{+\infty} e^{-\Delta\tau\sum_i x_{\ell,i}^2} e^{\Delta\tau\sum_i(2U)^{\frac{1}{2}}x_{\ell,i}n_{i+}} \\
&\quad e^{-\Delta\tau\sum_i(2U)^{\frac{1}{2}}x_{\ell,i}n_{i-}} dx_{\ell,i} \\
&\equiv (C_2)^N \int [\delta x] e^{-S_B(x)} e^{\Delta\tau\mathcal{H}_{V+}^\ell} e^{\Delta\tau\mathcal{H}_{V-}^\ell},
\end{aligned}
$$

where

$$
S_B(x) = \Delta\tau \sum_{\ell,i} x_{\ell,i}^2,
$$
$$
\mathcal{H}_{V_\sigma}^\ell = \sum_i (2U)^{\frac{1}{2}} x_{\ell,i} n_{i\sigma} = \sigma(2U)^{\frac{1}{2}} c_\sigma^\dagger V_\ell(x_\ell) c_\sigma,
$$

and $V_\ell(x_\ell)$ is a diagonal matrix,

$$
V_\ell(x_\ell) = \mathrm{diag}(x_{\ell,1}, x_{\ell,2}, \cdots, x_{\ell,N}).
$$

By an analogous argument as in section 1.2.1, we have the following approximation of the partition function $\mathcal{Z}$:

$$
\begin{aligned}
\mathcal{Z} &= \mathrm{Tr}\left(\prod_{\ell=1}^L e^{-\Delta\tau\mathcal{H}_K} e^{-\Delta\tau\mathcal{H}_V}\right) \\
&= (C_2)^{NL} \int [\delta x] e^{-S_B(x)} \mathrm{Tr}\left(\prod_{\ell=1}^L e^{-\Delta\tau\mathcal{H}_{K+}} e^{\Delta\tau\mathcal{H}_{V+}^\ell}\right) \\
&\quad \times \mathrm{Tr}\left(\prod_{\ell=1}^L e^{-\Delta\tau\mathcal{H}_{K-}} e^{\Delta\tau\mathcal{H}_{V-}^\ell}\right).
\end{aligned}
$$

Note that all the operators $e^{-\Delta\tau\mathcal{H}_K}$, $e^{-\Delta\tau\mathcal{H}_{V+}^\ell}$ and $e^{-\Delta\tau\mathcal{H}_{V-}^\ell}$ are quadratic in the fermion operators. By argument (1.16), we derive the following path integral expression for the partition function $\mathcal{Z}$:

$$
\begin{aligned}
Z_x &= (C_2)^{NL} \int [\delta x] e^{-S_B(x)} \det\left(I + \prod_{\ell=1}^L e^{t\Delta\tau K} e^{(2U)^{\frac{1}{2}}\Delta\tau V_l(x_\ell)}\right) \\
&\quad \times \det\left(I + \prod_{\ell=1}^L e^{t\Delta\tau K} e^{-(2U)^{\frac{1}{2}}\Delta\tau V_\ell(x_\ell)}\right) \\
&= (C_2)^{NL} \int [\delta x] e^{-S_B(x)} \det[M_+(x)] \det[M_-(x)], \qquad (1.23)
\end{aligned}
$$

where for $\sigma = \pm$, the fermion matrices

$$
M_\sigma(x) = I + B_{L,\sigma}(x_L) B_{L-1,\sigma}(x_{L-1}) \cdots B_{1,\sigma}(x_1), \qquad (1.24)
$$

and for $\ell = 1, 2, \cdots, L$,

$$B_{\ell,\sigma}(x_\ell) = e^{t \Delta \tau K} e^{\sigma (2U)^{\frac{1}{2}} \Delta \tau V_\ell(x_\ell)}. \qquad (1.25)$$

By a so-called particle-hole transformation (see Appendix B), we have

$$\det[M_-(x)] = e^{-\Delta \tau (2U)^{\frac{1}{2}} \sum_{\ell,i} x_{\ell,i}} \det[M_+(x)]. \qquad (1.26)$$

Therefore, the integrand of Z_x in (1.23) is positive definite.[1]

Consequently, a computable approximation of the distribution operator $\mathcal{P}$ is given by

$$P(x) = \frac{\eta_h'}{Z_x} e^{-S_B(x)} \det[M_+(x)] \det[M_-(x)], \qquad (1.27)$$

where $\eta_h' = (C_2)^{NL}$ is a normalizing constant.

1.3.2   Algorithm

At this point, our computational task becomes how to generate Hubbard-Stratonovich variables (configurations) x that follow the probability distribution function $P(x)$ defined as (1.27). To develop an efficient Monte Carlo method, we reformulate the function $P(x)$. First, let us recall the following two facts:

1. Let $M_\sigma(x)$ denote an $L \times L$ block matrix[2]

$$M_\sigma(x) = \begin{bmatrix} I & & & & B_{1,\sigma}(x_1) \\ -B_{2,\sigma}(x_2) & I & & & \\ & -B_{3,\sigma}(x_3) & I & & \\ & & \ddots & \ddots & \\ & & & -B_{L,\sigma}(x_L) & I \end{bmatrix}, \qquad (1.28)$$

where $B_{\ell,\sigma}(x_\ell)$ are $N \times N$ matrices as defined in (1.25). Then we have[3]

$$\det[M_\sigma(x)] = \det[I + B_{L,\sigma}(x_L) B_{L-1,\sigma}(x_{L-1}) \cdots B_{1,\sigma}(x_1)]. \quad (1.29)$$

---

[1] Note that we assume that $\mu = 0$ (half-filling case). Otherwise, there exists "sign problem": $P(x)$ may be negative and cannot be used as a probability distribution function.

[2] We use the same notation $M_\sigma(x)$ to denote the $N \times N$ matrix as defined in (1.24), and $NL \times NL$ matrix as defined in (1.28). It depends on the context which one we refer to.

[3] The identity can be easily derived based on the following observation. If A is a 2×2 block matrix,

$$A = \begin{bmatrix} A_{11} & A_{12} \\ A_{21} & A_{22} \end{bmatrix},$$

then $\det(A) = \det(A_{22}) \det(F_{11}) = \det(A_{11}) \det(F_{22})$, where $F_{11} = A_{11} - A_{12} A_{22}^{-1} A_{21}$ and $F_{22} = A_{22} - A_{21} A_{11}^{-1} A_{12}$.

2. If F is an $N \times N$ symmetric and positive definite matrix, then[1]

$$\int e^{-v^T F^{-1} v} dv = \pi^{\frac{N}{2}} \det[F^{\frac{1}{2}}]. \tag{1.30}$$

Now, by introducing two auxiliary fields Φ_σ, $\sigma = \pm$, we have

$$|\det[M_\sigma(x)]| = \det\left[M_\sigma^T(x)M_\sigma(x)\right]^{\frac{1}{2}} = \pi^{-\frac{NL}{2}} \int e^{-\Phi_\sigma^T A_\sigma^{-1}(x)\Phi_\sigma}, \tag{1.31}$$

where

$$A_\sigma(x) = M_\sigma(x)^T M_\sigma(x).$$

By combining (1.23) and (1.31), expression (1.23) of the partition function Z_x can be recast as the following[2]:

$$Z_x = (C_2)^{NL} \int [\delta x] e^{-S_B(x)} \det[M_+(x)] \det[M_-(x)]$$

$$= \left(\frac{C_2}{\pi}\right)^{NL} \int [\delta x \delta\Phi_+ \delta\Phi_-] e^{-\left(S_B(x) + \Phi_+^T A_+^{-1}(x)\Phi_+ + \Phi_-^T A_-^{-1}(x)\Phi_-\right)}$$

$$\equiv \left(\frac{C_2}{\pi}\right)^{NL} \int [\delta x \delta\Phi_\sigma] e^{-V(x,\Phi_\sigma)},$$

where

$$V(x, \Phi_\sigma) = S_B(x) + \Phi_+^T A_+^{-1}(x)\Phi_+ + \Phi_-^T A_-^{-1}(x)\Phi_-. \tag{1.32}$$

Now let us consider how to move the configuration x satisfying the distribution:

$$P(x, \Phi_\sigma) \propto \frac{1}{Z_x} e^{-V(x,\Phi_\sigma)}. \tag{1.33}$$

Similar to the DQMC method, at each Monte Carlo step, we can try to move $x = \{x_{\ell,i}\}$ with respect to each imaginary time ℓ and spatial site i. Alternatively, we can also move the entire configuration x by adding a Gaussian noise

$$x \longrightarrow x + \Delta x,$$

where

$$\Delta x = -\frac{\partial V(x, \Phi_\sigma)}{\partial x} \Delta t + \sqrt{\Delta t} \, w_t,$$

Δt is a parameter of step size, and w_t follows a Gaussian distribution $N(0,1)$ with mean 0 and variance 1. It is known as Langevin-Euler move, for example, see [10, p.192].

---

[1] The identity can be proven by using the eigen-decomposition of F. For example, see page 97 of [R. Bellman, Introduction to Matrix Analysis, SIAM Edition, 1997].

[2] Here we assume that $\det[M_\sigma(x)]$ is positive. Otherwise, we will have the so-called "sign problem".

Spurred by the popularity of the molecular dynamics (MD) method, Scalettar $et.al$ [15] proposed a hybrid method to move x by combining Monte Carlo and molecular dynamics and derived a so-called Hybrid Quantum Monte Carlo (HMQC) method. In the HQMC, an additional auxiliary momentum field $p = \{p_{\ell,i}\}$ is introduced. By the identity

$$\int_{-\infty}^{\infty} e^{-z^2} dz = \sqrt{\pi},$$

we see that the partition function Z_x can be rewritten as

$$Z_x = (C_2)^{NL} \pi^{-NL} \int [\delta x \delta \Phi_\sigma] e^{-V(x,\Phi_\sigma)}$$

$$= (C_2)^{NL} \pi^{-\frac{3NL}{2}} \int [\delta x \delta p \delta \Phi_\sigma] e^{-\left[\sum_{\ell,i} p_{\ell,i}^2 + V(x,\Phi_\sigma)\right]}$$

$$= (C_2)^{NL} \pi^{-\frac{3NL}{2}} \int [\delta x \delta p \delta \Phi_\sigma] e^{-H(x,p,\Phi_\sigma)}$$

$$\equiv Z_H,$$

where

$$H(x,p,\Phi_\sigma) = p^T p + V(x,\Phi_\sigma)$$

$$= p^T p + S_B(x) + \Phi_+^T A_+^{-1}(x)\Phi_+ + \Phi_-^T A_+^{-1}(x)\Phi_-. \quad (1.34)$$

At this point, the computational task becomes a classical Monte Carlo problem: seek the configurations $\{x, p, \Phi_\sigma\}$ that obey the following probability distribution function:

$$P(x,p,\Phi_\sigma) = \frac{\eta_h}{Z_H} e^{-H(x,p,\Phi_\sigma)}, \quad (1.35)$$

where $\eta_h = (C_2)^{NL} \pi^{-\frac{3NL}{2}}$ is a normalizing constant.

Recall that initially we are interested in drawing samples x from the distribution $P(x,\Phi_\sigma)$ defined in (1.33). The motivation of introducing the auxiliary momentum variable p is as the following. If we draw samples (x,p) from the distribution $P(x,p,\Phi_\sigma)$ defined in (1.35), then marginally, it can be shown that x follows the desired distribution $P(x,\Phi_\sigma)$ and entries of p follow the Gaussian distribution $N(0,\frac{1}{2})$ with mean 0 and variance $\frac{1}{2}$ (the standard deviation $\frac{1}{\sqrt{2}}$).

If the configurations (x,p) are moved satisfying the Hamiltonian equations

$$\dot{p}_{\ell,i} = -\frac{\partial H}{\partial x_{\ell,i}} = -\frac{\partial V}{\partial x_{\ell,i}}, \quad (1.36)$$

$$\dot{x}_{\ell,i} = \frac{\partial H}{\partial p_{\ell,i}} = 2p_{\ell,i}, \quad (1.37)$$

then by Liouville's theorem[①], (x, p) is moved along a trajectory in which both H and the differential volume element in phase space are constant and the system is preserved in equilibrium.

In our current implementation of the HQMC method, Hamiltonian equations (1.36) and (1.37) are solved by Verlet (leap-frog) method, see [4]. It is a simple and efficient numerical method to move the configuration from $(x(0), p(0))$ to $(x(T), p(T))$. Since the MD algorithm is "time reversible", HQMC transition guarantees the invariance of the target distribution. This is to say that if the starting x follows the target distribution $P(x, \Phi_\sigma)$, then the new configuration $x(T)$ still follows the same distribution [10, Chap. 9].

Finally, the update of the auxiliary field Φ_σ is simply done by the MC strategy:

$$\Phi_\sigma = M_\sigma^T R_\sigma,$$

where the entries of the vector R_σ are chosen from the distribution $N(0, \frac{1}{2})$. Note that the fields Φ_σ and p are both auxiliary. We first fix the fields Φ_σ, vary the fields p, and move (x, p) together. It is possible to use a different moving order.

The following is an outline of the Hybrid Quantum Monte Carlo (HQMC) method.

HQMC
- Initialize
  - $x(0) = 0$,
  - $\Phi_\sigma = M_\sigma^T(x(0))R_\sigma = M_\sigma^T(x(0))(R_{\ell,i}^\sigma)$, $R_{\ell,i}^\sigma \sim N(0, \frac{1}{2})$.
- MC loop (total steps = warm up + measurement)
  1. MD steps $(x(0), p(0)) \longrightarrow (x(T), p(T))$ with step size Δt:
     (a) Initialize $p(0) = (p_{\ell,i})$, $p_{\ell,i} \sim N(0, \frac{1}{2})$.
     (b) p is evolved in a half time step $\frac{1}{2}\Delta t$:

$$p_{\ell,i}\left(\frac{1}{2}\Delta t\right) = p_{\ell,i}(0) - \left[\frac{\partial V(x(0), \Phi_\sigma)}{\partial x_{\ell,i}}\right]\left(\frac{1}{2}\Delta t\right).$$

     (c) For $n = 0, 1, \cdots, N_T - 2$,

$$x_{\ell,i}(t_n + \Delta t) - x_{\ell,i}(t_n) = 2p_{\ell,i}\left(t_n + \frac{1}{2}\Delta t\right)\Delta t,$$

$$p_{\ell,i}\left(t_n + \frac{3}{2}\Delta t\right) - p_{\ell,i}\left(t_n + \frac{1}{2}\Delta t\right)$$
$$= -\left[\frac{\partial V(x(t_n + \Delta t), \Phi_\sigma)}{\partial x_{\ell,i}}\right]\Delta t,$$

[①]For example, see [1, Chap.3] or [12, Chap.2].

where $N_T = T/\Delta t$, $t_n = n\Delta t$ and $t_0 = 0$.

(d) For $n = N_T - 1$,

$$x_{\ell,i}(T) - x_{\ell,i}(t_n) = 2p_{\ell,i}\left(t_n + \frac{1}{2}\Delta t\right)\Delta t,$$

$$p_{\ell,i}(T) - p_{\ell,i}\left(t_n + \frac{1}{2}\Delta t\right) = -\left[\frac{\partial V(x(T), \Phi_\sigma)}{\partial x_{\ell,i}}\right]\Delta t.$$

2. Metropolis acceptance-rejection:

$$x(T) = \begin{cases} x(T), \text{ if } r \leqslant \min\left\{1, \frac{e^{-H(x(T),p(T),\Phi_\sigma)}}{e^{-H(x(0),p(0),\Phi_\sigma)}}\right\}, \\ x(0), \text{ otherwise,} \end{cases}$$

where r is a random number chosen from Uniform$[0, 1]$.

3. Perform the "heat-bath" step

$$\Phi_\sigma = M_\sigma^T(x(T))R_\sigma = M_\sigma^T(x(T))(R_{\ell,i}^\sigma),$$

where $R_{\ell,i}^\sigma \sim N(0, \frac{1}{2})$.

4. Perform physical measurements after warm up steps.

5. Set $x(0) := x(T)$, go to MD step 1.

Remark 1.8. The Langevin-Euler update is equivalent to a single-step HQMC move [10]. The MD and Langevin-Euler are two alternate methods which both have the virtue of moving all the variables together. Which is better depends basically on which allows the larger step size, the fastest evolution of the Hubbard-Stratonovich fields to new values.

Remark 1.9. If Δt is sufficient small,

$$H(x(T), p(T), \Phi_\sigma) = H(x(0), p(0), \Phi_\sigma)$$

since the MD conserves H. Then at step 2 of the MC loop, all moves will be accepted. However, in practice, for finite Δt the integrator does not conserve H, so step 2 of the MC loop is needed to keep the algorithm exact.

To this end, let us consider how to compute the force term $\frac{\partial V(x,\Phi_\sigma)}{\partial x_{\ell,i}}$. First, we note the matrix $M_\sigma(x)$ defined in (1.28) can be compactly written as

$$M_\sigma(x) = I - K_{[L]}D_{[L]}^\sigma(x)\Pi, \tag{1.38}$$

where

$$K_{[L]} = \text{diag}\left(e^{t\Delta_\tau K}, \cdots, e^{t\Delta_\tau K}\right),$$

$$D_{[L]}^\sigma(x) = \text{diag}\left(e^{\sigma(2U)^{\frac{1}{2}}\Delta_\tau V_1(x_1)}, \cdots, e^{\sigma(2U)^{\frac{1}{2}}\Delta_\tau V_L(x_L)}\right)$$

and

$$\Pi = \begin{bmatrix} 0 & & -I \\ I & 0 & \\ & \ddots & \ddots \\ & & I & 0 \end{bmatrix}.$$

By the definition of $V(x, \Phi_\sigma)$ in (1.32), and recalling that

$$A_\sigma(x) = M_\sigma^T(x) M_\sigma(x)$$

and

$$X_\sigma = A_\sigma^{-1}(x) \Phi_\sigma,$$

we have

$$\frac{\partial [\Phi_\sigma^T A_\sigma^{-1}(x) \Phi_\sigma]}{\partial x_{\ell,i}} = -\Phi_\sigma^T A_\sigma^{-1}(x) \frac{\partial A_\sigma(x)}{\partial x_{\ell,i}} A_\sigma^{-1}(x) \Phi_\sigma$$

$$= -X_\sigma^T \frac{\partial A_\sigma(x)}{\partial x_{\ell,i}} X_\sigma$$

$$= 2 \left[M_\sigma(x) X_\sigma \right]^T \frac{\partial M_\sigma(x)}{\partial x_{\ell,i}} X_\sigma.$$

By expression (1.38) of $M_\sigma(x)$, we have

$$\frac{\partial M_\sigma(x)}{\partial x_{\ell,i}} = -K_{[L]} \frac{\partial D_{[L]}^\sigma(x)}{\partial x_{\ell,i}} \Pi.$$

Note that $D_{[L]}^\sigma(x)$ is a diagonal matrix, and only the (ℓ, i)-diagonal element

$$d_{\ell,i}^\sigma = \exp(\sigma(2U)^{\frac{1}{2}} \Delta \tau x_{\ell,i})$$

depends on $x_{\ell,i}$. Therefore,

$$\frac{\partial M_\sigma(x)}{\partial x_{\ell,i}} = -K_{[L]} \frac{\partial D_{[L]}^\sigma(x)}{\partial x_{\ell,i}} \Pi.$$

We have

$$\frac{\partial [\Phi_\sigma^T A_\sigma^{-1}(x) \Phi_\sigma]}{\partial x_{\ell,i}} = -2 \frac{\partial d_{\ell,i}^\sigma}{\partial x_{\ell,i}} \left[K_{[L]}^T M_\sigma(x) X_\sigma \right]_{\ell,i} (\Pi X_\sigma)_{\ell,i}.$$

Subsequently, the force term is computed by

$$\frac{\partial V(x, \Phi_\sigma)}{\partial x_{\ell,i}} = 2\Delta \tau x_{\ell,i} - 2(2U)^{\frac{1}{2}} \Delta \tau d_{\ell,i}^+ \left[K_{[L]}^T M_+(x) X_+ \right]_{\ell,i} [\Pi X_+]_{\ell,i}$$

$$+ 2(2U)^{\frac{1}{2}} \Delta \tau d_{\ell,i}^- \left[K_{[L]}^T M_-(x) X_- \right]_{\ell,i} [\Pi X_-]_{\ell,i}.$$

Therefore, the main cost of the HQMC is on computing the force term, which in turn is on solving the symmetric positive definite linear system

$$A_\sigma(x)X_\sigma = \Phi_\sigma,$$

where $\sigma = \pm$. In sections 3 and 4, we will discuss direct and iterative methods to such linear system of equations.

1.3.3 Physical measurements

In section 1.2.3 we described how measurements are made in a determinant QMC code. The procedure in the hybrid QMC is identical in that one expresses the desired quantities in terms of precisely the same products of Green's functions. The only difference is that these Green's functions are obtained from matrix-vector products instead of the entries of the Green's function. The basic identity is this:

$$2\langle X_{\sigma,i}R_{\sigma,j}\rangle \leftrightarrow (M_\sigma)_{i,j}^{-1} = G_{ij}^\sigma.$$

This follows from the fact that

$$X_\sigma = A_\sigma^{-1}(x)\Phi_\sigma = M_\sigma^{-1}(x)R_\sigma$$

and that the components R_i of R_σ are independently distributed Gaussian random numbers satisfying $\langle R_i R_j\rangle = \frac{1}{2}\delta_{i,j}$.

Hence, the expression for the spin-spin correlation function would become

$$\langle c(l)\rangle = \langle c_{i+l,\downarrow}^\dagger c_{i+l,\uparrow} c_{i,\uparrow}^\dagger c_{i,\downarrow}\rangle$$
$$= G_{i+l,i}^\uparrow G_{i,i+l}^\downarrow \leftrightarrow 4\langle R_{\uparrow,i+l}X_{\uparrow,i}R_{\downarrow,i}X_{\downarrow,i+l}\rangle.$$

The only point should be to be cautious of concerns the evaluation of expectation values of four fermion operators if the operators have the same spin index. There it is important that two different vectors of random numbers are used: R_σ, X_σ and R'_σ, X'_σ. Otherwise the averaging over the Gaussian random numbers generates additional, unwanted, values:

$$\langle R_i R_j R_k R_l\rangle = \frac{1}{4}(\delta_{i,j}\delta_{k,l} + \delta_{i,k}\delta_{j,l} + \delta_{i,l}\delta_{j,k}),$$

whereas

$$\langle R'_i R'_j R_k R_l\rangle = \frac{1}{4}\delta_{i,j}\delta_{k,l}.$$

It should be apparent that if the indices i and j are in the same N-dimensional block, we get the equal-time Green's function

$$G^\sigma = M_\sigma^{-1}$$
$$= [I + B_{L,\sigma}B_{L-1,\sigma}\cdots B_{1,\sigma}]^{-1},$$

which is the quantity used in traditional determinant QMC.

However, choosing i, j in different N-dimensional blocks, we can also access the nonequal-time Green's function,

$$G_{\ell_1, i; \ell_2, j} = \langle c_i(\ell_1) c_j^\dagger(\ell_2) \rangle$$
$$= \left(B_{\ell_1} B_{\ell_1 - 1} \cdots B_{\ell_2 + 1} (I + B_{\ell_2} \cdots B_1 B_L \cdots B_{\ell_2 + 1})^{-1} \right)_{ij}.$$

At every measurement step in the HQMC simulation, the equal-time Green's function G_{ij} can be obtained from the diagonal block of M_σ^{-1} and the unequal-time Green's function $G_{\ell_1, i; \ell_2, j}$ can be computed from the (ℓ_1, ℓ_2) block submatrix of M_σ^{-1}.

As we already remarked in describing the measurements in determinant QMC, it is often useful to generalize the definition of correlation functions so that the pair of operators are separated in imaginary time as well as spatially. The values of the non-equal time Green's function allow us to evaluate these more general correlation functions $c(l, \Delta\tau)$. To motivate their importance we comment that just as the structure factor $S(q)$, the Fourier transform of the real space correlation function $c(l)$, describes the scattering of particles with change in momentum q, the Fourier transform of $c(l, \Delta\tau)$ into what is often called the susceptibility $\chi(q, \omega)$, tells us about scattering events where the momentum changes by q and the energy changes by ω.

2 Hubbard matrix analysis

For developing robust and efficient algorithmic techniques and high performance software for the QMC simulations described in section 1, it is important to understand mathematical and numerical properties of the underlying matrix computation problems. In this section, we study the dynamics and transitional behaviors of these properties as functions of multi-length scale parameters.

2.1 Hubbard matrix

To simplify the notation, we write the matrix M introduced in (1.28) as

$$M = \begin{bmatrix} I & & & & B_1 \\ -B_2 & I & & & \\ & -B_3 & I & & \\ & & \ddots & \ddots & \\ & & & -B_L & I \end{bmatrix}, \tag{2.1}$$

and refer to it as the *Hubbard matrix*, where

- I is an $N \times N$ identity matrix,
- B_ℓ is an $N \times N$ matrix of the form

$$B_\ell = BD_\ell, \qquad (2.2)$$

and

$$B = e^{t \Delta \tau K} \quad \text{and} \quad D_\ell = e^{\sigma \nu V_\ell(h_\ell)},$$

- t is a hopping parameter,
- $\Delta \tau = \beta/L$, β is the inverse temperature, and L is the number of imaginary time slices,
- $\sigma = +$ or $-$,
- ν is a parameter related to the interacting energy parameter U,

$$\nu = \cosh^{-1} e^{\frac{U \Delta \tau}{2}} = (\Delta \tau U)^{\frac{1}{2}} + \frac{1}{12}(\Delta \tau U)^{\frac{3}{2}} + O((\Delta \tau U)^2)$$

in the DQMC and

$$\nu = (2U)^{\frac{1}{2}} \Delta \tau$$

in the HQMC,

- $K = (k_{ij})$ is a matrix describing lattice structure:

$$k_{ij} = \begin{cases} 1, & \text{if } i \text{ and } j \text{ are nearest neighbor,} \\ 0, & \text{otherwise,} \end{cases}$$

- $V_\ell(h_\ell)$ is an $N \times N$ diagonal matrix,

$$V_\ell(h_\ell) = \text{diag}(h_{\ell,1}, h_{\ell,2}, \cdots, h_{\ell,N}),$$

- $h_{\ell,i}$ are random variables, referred to as Hubbard-Stratonovich field or configurations, $h_{\ell,i} = 1$ or -1 in the DQMC. In the HQMC, $h_{\ell,i}$ are obtained from the MD move.

The Hubbard matrix M displays the property of *multi-length scaling*, since the dimensions and numerical properties of M are characterized by multiple length and energy parameters, and random variables. Specifically, we have

1. Length parameters: N and L.
   - N is the spatial size. If the density ρ is given, N also measures the number of electrons being simulated.
   - L is the number of blocks related to the inverse temperature $\beta = L \Delta \tau$.

2. Energy-scale parameters: t, U and β.

- t determines the hopping of electrons between different atoms in the solid and thus measures the material's kinetic energy.

- U measures the strength of the interactions between the electrons, that is the potential energy.

- β is the inverse temperature, $\beta = \frac{1}{k_B T}$, where k_B is the Boltzmann's constant and T is the temperature.

3. The parameter connecting length and energy scales: $\Delta\tau = \beta/L$.

- $\Delta\tau$ is a discretization parameter, a measure of the accuracy of the Trotter-Suzuki decomposition, see (1.11).

In more complex situations other energy scales also enter, such as the frequency of ionic vibrations (phonons) and the strength of the coupling of electrons to those vibrations.

Under these parameters of multi-length scaling, the Hubbard matrix M has the following features:

- M incorporates multiple structural scales: The inverse temperature β determines the number of blocks $L = \beta/\Delta\tau$, where $\Delta\tau$ is a discretization step size. Typically $L = O(10^2)$. The dimension of the individual blocks is set by N the number of spatial sites. In a typical 2D simulations $N = N_x \times N_y = O(10^3)$. Thus the order of the Hubbard matrix M is about $O(10^5)$.

- M incorporates multiple energy scales: The parameter t determines the kinetic energy of the electrons, and the interaction energy scale U determines the potential energy. We will see that the condition number and eigenvalue distribution of the Hubbard matrix M are strongly influenced by these energy parameters.

- M is a function of NL random variables $h_{\ell i}$, the so-called Hubbard-Stratonovich field or configurations. The goal of the simulation is to sample these configurations which make major contributions to physical measurements. Therefore, the underlying matrix computation problems need to be solved several thousand times in a full simulation. Figure 2.1 shows a typical MD trajectory in an HQMC simulation, where at every step, we need to a linear system of equations associated with the Hubbard matrix M.

The matrix computation problems arising from the quantum Monte Carlo simulations include

1. Computation of the ratio of the determinants $\dfrac{\det[\widehat{M}]}{\det[M]}$, where $\widehat{M}$ is a low-rank update of M, see the Metropolis acceptance-rejection step in the DQMC algorithm, section 1.2.2 and Appendix A.

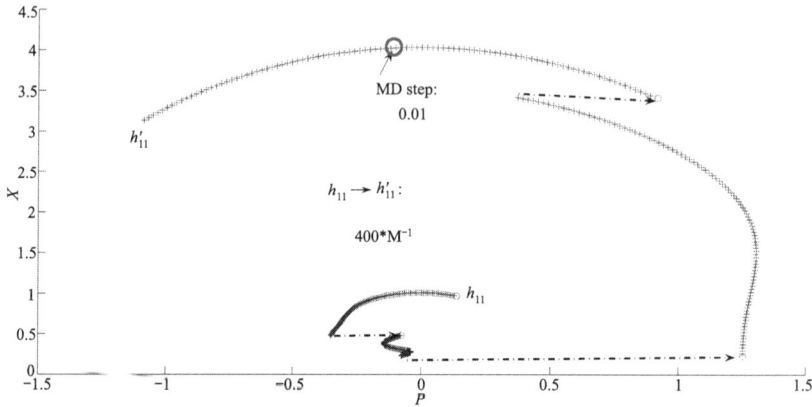

Figure 2.1. A typical MD trajectory in an HQMC simulation.

2. Solution of linear systems of the form $M^T M x = b$, see the HQMC algorithm for computing the force term in the molecular dynamics step, section 1.3.2.

3. Computation of certain entries and the traces of the inverse of the matrix M for physical observables, such as energy, density, moments, magnetism and superconductivity, see sections 1.2.3 and 1.3.3.

One of computational challenges associated with the QMC simulation of the Hubbard model is the wide range of values of parameters. For example, the spatial dimension is $N = N_x \times N_y$. When N is increased from $O(10^2)$ to $O(10^4)$, that is, to do a 10000 electron QMC simulation, it would have a tremendous impact on our understanding of strongly interacting materials. It would allow for the first time the simulation of systems incorporating a reasonable number of mesoscopic structures, such as a "checkerboard" electronic crystal [18], and stripe structure arising from removing electrons from the filling of one electron per site [21].

Another computational challenge is on the ill-conditioning of the underlying matrix computation problems when the energy scale parameters U and β are in certain ranges. In the rest of this section, we will illustrate these challenges in detail.

2.2   Basic properties

Let us first study some basic properties of the Hubbard matrix M as defined in (2.1).

1. The Hubbard matrix M can be compactly written as

$$M = I_{NL} - \text{diag}(B_1, B_2, \cdots, B_L)\Pi \tag{2.3}$$

or

$$M = I_{NL} - (I_N \otimes B)D_{[L]}(P \otimes I_N), \tag{2.4}$$

where I_N and I_{NL} are $N \times N$ and $NL \times NL$ unit matrices, respectively,

$$P = \begin{bmatrix} 0 & & & -1 \\ 1 & 0 & & \\ & \ddots & \ddots & \\ & & 1 & 0 \end{bmatrix}, \quad \Pi = P \otimes I_N = \begin{bmatrix} 0 & & & -I_N \\ I_N & 0 & & \\ & \ddots & \ddots & \\ & & I_N & 0 \end{bmatrix}$$

and

$$B = e^{t\Delta\tau K},$$
$$D_{[L]} = \text{diag}\left(D_1, D_2, \cdots, D_L\right).$$

2. A block LU factorization of M is given by

$$M = LU, \tag{2.5}$$

where

$$L = \begin{bmatrix} I & & & & \\ -B_2 & I & & & \\ & -B_3 & I & & \\ & & \ddots & \ddots & \\ & & & -B_L & I \end{bmatrix}$$

and

$$U = \begin{bmatrix} I & & & & B_1 \\ & I & & & B_2 B_1 \\ & & \ddots & & \vdots \\ & & & I & B_{L-1}B_{L-2}\cdots B_1 \\ & & & & I + B_L B_{L-1} \cdots B_1 \end{bmatrix}.$$

3. The inverses of the factors L and U are given by

$$L^{-1} = \begin{bmatrix} I & & & & \\ B_2 & I & & & \\ B_3 B_2 & B_3 & I & & \\ \vdots & & \ddots & \ddots & \\ B_L \cdots B_2 & B_L \cdots B_3 & \cdots & B_L & I \end{bmatrix}$$

and

$$U^{-1} = \begin{bmatrix} I & & & & -B_1F \\ & I & & & -B_2B_1F \\ & & \ddots & & \vdots \\ & & & I & -B_{L-1}B_{L-2}\cdots B_1F \\ & & & & F \end{bmatrix},$$

where $F = (I + B_LB_{L-1}\cdots B_2B_1)^{-1}$.

4. The inverse of M is explicitly given by

$$M^{-1} = U^{-1}L^{-1} = W^{-1}Z, \qquad (2.6)$$

where

$$W = \begin{bmatrix} I + B_1B_L\cdots B_2 & & & \\ & I + B_2B_1B_L\cdots B_3 & & \\ & & \ddots & \\ & & & I + B_LB_{L-1}\cdots B_1 \end{bmatrix}$$

and

$$Z = \begin{bmatrix} I & -B_1B_L\cdots B_3 & \cdots & -B_1 \\ B_2 & I & \cdots & -B_2B_1 \\ B_3B_2 & B_3 & \cdots & -B_3B_2B_1 \\ \vdots & \vdots & \vdots & \vdots \\ B_{L-1}\cdots B_2 & B_{L-1}\cdots B_3 & \cdots & -B_{L-1}\cdots B_2B_1 \\ B_L\cdots B_2 & B_L\cdots B_3 & \cdots & I \end{bmatrix}.$$

In other words, the (i,j) block submatrix of M^{-1} is given by

$$\{M^{-1}\}_{i,j} = (I + B_i\cdots B_1B_L\cdots B_{i+1})^{-1}Z_{ij}$$

where

$$Z_{ij} = \begin{cases} -B_iB_{i-1}\cdots B_1B_LB_{L-1}\cdots B_{j+1}, & i < j, \\ I, & i = j, \\ B_iB_{i-1}\cdots B_{j+1}, & i > j. \end{cases}$$

5. By the LU factorization (2.5), we immediately have the following determinant identity:

$$\det[M] = \det[I + B_LB_{L-1}\cdots B_1]. \qquad (2.7)$$

2.3   Matrix exponential $B = e^{t\Delta_\tau K}$

In this section, we discuss how to compute the matrix exponential

$$B = e^{t\Delta_\tau K},$$

where K defines a 2-D $N_x \times N_y$ rectangle spatial lattice:

$$K = I_y \otimes K_x + K_y \otimes I_x,$$

I_x and I_y are unit matrices of dimensions N_x and N_y, respectively, and K_x and K_y are $N_x \times N_x$ and $N_y \times N_y$ matrices of the form

$$K_x, K_y = \begin{bmatrix} 0 & 1 & & & & 1 \\ 1 & 0 & 1 & & & \\ & \ddots & \ddots & \ddots & & \\ & & 1 & 0 & 1 \\ 1 & & & & 1 & 0 \end{bmatrix},$$

and $\otimes$ is the Kronecker product.

A survey of numerical methods for computing the matrix exponential can be found in [19]. A simple approach is to use the eigen-decomposition of the matrix K. First, by the definition of K and the property of the Kronecker product (see Appendix B.4), the matrix exponential $e^{t\Delta_\tau K}$ can be written as the product of two matrix exponentials:

$$B = e^{t\Delta_\tau K} = (I_y \otimes e^{t\Delta_\tau K_x})(e^{t\Delta_\tau K_y} \otimes I_x) = e^{t\Delta_\tau K_y} \otimes e^{t\Delta_\tau K_x}.$$

By straightforward calculation, we can verify the following lemma.

Lemma 2.1. *The eigenvalues of K are*

$$\kappa_{ij} = 2(\cos\theta_i + \cos\theta_j), \tag{2.8}$$

where

$$\theta_i = \frac{2i\pi}{N_x}, \quad for \quad i = 0, 1, 2, \cdots, N_x - 1,$$

$$\theta_j = \frac{2j\pi}{N_y}, \quad for \quad j = 0, 1, 2, \cdots, N_y - 1.$$

The corresponding eigenvectors are

$$v_{ij} = u_j \otimes u_i,$$

where

$$u_i = \frac{1}{\sqrt{N_x}}[1, e^{i\theta_i}, e^{i2\theta_i}, \cdots, e^{i(N_x-1)\theta_i}]^T,$$

$$u_j = \frac{1}{\sqrt{N_y}}[1, e^{i\theta_j}, e^{i2\theta_j}, \cdots, e^{i(N_y-1)\theta_j}]^T.$$

By Lemma 2.1, we can use the FFT to compute B. The computational complexity of formulating the matrix B explicitly is $O(N^2)$. The cost of the matrix-vector multiplication is $O(N \log N)$.

We now consider a computational technique referred to as the *"checkerboard method"* in computational physics. The method is particularly useful when the hopping parameter t depends on the location (i, j) on the lattice, i.e., t is not a constant. The checkerboard method only costs $O(N)$. Let us describe this method in detail. For simplicity, assume that N_x and N_y are even. Write K_x as

$$K_x = K_x^{(1)} + K_x^{(2)},$$

where

$$K_x^{(1)} = \begin{bmatrix} D & & & \\ & D & & \\ & & \ddots & \\ & & & D \end{bmatrix}, \quad K_x^{(2)} = \begin{bmatrix} 0 & & & 1 \\ & D & & \\ & & \ddots & \\ & & D & \\ 1 & & & 0 \end{bmatrix}, \quad D = \begin{bmatrix} 0 & 1 \\ 1 & 0 \end{bmatrix}.$$

Note that for any scalar $\alpha \neq 0$, the matrix exponential $e^{\alpha D}$ is given by

$$e^{\alpha D} = \begin{bmatrix} \cosh \alpha & \sinh \alpha \\ \sinh \alpha & \cosh \alpha \end{bmatrix}.$$

Therefore, we have

$$e^{\alpha K_x^{(1)}} = \begin{bmatrix} e^{\alpha D} & & & \\ & e^{\alpha D} & & \\ & & \ddots & \\ & & & e^{\alpha D} \end{bmatrix}, \quad e^{\alpha K_x^{(2)}} = \begin{bmatrix} \cosh \alpha & & & \sinh \alpha \\ & e^{\alpha D} & & \\ & & \ddots & \\ & & e^{\alpha D} & \\ \sinh \alpha & & & \cosh \alpha \end{bmatrix}.$$

Since $K_x^{(1)}$ does not commute with $K_x^{(2)}$, we use the Trotter-Suzuki approximation

$$e^{\alpha K_x} = e^{\alpha K_x^{(1)}} e^{\alpha K_x^{(2)}} + O(\alpha^2).$$

By an exactly analogous calculation, we have the approximation

$$e^{\alpha K_y} = e^{\alpha K_y^{(1)}} e^{\alpha K_y^{(2)}} + O(\alpha^2).$$

Subsequently, we have

$$B = e^{t\Delta_\tau K_y} \otimes e^{t\Delta_\tau K_x}$$
$$= (e^{t\Delta_\tau K_y^{(1)}} e^{t\Delta_\tau K_y^{(2)}}) \otimes (e^{t\Delta_\tau K_x^{(1)}} e^{t\Delta_\tau K_x^{(2)}}) + O((t\Delta_\tau)^2).$$

Therefore, when $t\Delta\tau$ is small, the matrix B can be approximated by

$$\widehat{B} = (e^{t\Delta\tau K_y^{(1)}} e^{t\Delta\tau K_y^{(2)}}) \otimes (e^{t\Delta\tau K_x^{(1)}} e^{t\Delta\tau K_x^{(2)}}). \qquad (2.9)$$

There are no more than 16 nonzero elements in each row and column of the matrix $\widehat{B}$. If $\cosh\alpha$ and $\sinh\alpha$ are computed in advance, the cost of constructing the matrix $\widehat{B}$ is $16N$.

Note the the approximation $\widehat{B}$ is not symmetric. A symmetric approximation of B is given by

$$\widehat{B} = (e^{\frac{t\Delta\tau}{2} K_y^{(2)}} e^{t\Delta\tau K_y^{(1)}} e^{\frac{t\Delta\tau}{2} K_y^{(2)}}) \otimes (e^{\frac{t\Delta\tau}{2} K_x^{(2)}} e^{t\Delta\tau K_x^{(1)}} e^{\frac{t\Delta\tau}{2} K_x^{(2)}}).$$

In this case, there are no more than 36 nonzero elements in each row and column.

Sometimes, it is necessary to compute the matrix-vector multiplication $\widehat{B}w$. Let w be a vector of the dimension $N = N_x \times N_y$ obtained by stacking the columns of an $N_x \times N_y$ matrix W into one long vector:

$$w = \text{vec}(W).$$

Then it can be verified that

$$\widehat{B}w = \text{vec}(e^{t\Delta\tau K_x^{(2)}} e^{t\Delta\tau K_x^{(1)}} W e^{t\Delta\tau K_y^{(1)}} e^{t\Delta\tau K_y^{(2)}}). \qquad (2.10)$$

As a result, the cost of the matrix-vector multiplication $\widehat{B}w$ is $12N$ flops. It can be further reduced by rewriting the block $e^{\alpha D}$ as

$$e^{\alpha D} = \cosh\alpha \begin{bmatrix} 1 & \tanh\alpha \\ \tanh\alpha & 1 \end{bmatrix}.$$

Using this trick, the cost of the matrix-vector multiplication $\widehat{B}w$ is $9N$ flops.

2.4   Eigenvalue distribution of M

The study of eigenvalues of a cyclic matrix of form (2.1) can be traced back to the work of Frobenius, Romanovsky and Varga, see [20]. The following theorem characterizes the eigenvalues of the Hubbard matrix M defined in (2.1).

Theorem 2.1. *For each eigenpair (θ, z) of the matrix $B_L \cdots B_2 B_1$:*

$$(B_L \cdots B_2 B_1)z = \theta z,$$

there are L corresponding eigenpairs (λ_ℓ, v_ℓ) of the matrix M:

$$M v_\ell = \lambda_\ell v_\ell \quad \text{for } \ell = 0, 1, \cdots, L-1,$$

where

$$\lambda_\ell = 1 - \mu_\ell \quad and \quad v_\ell = \begin{bmatrix} (B_L \cdots B_3 B_2)^{-1} \mu_\ell^{L-1} z \\ (B_L \cdots B_3)^{-1} \mu_\ell^{L-2} z \\ \vdots \\ B_L^{-1} \mu_\ell z \\ z \end{bmatrix}$$

and $\mu_\ell = \theta^{\frac{1}{L}} e^{\mathrm{i} \frac{(2\ell+1)\pi}{L}}$.

Proof. By verifying $(I - M)v_\ell = \mu_\ell v_\ell$.   □

2.4.1   The case $U = 0$

When the Hubbard system without Coulomb interaction, $U = 0$, we have

$$B_1 = B_2 = \cdots = B_L = B = e^{t\Delta_\tau K}.$$

In this case, the eigenvalues of the matrix M are known explicitly.

Theorem 2.2. *When $U = 0$, the eigenvalues of the matrix M are*

$$\lambda(M) = 1 - e^{t\Delta_\tau \kappa_{ij}} e^{\mathrm{i} \frac{(2\ell+1)\pi}{L}} \tag{2.11}$$

for $i = 0, 1, \cdots, N_x - 1$, $j = 0, 1, \cdots, N_y - 1$ and $\ell = 0, 1, \cdots, L - 1$, where κ_{ij} is defined in (2.8). Furthermore,

$$\max |1 - \lambda(M)| = e^{4t\Delta_\tau} \quad and \quad \min |1 - \lambda(M)| = e^{-4t\Delta_\tau}.$$

Figure 2.2 shows the eigenvalue distribution of the matrix M with the setting $(N, L, U, \beta, t) = (4 \times 4, 8, 0, 1, 1)$. In this case, the order of the matrix M is $NL = 4 \times 4 \times 8 = 128$. Theorem 2.2 can be used to interpret the distribution. It has a ring structure, centered at $(1, 0)$. On every ring there are $L = 8$ circles. Alternatively, we can also view that the eigenvalues are distributed on $L = 8$ rays, originated from the point $(1, 0)$. The eigenvalues κ_{ij} of the matrix K only have 5 different values. There are total 40 circles, which indicate the multiplicity of some eigenvalues.

Let us examine the dynamics of eigenvalue distributions of M under the variation of the parameters N, L, U and t.

1. Lattice size N: Figure 2.3 shows the eigenvalue distributions for $N = 4 \times 4$ and $N = 16 \times 16$. Other parameters are set as $(L, U, \beta, t) = (8, 0, 1, 1)$. Since L is fixed, the number of rays does not change. When N is increased from 4×4 to 16×16, there are more points (eigenvalues) on each ray. Note that the range of eigenvalue distribution on each ray stays the same.

2. Block number L: Figure 2.4 shows the eigenvalue distribution for block numbers $L = 8$ and $L = 64$. Other parameters are set as $(N, U, \beta, t) = (4 \times 4, 0, 1, 1)$. As we observe that as L increases, the number of rays increases, and the range of the eigenvalue distribution on each ray shrinks and becomes more clustered.

3. Block number L and hopping parameter t: Figure 2.5 shows the eigenvalue distributions for pairs $(L, t) = (8, 1)$ and $(64, 8)$. Other parameters are set as $(N, U, \beta) = (4 \times 4, 0, 1)$. By Theorem 2.2, we know that the points (eigenvalues) on each ring are L. When L increases, the points on each ring increase. At the same time, since $\Delta\tau = \frac{1}{L}$, the range of $|1 - \lambda(M)|$ is $[e^{-\frac{4t}{L}}, e^{\frac{4t}{L}}]$, the bandwidth of the range will shrink when L increases. Since the ratio $\frac{t}{L}$ is fixed, the bandwidth of the ring keeps the same.

2.4.2   The case $U \neq 0$

Unfortunately, there is no explicit expression for the eigenvalues of the matrix M when $U \neq 0$. Figure 2.6 shows that as U increases, the eigenvalues on each ray tend to spread out. Other parameters are set as $(N, L, \beta, t) = (4 \times 4, 8, 1, 1)$.

2.5   Condition number of M

In this section, we study the condition number of the Hubbard matrix M defined in (2.1).

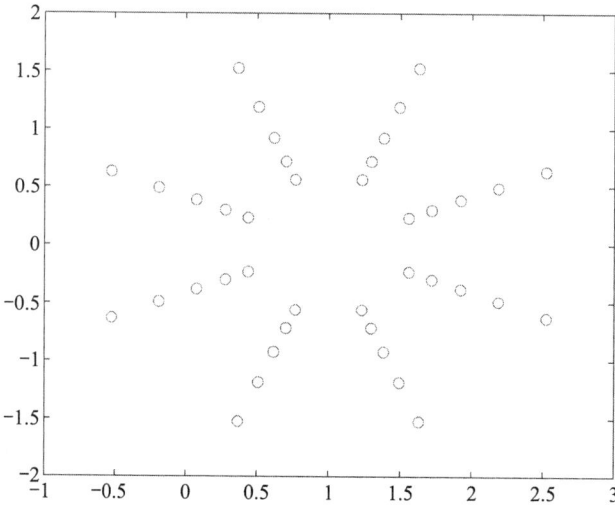

Figure 2.2. Eigenvalue distribution of M.

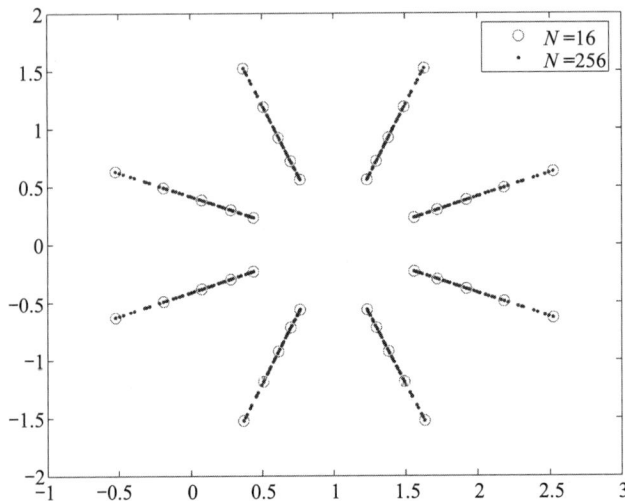

Figure 2.3. Eigenvalue distributions of M for different N.

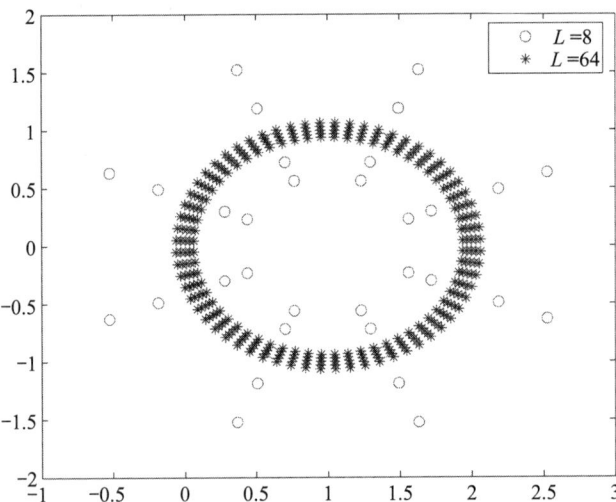

Figure 2.4. Eigenvalue distributions of M for different L.

2.5.1　The case $U = 0$

When $U = 0$, M is a deterministic matrix and $B_1 = \cdots = B_L = B = e^{t\Delta\tau K}$. First, we have the following lemma about the eigenvalues of the symmetric matrix $M^T M$.

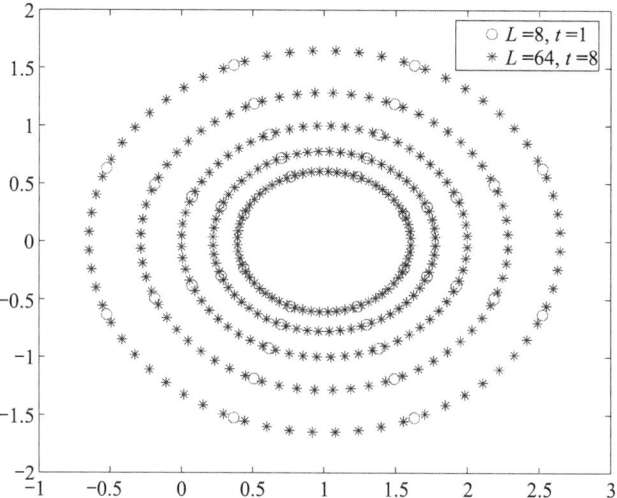

Figure 2.5. Eigenvalue distributions of M for different (L, t).

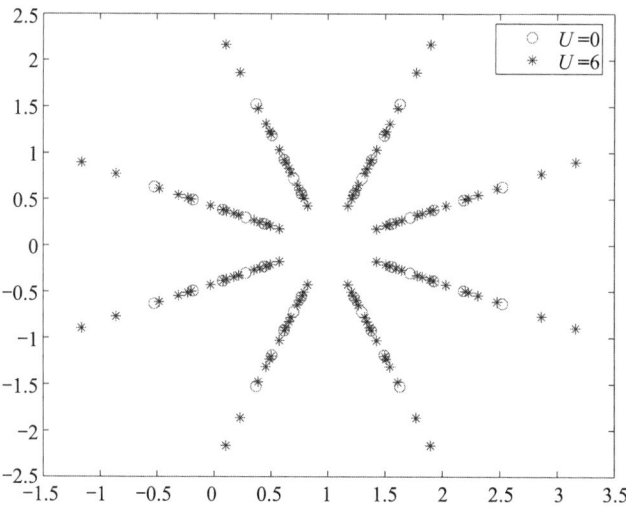

Figure 2.6. Eigenvalue distributions of M for different U.

Lemma 2.2. *When $U = 0$, the eigenvalues of $M^T M$ are*

$$\lambda_\ell(M^T M) = 1 + 2\lambda(B)\cos\theta_\ell + (\lambda(B))^2, \qquad (2.12)$$

where

$$\theta_\ell = \frac{(2\ell + 1)\pi}{L}$$

for $\ell = 0, 1, \cdots, L - 1$.

Proof. The lemma is based on the following fact. For any real number a, the eigenvalues of the matrix

$$A(a) = \begin{bmatrix} 1+a^2 & -a & & & a \\ -a & 1+a^2 & -a & & \\ & \ddots & \ddots & \ddots & \\ a & & & -a & 1+a^2 \end{bmatrix}$$

are $\lambda(A(a)) = 1 - 2a\cos\theta_\ell + a^2$. $\qquad\qquad\square$

We note that when a is a real number,

$$\sin^2\theta \leqslant 1 - 2a\cos\theta + a^2 \leqslant (1+|a|)^2.$$

Therefore, by expression (2.12), we have the following inequalities to bound the largest and smallest eigenvalues of the matrix $M^T M$:

$$\max \lambda_\ell(M^T M) \leqslant (1 + \max|\lambda(B)|)^2$$

and

$$\min \lambda_\ell(M^T M) \geqslant \sin^2 \frac{\pi}{L}.$$

By these inequalities, the norms of M and M^{-1} are bounded by

$$\|M\| = \max \lambda_\ell(M^T M)^{\frac{1}{2}} \leqslant 1 + \max|\lambda(B)|,$$

and

$$\|M^{-1}\| = \frac{1}{\min \lambda_\ell(M^T M)^{\frac{1}{2}}} \leqslant \frac{1}{\sin \frac{\pi}{L}}. \qquad (2.13)$$

Note that $B = e^{t\Delta\tau K}$ and $\lambda_{\max}(K) = 4$, we have the following upper bound of the condition number $\kappa(M)$ of M.

Theorem 2.3. *When $U = 0$,*

$$\kappa(M) = \|M\|\,\|M^{-1}\| \leqslant \frac{1 + e^{4t\Delta\tau}}{\sin \frac{\pi}{L}} = O(L). \qquad (2.14)$$

By the theorem, we conclude that when $U = 0$, the matrix M is well-conditioned.

2.5.2 The case $U \neq 0$

We now consider the situation when $U \neq 0$. By the representation of M in (2.3), we have

$$\|M\| \leqslant 1 + \max_\ell \|B_\ell\| \, \|\Pi\| = 1 + \max_\ell \|B_\ell\| \leqslant 1 + e^{4t\Delta\tau + \nu}. \qquad (2.15)$$

To bound $\|M^{-1}\|$, we first consider the case where U is small. In this case, we can treat it as a small perturbation of the case $U = 0$. We have the following result.

Theorem 2.4. *If U is sufficient small such that*

$$e^{\nu} < 1 + \sin \frac{\pi}{L}, \tag{2.16}$$

then

$$\kappa(M) = \|M\| \, \|M^{-1}\| \leqslant \frac{1 + e^{4t\Delta\tau + \nu}}{\sin \frac{\pi}{L} + 1 - e^{\nu}}.$$

Proof: First we note that M can be written as a perturbation of M at $U = 0$:

$$M = M_0 + \text{diag}(B - B_{\ell})\Pi,$$

where M_0 denotes the matrix M when $U = 0$. Therefore, if

$$\|M_0^{-1}\text{diag}(B - B_{\ell})\| < 1,$$

then we have

$$\|M^{-1}\| \leqslant \frac{\|M_0^{-1}\|}{1 - \|M_0^{-1}\text{diag}(B - B_{\ell})\|}. \tag{2.17}$$

Note that $\|\Pi\| = 1$.

Since the block elements of the matrix M_0 are B or I, and the eigenvalues of the matrix B and B^{-1} are the same, by following the proof of Lemma 2.2, we have

$$\|M_0^{-1}\text{diag}(B)\| \leqslant \frac{1}{\sin \frac{\pi}{L}}.$$

Hence

$$\begin{aligned} \|M_0^{-1}\text{diag}(B - B_{\ell})\| &\leqslant \|M_0^{-1}\text{diag}(B)\| \cdot \|\text{diag}(I - D_{\ell})\| \\ &\leqslant \frac{e^{\nu} - 1}{\sin \frac{\pi}{L}}. \end{aligned}$$

If

$$\frac{e^{\nu} - 1}{\sin \frac{\pi}{L}} < 1,$$

then by (2.17) and (2.13), we have

$$\|M^{-1}\| \leqslant \frac{\frac{1}{\sin \frac{\pi}{L}}}{1 - \frac{e^{\nu} - 1}{\sin \frac{\pi}{L}}} = \frac{1}{\sin \frac{\pi}{L} + 1 - e^{\nu}}. \tag{2.18}$$

The theorem is proven by combining inequalities (2.15) and (2.18). $\square$

Note that the Taylor expansion of ν gives the expression

$$\nu = \sqrt{U\Delta\tau} + \frac{(U\Delta\tau)^{\frac{3}{2}}}{12} + O(U^2\Delta\tau^2). \tag{2.19}$$

Therefore, to the first-order approximation, condition (2.16) is equivalent to

$$\sqrt{U} \leqslant \frac{\pi}{\beta}\sqrt{\Delta\tau} + O(\Delta\tau). \qquad (2.20)$$

Consequently, to the first order approximation, we have

$$\kappa(M) \leqslant \frac{L(1 + e^{4t\Delta\tau+\nu})}{\pi - \beta\sqrt{U\Delta\tau} - U\beta/2} + O(U^{\frac{3}{2}}\beta\Delta\tau^{\frac{1}{2}}).$$

By the inequality, we conclude that when U is sufficient small enough, M is well-conditioned and $\kappa(M) = O(L)$.

It is an open problem to find a rigorous sharp bound of $\kappa(M)$ when $U \neq 0$. Figure 2.7 shows the averages of the condition numbers of M for 100 Hubbard-Stratonovich configurations $h_{\ell,i} = \pm 1$, where $N = 16$, $L = 8\beta$ with $\beta = [1 : 10]$, and $t = 1$.

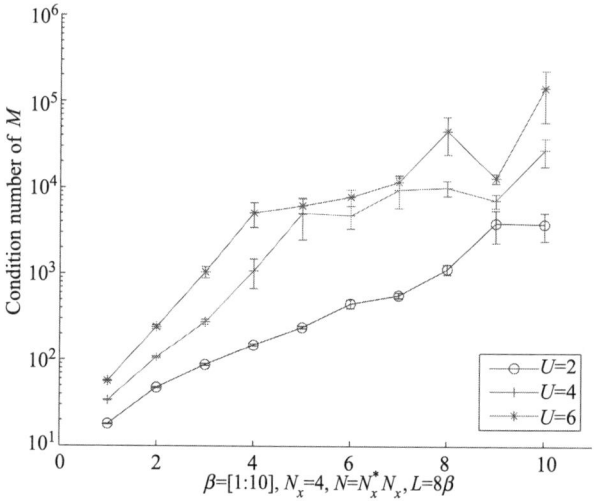

Figure 2.7. Condition numbers $\kappa(M)$ for different U.

Figure 2.7 reveals several key issues concerning the transitional behaviors of the condition number of $\kappa(M)$:

1. The condition number increases much more rapidly than the linear rise which we know analytically at $U = 0$.

2. Not only does the condition number increase with U, but also so do its fluctuations over the 100 chosen field configurations.

3. When the inverse temperature β increases, the conditioning of M becomes worse.

These observations tell us that the energy parameters U and β, which in turn determines L, are two critical parameters for the conditioning of the underlying matrix problems. The matrix M becomes ill-conditioned at low temperature (large β) or strong coupling (large U). It suggests that widely varying conditioning of the matrix problems is encountered in the course of a simulation, robust and efficient matrix solvers need to adopt different solution strategies depending on the conditioning of the underlying matrix problems.

2.6 Condition number of $M^{(k)}$

For an integer $k \leqslant L$, a structure-preserving factor-of-k reduction of the matrix M leads a matrix $M^{(k)}$ of the same block form

$$
M^{(k)} = \begin{bmatrix}
I & & & & B_1^{(k)} \\
-B_2^{(k)} & I & & & \\
& -B_3^{(k)} & I & & \\
& & \ddots & \ddots & \\
& & & -B_{L_k}^{(k)} & I
\end{bmatrix},
$$

where $L_k = \lceil \frac{L}{k} \rceil$ is the number of blocks and

$$
\begin{aligned}
B_1^{(k)} &= B_k B_{k-1} \cdots B_2 B_1, \\
B_2^{(k)} &= B_{2k} B_{2k-1} \cdots B_{k+2} B_{k+1}, \\
&\cdots \\
B_{L_k}^{(k)} &= B_L B_{L-1} \cdots B_{(L_k-1)k+1}.
\end{aligned}
$$

First we have the following observation. The inverse of $M^{(k)}$ is a "submatrix" of the inverse of M. Specifically, since M and $M^{(k)}$ have the same block cyclic structure, by expression (2.6), we immediately have the following expression for the (i,j) block of $\{M^{(k)}\}_{i,j}^{-1}$:

$$
\{M^{(k)}\}_{i,j}^{-1} = (I + B_i^{(k)} \cdots B_1^{(k)} B_L^{(k)} \cdots B_{i+1}^{(k)})^{-1} Z_{i,j}^{(k)},
$$

where

$$
Z_{i,j}^{(k)} = \begin{cases}
-B_i^{(k)} \cdots B_1^{(k)} B_L^{(k)} \cdots B_{j+1}^{(k)}, & i < j, \\
I, & i = j, \\
B_i^{(k)} B_{i-1}^{(k)} \cdots B_{j+1}^{(k)}, & i > j.
\end{cases}
$$

By the definition of $B_i^{(k)}$, and if $i \neq L^{(k)}$,

$$
\{M^{(k)}\}_{i,j}^{-1} = (I + B_{ik} \cdots B_1 B_L \cdots B_{i*k+1})^{-1}
$$

$$
\times \begin{cases}
-B_{i*k} \cdots B_1 B_L \cdots B_{j*k+1}, & i < j, \\
I, & i = j, \\
B_{i*k} \cdots B_{j*k+1}, & i > j.
\end{cases}
$$

Hence if $i \neq L^{(k)}$,

$$\{M^{(k)}\}^{-1}_{i,j} = \{M^{-1}\}_{i*k, j*k}.$$

If $i = L^{(k)}$, we have

$$\{M^{(k)}\}^{-1}_{i,j} = (I + B_L \cdots B_1)^{-1} \begin{cases} B_L \cdots B_{j*k+1}, & j < L, \\ I, & j = L. \end{cases}$$

Hence if $i = L^{(k)}$,

$$\{M^{(k)}\}^{-1}_{L^{(k)}, j} = \{M^{-1}\}_{L, j*k}.$$

We now turn to the discussion of the condition number of the matrix $M^{(k)}$. We have the following two immediate results:

1. The upper bound of the norm of the matrix $M^{(k)}$ is given by

$$\|M^{(k)}\| \leqslant 1 + e^{(4t\Delta\tau + \nu)k}. \tag{2.21}$$

   This is due to the fact that $\|B^{(k)}_\ell\| \leqslant e^{(4t\Delta\tau + \nu)k}$.

2. The norm of the inverse of the matrix $M^{(k)}$ is bounded by

$$\|(M^{(k)})^{-1}\| \leqslant \|M^{-1}\|. \tag{2.22}$$

   This is due to the fact that $(M^{(k)})^{-1}$ is a "submatrix" of M^{-1}.

By combining (2.21) and (2.22), we have an upper bound of the condition number of the reduced matrix $M^{(k)}$ in terms of the condition number of the original matrix M:

$$\kappa(M^{(k)}) = \|M^{(k)}\| \, \|(M^{(k)})^{-1}\|$$

$$\leqslant \frac{\|M^{(k)}\|}{\|M\|} \kappa(M)$$

$$\leqslant \frac{1 + e^{(4t\Delta\tau + \nu)k}}{\|M\|} \kappa(M)$$

$$\leqslant c \, e^{(4t\Delta\tau + \nu)k} \kappa(M), \tag{2.23}$$

where c is some constant. Inequality (2.23) shows that comparing to the condition number of M, the condition number of $M^{(k)}$ is amplified by a factor proportional to the reduction factor k.

For further details, we consider the following three cases:

1. $U = 0$ and $k = L$. In this case, the matrix M is reduced to a single block

$$
\begin{aligned}
M^{(L)} &= I + B_L \cdots B_2 B_1 \\
&= I + B \cdots BB \\
&= I + B^L \\
&= I + (e^{t\Delta\tau K})^L \\
&= I + e^{t\beta K}.
\end{aligned}
$$

By the eigen-decomposition of the matrix K, see Lemma 2.1, the condition number of $M^{(L)}$ is given by

$$
\kappa(M^{(L)}) = \frac{1 + e^{4t\beta}}{1 + e^{-4t\beta}}.
$$

Therefore, for large β, $M^{(L)}$ is extremely ill-conditioned.

2. $U = 0$ and $k < L$ and $L_k = \frac{L}{k}$ is an integer. In this case, we have

$$
\kappa(M^{(k)}) \leqslant \frac{1 + e^{4t\Delta\tau k}}{\sin\frac{\pi}{L_k}}.
$$

The inequality can be proven similarly as the proof of Lemma 2.2. It is anticipated that the bound is still true when L/k is not an integer. Figure 2.8 shows condition numbers of $M^{(k)}$ with respect to the reduction factor k when $U = 0$. The computational and estimated condition numbers fit well.

3. For the general case of $U \neq 0$, we have bound (2.23). Figure 2.9 shows the condition numbers of sample matrices $M^{(k)}$ (solid lines) for $U = 4$ and $U = 6$. The upper bound (2.23) are dashed line with circles. The mean of condition numbers $\kappa(M)$ of sample matrices M are used in bound (2.23). It is clear that the upper bound (2.23) of the condition numbers of $M^{(k)}$ is an over estimation of the actual conditioning of $M^{(k)}$. This may partially due to the over-estimation of the norm of $M^{(k)}$. We observe that the condition number of $M^{(k)}$ is closer to $e^{\frac{1}{2}k(4t\Delta\tau+\nu)}\kappa(M)$. This is shown as the dashed lines with diamonds in the following plots. It is a subject of further study.

3 Self-adaptive direct linear solvers

In this section, we consider one of the computational kernels of the QMC simulations discussed in sections 1 and 2, namely solving the linear system of equations

$$
Mx = b, \tag{3.1}
$$

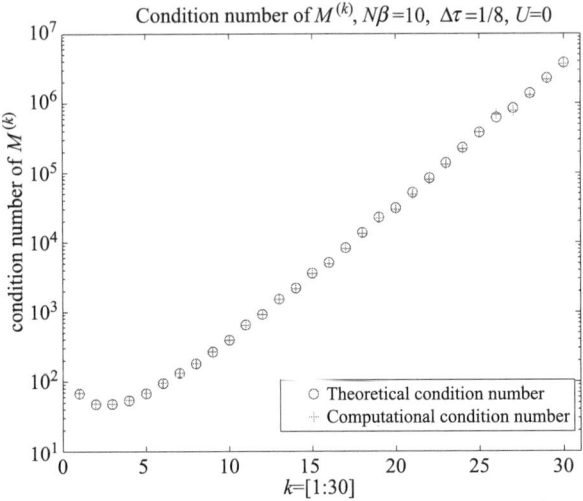

Figure 2.8. Condition numbers of $M^{(k)}$, $U = 0$.

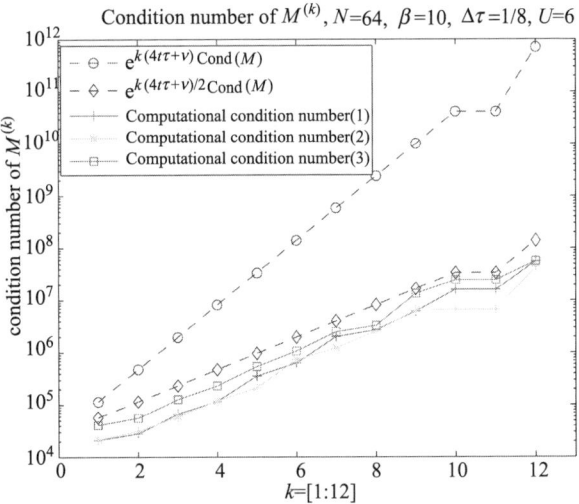

Figure 2.9. Condition numbers of $M^{(k)}$, $U = 6$.

where the coefficient matrix M is the Hubbard matrix as defined in (2.1) of section 2. One of challenges in QMC simulations is to develop algorithmic techniques that can robustly and efficiently solve numerical linear algebra problems with underlying multi-length scale coefficient matrices in a self-adapting fashion. In this section, we present such an approach for solving the linear system (3.1).

The structure of Hubbard matrix M exhibits the form of so-called block p-cyclic consistently ordered matrix [20]. p-cyclic matrices arise in a number of contexts in applied mathematics, such as numerical solution of boundary value problems for ordinary differential equations [28], finite-difference equations for the steady-state solution of parabolic equations with periodic boundary conditions [27], and the stationary solution of Markov chains with periodic graph structure [26]. It is known that the block Gaussian elimination with and without pivoting for solving p-cyclic linear systems could be numerically unstable, as shown in the cases arising from the multiple shooting method for solving two-point boundary value problems [23,30] and Markov chain modeling [25]. Block cyclic reduction [22] is a powerful idea to solve such p-cyclic system since the number of unknowns is reduced exponentially. However, a full block cyclic reduction is inherently unstable and is only applicable for some energy scales, namely $U \leqslant 1$, due to ill-conditioning of the reduced system. A stable p-cyclic linear system solver is based on the structural orthogonal factorization [23,29]. However, the costs of memory requirements and flops are prohibitively expensive when the length scale parameters N and L increase.

To take advantages of the block cyclic reduction and the structural orthogonal factorization method, in this section, we present a hybrid method to solve the linear system (3.1). The hybrid method performs a partial cyclic reduction in a self-adaptive way depending on system parameters such that the reduced system is still sufficiently well-conditioned to give rise a solution of the desired accuracy computed by the structural orthogonal factorization method. The hybrid method is called *self-adaptive block cyclic reduction*, or SABCR in short.

3.1   Block cyclic reduction

Consider the following 16×16 block cyclic linear system (3.1):

$$Mx = b,$$

where

$$M = \begin{bmatrix} I & & & & & & & B_1 \\ -B_2 & I & & & & & & \\ & -B_3 & I & & & & & \\ & & -B_4 & I & & & & \\ & & & \ddots & \ddots & & & \\ & & & & & -B_{15} & I & \\ & & & & & & -B_{16} & I \end{bmatrix}.$$

Correspondingly, partitions of the vectors x and b are conformed to the blocks of M:

$$x = \begin{bmatrix} x_1 \\ x_2 \\ x_3 \\ x_4 \\ \vdots \\ x_{15} \\ x_{16} \end{bmatrix}, \quad b = \begin{bmatrix} b_1 \\ b_2 \\ b_3 \\ b_4 \\ \vdots \\ b_{15} \\ b_{16} \end{bmatrix}.$$

A factor-of-four block cyclic reduction (BCR) leads to the following 4×4 block cycle system:

$$M^{(4)} x^{(4)} = b^{(4)}, \tag{3.2}$$

where

$$M^{(4)} = \begin{bmatrix} I & & & B_1^{(4)} \\ -B_2^{(4)} & I & & \\ & -B_3^{(4)} & I & \\ & & -B_4^{(4)} & I \end{bmatrix}, \quad x^{(4)} = \begin{bmatrix} x_4 \\ x_8 \\ x_{12} \\ x_{16} \end{bmatrix}, \quad b^{(4)} = \begin{bmatrix} b_1^{(4)} \\ b_2^{(4)} \\ b_3^{(4)} \\ b_4^{(4)} \end{bmatrix}$$

and

$$\begin{aligned} B_1^{(4)} &= B_4 B_3 B_2 B_1, \\ B_2^{(4)} &= B_8 B_7 B_6 B_5, \\ B_3^{(4)} &= B_{12} B_{11} B_{10} B_9, \\ B_4^{(4)} &= B_{16} B_{15} B_{14} B_{13}. \end{aligned}$$

and

$$\begin{aligned} b_1^{(4)} &= b_4 + B_4 b_3 + B_4 B_3 b_2 + B_4 B_3 B_2 b_1, \\ b_2^{(4)} &= b_8 + B_8 b_7 + B_8 B_7 b_6 + B_8 B_7 B_6 b_5, \\ b_3^{(4)} &= b_{12} + B_{12} b_{11} + B_{12} B_{11} b_{10} + B_{12} B_{11} B_{10} b_9, \\ b_4^{(4)} &= b_{16} + B_{16} b_{15} + B_{16} B_{15} b_{14} + B_{16} B_{15} B_{14} b_{13}. \end{aligned}$$

Therefore, to solve the original linear system (3.1), one can first solve the reduced system (3.2). Once the vector $x^{(4)}$ is computed, i.e., the block components x_4, x_8, x_{12} and x_{16} of the solution x are computed, the rest of block components x_i of the solution x can be computed by the following forward and back substitutions:

- Forward substitution:

$$\begin{aligned} x_1 &= b_1 - B_1 x_{16}, & x_2 &= b_2 + B_2 x_1, \\ x_5 &= b_5 + B_5 x_4, & x_6 &= b_6 + B_6 x_5, \\ x_9 &= b_9 + B_9 x_8, & x_{10} &= b_{10} + B_{10} x_9, \\ x_{13} &= b_{13} + B_{13} x_{12}, & x_{14} &= b_{14} + B_{14} x_{13}. \end{aligned}$$

- Back substitution:

$$x_3 = B_4^{-1}(x_4 - b_4), \qquad x_7 = B_8^{-1}(x_8 - b_8),$$
$$x_{11} = B_{12}^{-1}(x_{12} - b_{12}), \qquad x_{15} = B_{16}^{-1}(x_{16} - b_{16}).$$

The use of both forward and back substitutions is intended to minimize the propagation of rounding errors in the computed components x_4, x_8, x_{12} and x_{16}.

The pattern for a general factor-of-k reduction is clear. Given an integer $k \leqslant L$, a-factor-of-k BCR leads to an $L_k \times L_k$ block cycle linear system:

$$M^{(k)} x^{(k)} = b^{(k)}, \tag{3.3}$$

where $L_k = \lceil \frac{L}{k} \rceil$,

$$M^{(k)} = \begin{bmatrix} I & & & & B_1^{(k)} \\ -B_2^{(k)} & I & & & \\ & -B_3^{(k)} & I & & \\ & & \ddots & \ddots & \\ & & & -B_{L_k}^{(k)} & I \end{bmatrix},$$

with

$$B_1^{(k)} = B_k B_{k-1} \cdots B_2 B_1,$$
$$B_2^{(k)} = B_{2k} B_{2k-1} \cdots B_{k+2} B_{k+1},$$
$$\cdots$$
$$B_{L_k}^{(k)} = B_L B_{L-1} \cdots B_{(L_k-1)k+1},$$

and vectors $x^{(k)}$ and $b^{(k)}$ are

$$x^{(k)} = \begin{bmatrix} x_k \\ x_{2k} \\ \vdots \\ x_{(L_k-1)k} \\ x_L \end{bmatrix}$$

and

$$b^{(k)} = \begin{bmatrix} b_k + \sum_{t=1}^{k-1} B_k \cdots B_{t+1} b_t \\ b_{2k} + \sum_{t=k+1}^{2k-1} B_{2k} \cdots B_{t+1} b_t \\ \vdots \\ b_{(L_k-1)k} + \sum_{t=(L_k-2)k+1}^{(L_k-1)k-1} B_{(L_k-1)k} \cdots B_{t+1} b_t \\ b_L + \sum_{t=(L_k-1)k+1}^{L-1} B_L \cdots B_{t+1} b_t \end{bmatrix}.$$

After the solution vector $x^{(k)}$ of the reduced system (3.3) is computed, the rest of block components x_i of the solution vector x are obtained by forward and back substitutions as shown in the following:

FORWARD AND BACK SUBSTITUTIONS
1. Let $\ell = [k, 2k, \cdots, (L_k - 1)k, L]$
2. For $j = 1, 2, \cdots, L_k$
   (a) $x_{\ell(j)} = x_j^{(k)}$
   (b) forward substitution
      For $i = \ell(j-1) + 1, \cdots, \ell(j-1) + \lceil \frac{1}{2}(\ell(j) - \ell(j-1) - 1) \rceil$ with $\ell(0) = 0$:
         If $i = 1$, $x_1 = b_1 - B_1 x_L$
         else $x_i = b_i + B_i x_{i-1}$
   (c) back substitution
      For $i = \ell(j) - 1, \ell(j) - 2, \cdots, \ell(j) - \lfloor \frac{1}{2}(\ell(j) - \ell(j-1) - 1) \rfloor$,
         $x_i = B_{i+1}^{-1}(x_{i+1} - b_{i+1})$.

Remark 3.1. If the reduction factor $k = L$, then $L_k = 1$. The reduced system is
$$M^{(L)} x_L = b^{(L)},$$
i.e.,
$$(I + B_L B_{L-1} \cdots B_1) x_L = b_L + \sum_{\ell=1}^{L-1} (B_L B_{L-1} \cdots B_{\ell+1}) b_\ell.$$

Remark 3.2. There are a number of ways to derive the reduced system (3.3). For example, we can use the block Gaussian elimination. Writing the matrix M of the original system (3.1) as an L_k by L_k block matrix:

$$M = \begin{bmatrix} D_1 & & & & \widehat{B}_1 \\ -\widehat{B}_2 & D_2 & & & \\ & -\widehat{B}_3 & D_3 & & \\ & & \ddots & \ddots & \\ & & & -\widehat{B}_{L_k} & D_{L_k} \end{bmatrix},$$

where D_i are $k \times k$ black matrices defined as

$$D_i = \begin{bmatrix} I & & & & \\ -B_{(i-1)k+2} & I & & & \\ & -B_{(i-1)k+3} & I & & \\ & & & \ddots & \ddots \\ & & & & -B_{ik} & I \end{bmatrix},$$

and $\widehat{B}_i$ are $k \times k$ block matrices defined as

$$
\widehat{B}_i = \begin{bmatrix} 0\,0\cdots & B_{(i-1)k+1} \\ 0\,0\cdots & 0 \\ \vdots\ \vdots\ \vdots & \vdots \\ 0\,0\cdots & 0 \end{bmatrix}.
$$

Define $\widehat{D} = \operatorname{diag}(D_1, D_2, \cdots, D_{L_k})$. Then

$$
\widehat{D}^{-1}M = \begin{bmatrix} I & & & & D_1^{-1}\widehat{B}_1 \\ -D_2^{-1}\widehat{B}_2 & I & & & \\ & -D_3^{-1}\widehat{B}_3 & I & & \\ & & & \ddots & \ddots \\ & & & -D_{L_k}^{-1}\widehat{B}_{L_k} & I \end{bmatrix},
$$

where the matrix $D_i^{-1}\widehat{B}_i$ is given by

$$
D_i^{-1}\widehat{B}_i = \begin{bmatrix} 0\,0\cdots & B_{(i-1)k+2}B_{(i-1)k+1} \\ 0\,0\cdots & B_{(i-1)k+3}B_{(i-1)k+2}B_{(i-1)k+1} \\ \vdots\ \vdots\ \vdots\ \vdots \\ 0\,0\cdots & B_{(i-1)k+k}\cdots B_{(i-1)k+2}B_{(i-1)k+1} \end{bmatrix}.
$$

Therefore, we immediately see that $M^{(k)}$ is a submatrix of $\widehat{D}^{-1}M$, namely, there exists a matrix P, such that

$$
M^{(k)} = P^T\widehat{D}^{-1}MP,
$$

where the matrix P is $NL \times (NL/k)$ matrix, whose $(i-1)N+1$ to iN columns are the $(ik-1)N+1$ to ikN columns of the identity matrix I_{NL}.

3.2   Block structural orthogonal factorization

Comparing with the Gaussian elimination, the block structural orthogonal factorization (BSOF) method presented in this section is computationally more expensive, but numerically backward stable.

By multiplying a sequence of orthogonal transformation matrices Q_i, the block cyclic matrix M of system (3.1) is transformed to a block upper triangular matrix R:

$$
Q_{L-1}^T \cdots Q_2^T Q_1^T M = R, \tag{3.4}
$$

where

$$
R = \begin{bmatrix}
R_{11} & R_{12} & & & & R_{1L} \\
& R_{22} & R_{23} & & & R_{2L} \\
& & \ddots & \ddots & & \vdots \\
& & & R_{L-1,L-1} & R_{L-1,L} \\
& & & & R_{LL}
\end{bmatrix},
$$

and diagonal blocks $R_{\ell\ell}$ are upper triangular, Q_ℓ are orthogonal matrices of the form

$$
Q_\ell = \begin{bmatrix}
I \\
& \ddots \\
& & Q_{11}^{(\ell)} & Q_{12}^{(\ell)} \\
& & Q_{21}^{(\ell)} & Q_{22}^{(\ell)} \\
& & & & \ddots \\
& & & & & I
\end{bmatrix}.
$$

The subblocks $Q_{ij}^{(\ell)}$ are defined by the orthogonal factor of the QR decomposition:

$$
\begin{bmatrix} \widetilde{M_{\ell\ell}} \\ -B_{\ell+1} \end{bmatrix} = \begin{bmatrix} Q_{11}^{(\ell)} & Q_{12}^{(\ell)} \\ Q_{21}^{(\ell)} & Q_{22}^{(\ell)} \end{bmatrix} \begin{bmatrix} R_{\ell\ell} \\ 0 \end{bmatrix},
$$

where

$$
\widetilde{M_{\ell\ell}} = \begin{cases} I, & \text{for } \ell = 1, \\ (Q_{22}^{(\ell-1)})^T, & \text{for } \ell = 2, 3, \cdots, L-2, \end{cases}
$$

except for $\ell = L - 1$, it is defined by the QR decomposition:

$$
\begin{bmatrix} \widetilde{M}_{L-1,L-1} & R_{L-1,L} \\ -B_L & I \end{bmatrix} = \begin{bmatrix} Q_{11}^{(L-1)} & Q_{12}^{(L-1)} \\ Q_{21}^{(L-1)} & Q_{22}^{(L-1)} \end{bmatrix} \begin{bmatrix} R_{L-1,L-1} & \check{R}_{L-1,L} \\ 0 & R_{LL} \end{bmatrix}.
$$

The following is a pseudo-code for the BSOF method to solve the block cyclic system (3.1).

BSOF METHOD

1. Set $M_{11} = I$, $R_{1L} = B_1$ and $c_1 = b_1$
2. For $\ell = 1, 2, \cdots, L - 2$

   (a) Compute the QR decomposition

   $$
   \begin{bmatrix} M_{\ell\ell} \\ -B_{\ell+1} \end{bmatrix} = \begin{bmatrix} Q_{11}^{(\ell)} & Q_{12}^{(\ell)} \\ Q_{21}^{(\ell)} & Q_{22}^{(\ell)} \end{bmatrix} \begin{bmatrix} R_{\ell\ell} \\ 0 \end{bmatrix}
   $$

   (b) Set $\begin{bmatrix} R_{\ell,\ell+1} \\ M_{\ell+1,\ell+1} \end{bmatrix} = \begin{bmatrix} Q_{11}^{(\ell)} & Q_{12}^{(\ell)} \\ Q_{21}^{(\ell)} & Q_{22}^{(\ell)} \end{bmatrix}^T \begin{bmatrix} 0 \\ I \end{bmatrix}$

(c) Update $\begin{bmatrix} R_{\ell L} \\ R_{\ell+1,L} \end{bmatrix} := \begin{bmatrix} Q_{11}^{(\ell)} & Q_{12}^{(\ell)} \\ Q_{21}^{(\ell)} & Q_{22}^{(\ell)} \end{bmatrix}^{T} \begin{bmatrix} R_{\ell L} \\ 0 \end{bmatrix}$

(d) Set $\begin{bmatrix} c_{\ell} \\ c_{\ell+1} \end{bmatrix} := \begin{bmatrix} Q_{11}^{(\ell)} & Q_{12}^{(\ell)} \\ Q_{21}^{(\ell)} & Q_{22}^{(\ell)} \end{bmatrix}^{T} \begin{bmatrix} c_{\ell} \\ b_{\ell+1} \end{bmatrix}$

3. Compute the QR decomposition

$$\begin{bmatrix} M_{L-1,L-1} & R_{L-1,L} \\ -B_L & I \end{bmatrix} = \begin{bmatrix} Q_{11}^{(L-1)} & Q_{12}^{(L-1)} \\ Q_{21}^{(L-1)} & Q_{22}^{(L-1)} \end{bmatrix} \begin{bmatrix} R_{L-1,L-1} & R_{L-1,L} \\ 0 & R_{LL} \end{bmatrix}$$

4. Set $\begin{bmatrix} c_{L-1} \\ c_L \end{bmatrix} := \begin{bmatrix} Q_{11}^{(L-1)} & Q_{12}^{(L-1)} \\ Q_{21}^{(L-1)} & Q_{22}^{(L-1)} \end{bmatrix}^{T} \begin{bmatrix} c_{L-1} \\ b_L \end{bmatrix}$

5. Back substitution to solve the block triangular system
   $Rx = c$

   (a) Solve $R_{LL} x_L = c_L$ for x_L
   (b) Solve $R_{L-1,L-1} x_{L-1} = c_{L-1} - R_{L-1,L} x_L$ for x_{L-1}
   (c) For $\ell = L - 2, L - 3, \cdots, 1$,
       solve $R_{\ell\ell} x_\ell = c_\ell - R_{\ell,\ell+1} x_{\ell+1} - R_{\ell L} x_L$ for x_ℓ

Floating point operations of the BSOF method is about $15 N^3 L$. The memory requirement is $3 N^2 L$.

3.3   A hybrid method

To take advantages of block order reduction of the BCR method and numerical stability of the BSOF method, we propose a hybrid method:

STEP 1. Perform a factor-of-k BCR of the original system (3.1) to derive a reduced block cyclic system (3.3) of the block order L/k.

STEP 2. Solve the reduced block cyclic system (3.3) by using the BSOF method.

STEP 3. Forward and back substitute to find the rest of block components x_i of the solution x:

$$\{x_i\} \quad \longleftarrow \quad x^{(k)} \quad \longrightarrow \quad \{x_j\}.$$

Figure 3.1 is a schematic map of the hybrid method for a 16-block cyclic system with a reduction factor $k = 4$. We use both forward and back substitutions to minimize the propagation of rounding errors induced at Steps 1 and 2.

By Step 1, the order of the original M is reduced by a factor of k. Consequently, the computational cost of the BSOF method at Step 2 is

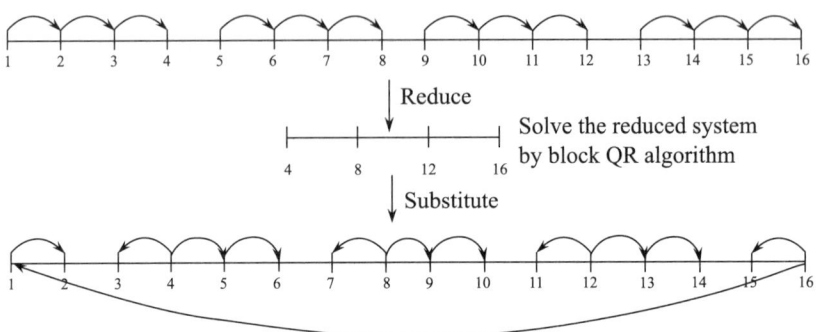

Figure 3.1. A schematic map of the hybrid method.

reduced to $O(N^3 \frac{L}{k})$, a factor of k speedup. Therefore, the larger k is, the better CPU performance is. However, the condition number of $M^{(k)}$ increases as k increases, see the analysis presented in section 2.6. As a result, the accuracy of the computed solution decreases. An interesting question is how to find a reduction factor k, such that the computed solution meets the required accuracy in QMC simulations. In practice, such a reduction factor k should be determined in a *self-adapting* fashion with respect to the changes of underlying multi-length and energy scales in QMC simulations. This is discussed in the next section.

3.4   Self-adaptive reduction factor k

Let us turn to the question of determining the reduction factor k for the BCR step of the proposed hybrid method. Since the BSOF method is backward stable, by well-established error analysis of the linear system, for example, see [24, p.120], we know that the relative error of the computed solution $\widehat{x}^{(k)}$ of the reduced system (3.3) is bounded by $\kappa(M^{(k)})\epsilon$, i.e.,

$$\frac{\|\delta x^{(k)}\|}{\|x^{(k)}\|} \equiv \frac{\|x^{(k)} - \widehat{x}^{(k)}\|}{\|x^{(k)}\|} \leqslant \kappa(M^{(k)})\epsilon, \tag{3.5}$$

where ϵ is the machine precision. For the clarity of notation, we only use the first-order upper bound and ignore the small constant coefficient.

To consider the propagation of the errors in the computed solution $\widehat{x}^{(k)}$ in the back and forward substitutions, let us start with the computed L-th block component $\widehat{x}_L$ of the solution vector x,

$$\widehat{x}_L = x_L + \delta x_L,$$

where δx_L is the computed error, with a related upper bound defined in (3.5). By the forward substitution, the computed first block component

$\widehat{x}_1$ of the solution x satisfies

$$
\begin{aligned}
\widehat{x}_1 &= b_1 - B_1 \widehat{x}_L \\
&= b_1 - B_1 (x_L + \delta x_L) \\
&= b_1 - B_1 x_L + B_1 \delta x_L \\
&= x_1 + \delta x_1,
\end{aligned}
$$

where $\delta x_1 = B_1 \delta x_L$ is the error propagated by the error in $\widehat{x}_L$. Note that the computational error induced in the substitution is ignored, since it is generally much smaller than the error in $\widehat{x}_L$.

By the relative error bound (3.5) of δx_L, it yields that the error in the computed $\widehat{x}_1$ could be amplified by the factor $\|B_1\|$, namely,

$$
\frac{\|\delta x_1\|}{\|x_1\|} \leqslant \|B_1\| \kappa(M^{(k)}) \epsilon.
$$

Subsequently, the relative error of the computed $\widehat{x}_2$ obtained by the forward substitution from x_1 is bounded by

$$
\frac{\|\delta x_2\|}{\|x_2\|} \leqslant \|B_2\| \|B_1\| \kappa(M^{(k)}) \epsilon.
$$

This process can be continued to bound the errors of δx_3 and so on until all that is left is $x_{\frac{k}{2}}$. The related error bound of computed $x_{\frac{k}{2}}$ is given by

$$
\frac{\|\delta x_{\frac{k}{2}}\|}{\|x_{\frac{k}{2}}\|} \leqslant \|B_{\frac{k}{2}}\| \cdots \|B_2\| \|B_1\| \kappa(M^{(k)}) \epsilon.
$$

By analogous calculations for the rest of substitutions, we conclude that for any computed block component $\widehat{x}_\ell$ of the solution x, where $\ell = 1, 2, \cdots, L$, the related error is bounded by

$$
\begin{aligned}
\frac{\|\delta x_\ell\|}{\|x_\ell\|} &\leqslant \|B_{\frac{k}{2}}\| \cdots \|B_2\| \|B_1\| \kappa(M^{(k)}) \epsilon \\
&\leqslant c\, e^{\frac{1}{2}k(4t\Delta\tau + \nu)} \cdot e^{k(4t\Delta\tau + \nu)} \kappa(M) \epsilon \\
&= c\, e^{\frac{3}{2}k(4t\Delta\tau + \nu)} \kappa(M) \epsilon,
\end{aligned} \tag{3.6}
$$

where for the second inequality we have used the upper bounds (2.15) and (2.23) for the norm of B_ℓ and the condition number of the matrix $M^{(k)}$.

Assume that the desired relative accuracy of the computed solution x is "tol", i.e.,

$$
\frac{\|\delta x\|}{\|x\|} \leqslant \text{tol}. \tag{3.7}
$$

Then by inequalities (3.6) and (3.7), a plausible choice of the reduction factor k is

$$k = \left\lfloor \frac{\frac{2}{3}\ln(\text{tol}/\epsilon)}{4t\Delta\tau + \nu} \right\rfloor. \tag{3.8}$$

To balance the number of the matrices B_ℓ in the product $B_\ell^{(k)}$, after k is computed, the final k is then slightly adjusted by $k = \left\lceil \frac{L}{L_k} \right\rceil$, where $L_k = \left\lceil \frac{L}{k} \right\rceil$.

We note that the factor of $\ln\kappa(M)$ is dropped in deciding reduction factor k. The reason is that as we discussed in section 2.5, $\kappa(M)$ grows slowly in the range of parameters of interest, $\ln\kappa(M)$ is small.

By expression (3.8), the reduction factor k is determined in a *self-adaptive* fashion. When energy parameters U and $\beta = L \cdot \Delta\tau$ change, k is determined adaptively to achieve the desired accuracy "tol". For example, let $t = 1$ and $\Delta\tau = \frac{1}{8}$. If the desired accuracy threshold is tol $=$ 10^{-8}, then with double precision arithmetic and machine precision $\epsilon = 10^{-16}$, the reduction factor k with respect to different energy parameter U is shown in the following table:

U	0	1	2	3	4	5	6
k	24	14	12	10	9	9	8

Figure 3.2 shows the reduced block sizes L_k with respect to different values of U, where $L = 8\beta$, $\beta = 1, 2, \cdots, 20$ and $t = 1$ The accuracy is set to be half of the machine precision, i.e., tol $= \sqrt{\epsilon} = 10^{-8}$.

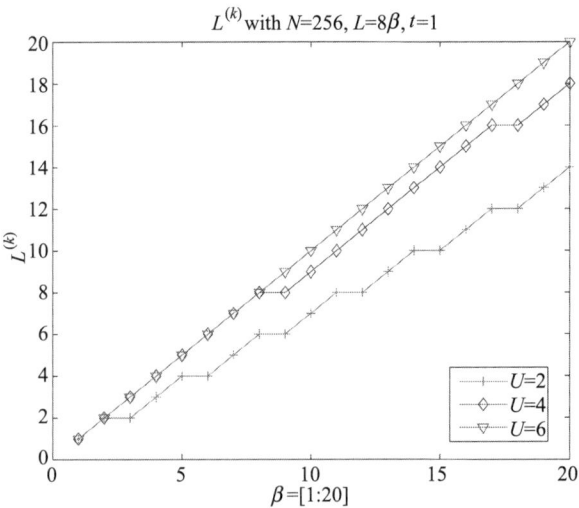

Figure 3.2. Reduced block size L_k.

3.5 Self-adaptive block cyclic reduction method

In summary, the self-adaptive block cyclic reduction method, SABCR in short, to solve the linear system (3.1) may be condensed as the following:

SABCR METHOD
1. Determine the reduction factor k by (3.8)
2. Reduce (M, b) to $(M^{(k)}, b^{(k)})$ by the BCR
3. Solve the reduced system $M^{(k)} x^{(k)} = b^{(k)}$ for $x^{(k)}$ by the BSOF method
4. Use forward and back substitutions to compute the remaining block components x_ℓ of x

3.6 Numerical experiments

In this section, we present numerical results of the SABCR method. SABCR is implemented in Fortran 90 using LAPACK and BLAS. All numerical results are performed on an HP Itanium2 workstation with 1.5GHz CPU and 2GB core memory. The threshold of desired relative accuracy of computed solution $\widehat{x}$ is set at the order of $\sqrt{\epsilon}$, namely,

$$\frac{\|\widehat{x} - x\|}{\|x\|} \leqslant \mathrm{tol} = 10^{-8}.$$

Example 1. In this example, we examine numerical accuracy and performance of the SABCR method when $U = 0$. The other parameters of the coefficient matrix M is set as $N = 16 \times 16$, $L = 8\beta$ for $\beta = 1, 2, \cdots, 20$, $t = 1$, $\Delta\tau = \frac{1}{8}$. Figure 3.3 shows the relative errors of computed solutions $\widehat{x}$ by the BSOF and SABCR methods. As we see, the BSOF method is of full machine precision $O(10^{-15})$. It indicates that the underlying linear system is well-conditioned and the BSOF method is backward stable. On the other hand, as shown in the figure, the relative errors of the SABCR method are at $O(10^{-8})$ as desired.

Table 1 shows the reduction factor k and the reduced block size L_k with respect to the inverse temperature β, CPU elapsed time and speedups of the SABCR method comparing to the BSOF method. Note that for $\beta \leqslant 3$, the reduction factor $k = L$ and the number of reduced block $L_k = 1$. As β increases, the reduction factor k decreases. For example, when $\beta = 20$, the reduction factor $k = 23$ and the number of reduced blocks $L_k = \lfloor \frac{160}{23} \rfloor + 1 = 7$.

Example 2. In this example, we examine numerical accuracy and performance of the SABCR method for $U = 2, 4, 6$. The other parameters of the coefficient matrix M are $N = 16 \times 16 = 256$, $L = 8\beta$ with $\beta = 1, 2, \cdots, 20$, $t = 1$ and $\Delta\tau = \frac{1}{8}$. Figure 3.4 shows that the relative errors

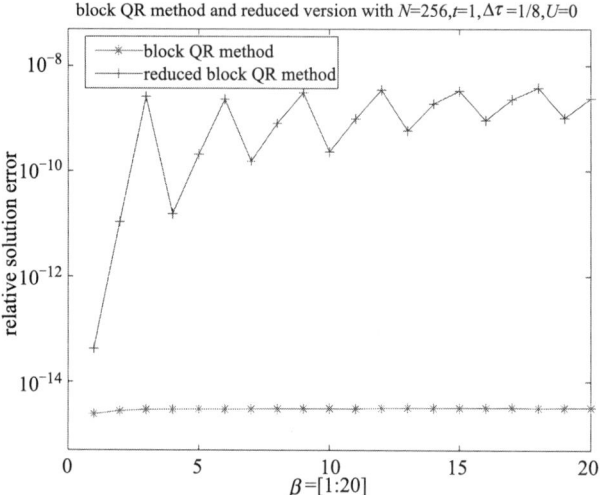

Figure 3.3. Relative errors, $U = 0$.

Table 1. Performance data of BSOF and SABCR methods, $U = 0$.

β	$L = 8\beta$	k	L_k	BSOF	SABCR	speedup
1	8	8	1	3.19	0.0293	108
2	16	16	1	7.06	0.042	168
3	24	24	1	10.8	0.0547	197
4	32	16	2	14.6	0.303	48
5	40	20	2	18.6	0.326	57
6	48	24	2	23.1	0.342	67
7	56	19	3	27.2	0.666	40
8	64	22	3	31.3	0.683	45
9	72	24	3	35.1	0.675	52
10	80	20	4	38.0	1.18	32
11	88	22	4	42.0	1.18	35
12	96	24	4	46.0	1.20	38
13	104	21	5	49.9	1.28	38
14	112	23	5	54.0	1.28	42
15	120	24	5	58.2	1.32	44
16	128	22	6	62.9	1.67	37
17	136	23	6	68.3	1.72	39
18	144	24	6	73.2	1.73	42
19	152	22	7	75.3	1.98	38
20	160	23	7	80.2	2.02	39

of the computed solution $\widehat{x}$ are under tol $= 10^{-8}$ as required. Table 2 shows the the reduced block size L_k, CPU elapsed time of the BSOF and SABCR methods with respect to U and β. The speedups of SABCR are shown in Figure 3.5.

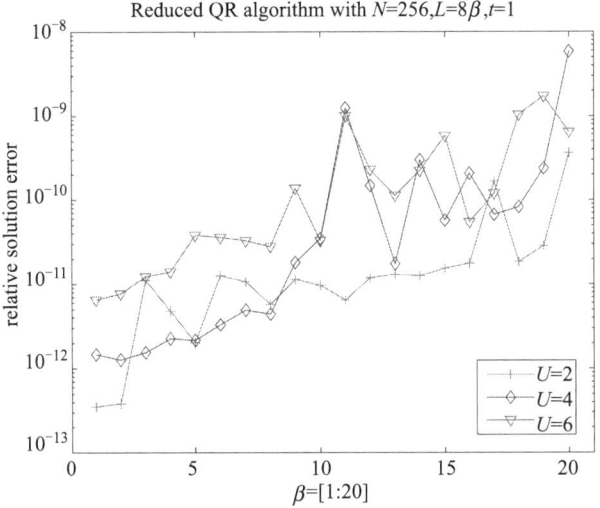

Figure 3.4. Relative errors, $U = 2, 4, 6$.

Table 2. Performance data of BSOF and SABCR methods, $U = 2, 4, 6$.

β	$U = 2$			$U = 4$			$U = 6$		
	L_k	BSOF	SABCR	L_k	BSOF	SABCR	L_k	BSOF	SABCR
1	1	3.08	0.0322	1	3.08	0.0312	1	3.08	0.0312
2	2	6.83	0.293	2	6.83	0.295	2	6.83	0.294
3	2	10.7	0.313	3	10.7	0.595	3	10.7	0.595
4	3	14.6	0.588	4	14.6	1.05	4	14.6	1.05
5	4	18.3	1.06	5	18.3	1.11	5	18.3	1.11
6	4	22.1	1.08	6	22.1	1.50	6	22.1	1.51
7	5	25.9	1.14	7	25.9	1.61	7	25.9	1.61
8	6	29.6	1.54	8	29.6	3.26	8	29.6	3.26
9	6	33.4	1.57	8	33.4	3.27	9	33.4	2.13
10	7	37.2	1.68	9	37.2	2.14	10	37.2	2.61
11	8	41.2	3.30	10	41.2	2.62	11	41.2	2.88
12	8	45.0	3.31	11	45.0	2.90	12	45.0	4.59
13	9	48.8	2.19	12	48.8	4.69	13	48.9	3.33
14	10	52.6	2.71	13	52.6	3.43	14	52.6	3.60
15	10	56.4	2.73	14	56.4	3.64	15	56.4	3.75
16	11	60.5	2.79	15	60.5	3.75	16	60.5	7.40
17	12	64.2	4.21	16	64.2	7.55	17	64.2	4.29
18	12	67.9	4.20	16	67.9	7.61	18	67.9	4.68
19	13	71.8	3.35	17	71.8	4.35	19	71.9	4.81
20	14	75.7	3.72	18	75.7	4.69	20	75.7	7.42

Example 3. In this example, we examine computational efficiency of the SABCR solver with respect to the number of time slices $L = \beta/\Delta\tau$ with $\beta = 1, 2, \cdots, 20$ and $\Delta\tau = \frac{1}{8}, \frac{1}{16}, \frac{1}{32}$. The dimensions of the coef-

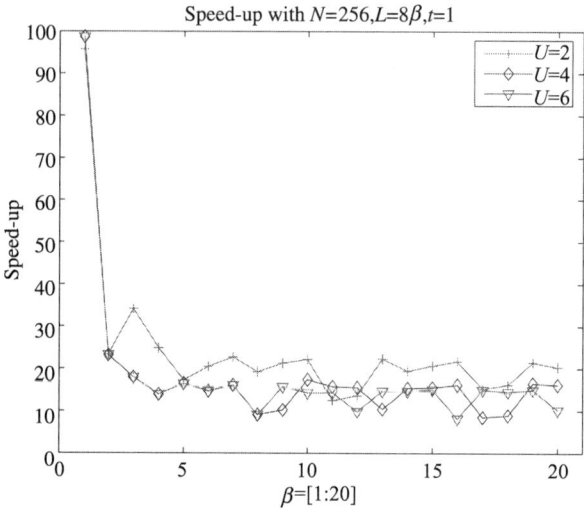

Figure 3.5. Speedups of SABCR for $U = 2, 4, 6$.

ficient matrices M vary from $NL = 2,048$ ($\beta = 1, \Delta\tau = \frac{1}{8}$) to $NL = 163,840$ ($\beta = 20, \Delta\tau = \frac{1}{32}$). The other parameters are $N = 16 \times 16$, $U = 6$ and $t = 1$.

Table 3 shows the reduced block size L_k with respect to β and $\Delta\tau$ ($L = \beta\Delta\tau$), CPU elapsed time in seconds of the BSOF and SABCR methods. Speedups are plotted in Figure 3.6.

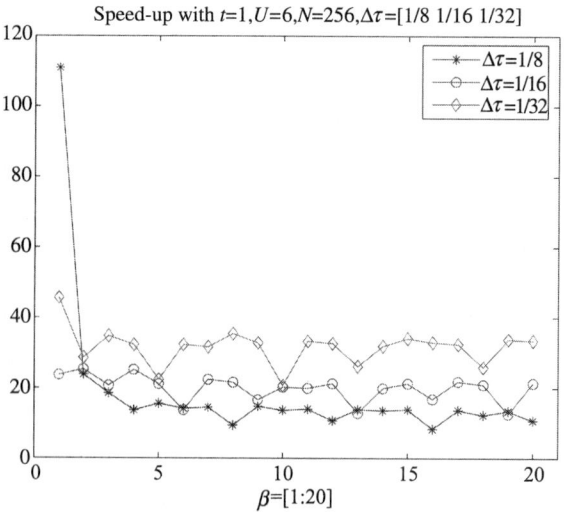

Figure 3.6. Speedups of SABCR for $L = \beta/\Delta\tau$.

Table 3. Performance data of BSOF and SABCR methods, $L = \beta/\Delta\tau$.

$\Delta\tau$	1/8			1/16			1/32		
β	L_k	BSOF	SABCR	L_k	BSOF	SABCR	L_k	BSOF	SABCR
1	1	3.25	0.0293	2	7.25	0.306	2	15.5	0.34
2	2	7.28	0.305	3	15.1	0.596	4	32.9	1.15
3	3	11.2	0.605	4	23.0	1.11	5	47.3	1.36
4	4	15.1	1.10	5	32.0	1.27	7	63.6	1.97
5	5	19.2	1.23	7	39.1	1.85	8	80.3	3.58
6	6	23.0	1.62	8	47.2	3.43	10	97.9	3.03
7	7	27.2	1.87	9	55.4	2.47	11	112	3.54
8	8	32.1	3.38	10	63.4	2.93	13	140	3.95
9	9	35.3	2.38	12	71.1	4.26	14	150	4.57
10	10	39.1	2.86	13	79.3	3.91	16	167	8.06
11	11	43.2	3.08	14	87.6	4.39	17	180	5.40
12	12	47.2	4.39	15	95.7	4.50	19	196	6.00
13	13	51.7	3.71	16	103	8.00	20	209	7.99
14	14	55.3	4.06	18	112	5.61	22	224	7.03
15	15	59.2	4.26	19	120	5.64	23	240	7.05
16	16	63.5	7.54	20	128	7.58	25	258	7.83
17	17	67.3	4.92	21	136	6.23	26	273	8.42
18	18	71.2	5.78	23	144	6.88	28	290	11.2
19	19	75.3	5.58	24	152	12.0	29	305	9.03
20	20	79.3	7.36	15	160	7.49	31	321	9.60

For large energy scale parameters t, β and U, small $\Delta\tau$ is necessary for the accuracy of the Trotter-Suzuki decomposition. Small $\Delta\tau$ implies large $L = \frac{\beta}{\Delta\tau}$. For the SABCR solver, small $\Delta\tau$ implies a large reduction factor k. The SABCR is more efficient for small $\Delta\tau$.

Example 4. Let us examine the memory limit with respect to the increase of the lattice size parameter N. The memory requirement of the BSOF method is $3N^2 L = 3N_x^4 L$. If $N_x = N_y = 32$, the memory storage of one $N \times N$ matrix is 8MB. Therefore for a 1.5GB memory machine, the number L of time slices is limited to $L < 63$. It implies that when $\Delta\tau = \frac{1}{8}$, the inverse temperature β must be smaller than 8. The BSOF method will run out of memory when $\beta \geqslant 8$. On the other hand, the SABCR solver should continue work for $L = 8\beta$ and $\beta = 1, 2, \cdots, 10$. Table 4 shows the memory limitation of the BSOF method, where $t = 1$ and $U = 6$.

4 Preconditioned iterative linear solvers

As discussed in sections 1 and 2, one of the computational kernels of the hybrid quantum Monte Carlo (HQMC) simulation is to repeatedly solve

Table 4. SABCR for large systems, $N = 32 \times 32$.

β	L	k	L_k	BSOF(sec.)	SABCR (sec.)	Speedup ($\times$)
1	8	8	1	148.00	2.10	70
2	16	8	2	322.00	17.8	18
3	24	8	3	509.00	40.1	12.7
4	32	8	4	689.00	64.5	10.6
5	40	8	5	875.00	88.6	9.8
6	48	8	6	1060.00	110.00	9.6
7	56	8	7	1250.00	131.00	9.5
8	64	8	8	out of memory	150.00	
9	72	8	9	out of memory	172.00	
10	80	8	10	out of memory	200.00	

the linear system of equations

$$Ax = b, \tag{4.1}$$

where A is a symmetric positive definite matrix of the form

$$A = M^T M$$

and M is the Hubbard matrix as defined in (2.1).

One can solve the linear system (4.1) by solving the coupled systems $M^T y = b$ for y and $Mx = y$ for x using the SABCR method described in section 3. However, the computational complexity will be $O(N^3 L/k)$, which is prohibitive for large lattice size N. In this section, we consider preconditioned iterative solvers. It is our goal to develop an efficient preconditioned iterative solver that exhibits an optimal linear-scaling, namely, the computational complexity scales linearly with respect to the lattice size N. At the end of this section, we will see that so far, we are only able to achieve the linear-scaling for moderately interacting systems, namely U is small.

4.1 Iterative solvers and preconditioning

We have conducted some numerical study of applying GMRES, QMR and Bi-CGSTAB methods to solve the p-cyclic system

$$Mx = b.$$

We observed that these methods suffer from slow convergence rates or erratic convergence behaviors. Figures 4.1 and 4.2 show the typical convergence behaviors of these methods. The parameters of the matrices M are set as $(N, L, U, t, \beta, \mu) = (8 \times 8, 24, 4, 1, 3, 0)$ and $h = \pm 1$ with equal probability, and the entries of the right-hand-side vector b are random numbers chosen from a uniform distribution on the interval $(0, 1)$.

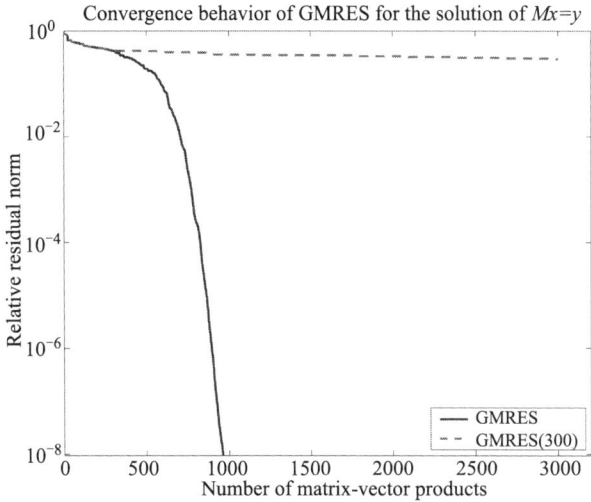

Figure 4.1. Relative residual norms of GMRES and GMRES(300).

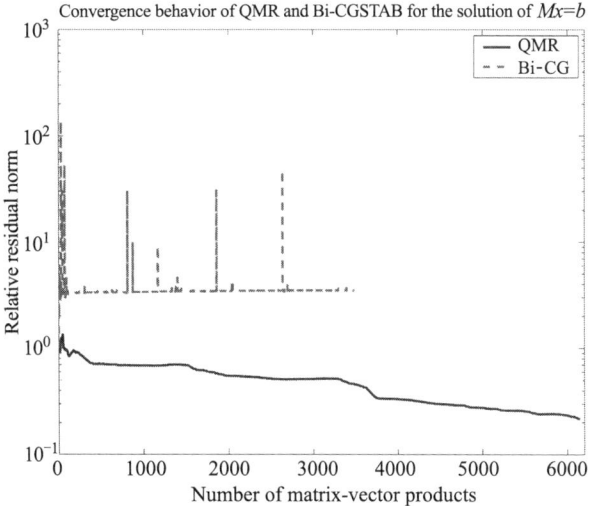

Figure 4.2. Relative residual norms of Bi-CGSTAB and QMR.

Although the convergence of conjugate gradient (CG) method to solve the symmetric positive definite system (4.1) is slow but robust in the sense that residual error decreases steadily as shown in Figure 4.3.

As it is well known, the convergence rate of CG could be improved dramatically by using a proper preconditioner R, which symmetrically

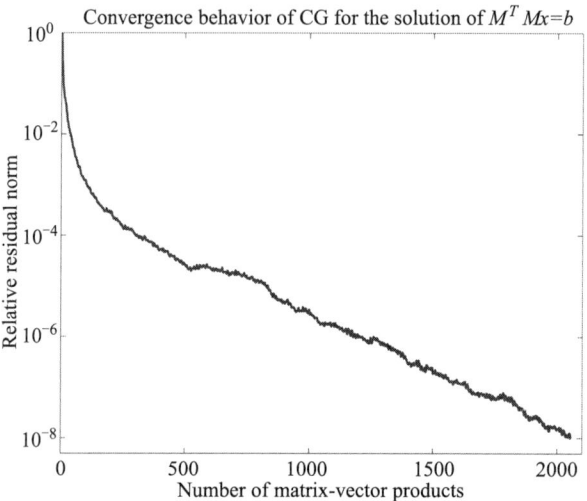

Figure 4.3. Relative residual norms of CG.

preconditions system (4.1):

$$R^{-1}AR^{-T} \cdot R^T x = R^{-1}b. \tag{4.2}$$

An ideal preconditioner R satisfies the following three conditions:

1) The cost of constructing R is cheap.

2) The application of R, i.e., solving $Rz = r$ for z, is not expensive.

3) RR^T is a good approximation of A.

However, in practice, there is a trade-off between costs 1) and 2) and quality 3). In this section, we focus on the development of robust and efficient preconditioning techniques for an optimal balance between costs and quality.

For all numerical results presented in this section, Hubbard matrices M are generated with the Hubbard-Stratonovich configurations $h_{\ell,i}$ such that the average condition numbers of the resulting test matrices are close to the ones arising in a full HQMC simulation. The right-hand-side vector b is set so that entries of the (exact) solution vector x are uniformly distributed on the interval $(0,1)$. The required accuracy of the computed solution $\widehat{x}$ is set as

$$\frac{\|x - \widehat{x}\|_2}{\|x\|_2} \leqslant 10^{-3}.$$

This is sufficient for the HQMC simulation.

All preconditioning algorithms presented in this section are implemented in Fortran 90. The numerical performance data are collected on an HP Itanium2 workstation with 1.5GHz CPU and 2GB of main memory. Intel Math Kernel Library (MKL) 7.2.1 and -O3 optimization option in `ifort` are used to compile the codes.

4.2   Previous work

There are a number of studies on preconditioning techniques to improve the convergence rate of PCG solver for the QMC simulations. One attempt is to precondition the system with $R = M_{(U=0)}$ [34, 49]. By using the Fast Fourier Transform, the computational cost of applying this preconditioner is $O(NL \log(NL))$. However, the quality of the preconditioner is poor for moderately and strongly interacting systems ($U \geqslant 3$), as shown in Figure 4.4. The results are the averages of 50 solutions of the systems $(N, L, t, \beta, \mu) = (8 \times 8, 40, 1, 5, 0)$.

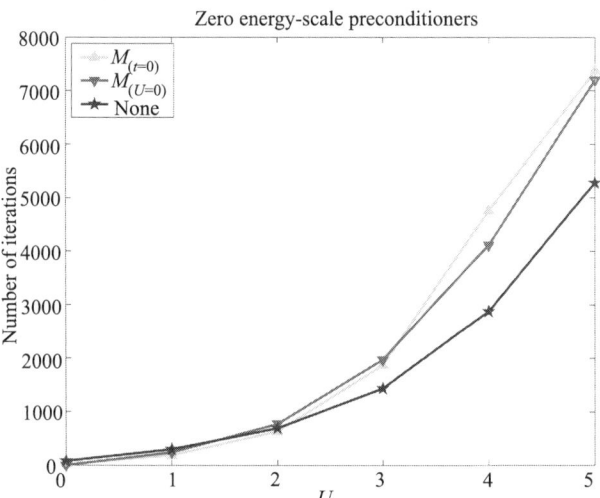

Figure 4.4. Number of PCG iterations using preconditioners $R = M_{(U=0)}$ and $R = M_{(t=0)}$.

It is suggested to use the preconditioner $R = M_{(t=0)}$ [34]. In this case, the submatrices B_ℓ are diagonal. The cost of applying the preconditioner R is $O(NL)$. However, the quality is poor, particularly for strongly interacting systems, as shown in Figure 4.4.

Jacobi preconditioner $R = \mathrm{diag}(a_{ii}^{1/2})$ is used in [42], where a_{ii} are the diagonal elements of A. The PCG convergence rate is improved consistently as shown in Figure 4.5. However, this is still insufficient for the full HQMC simulation. For example, for a small and moderately-interacting system $(N, L, U, t, \beta) = (8 \times 8, 40, 4, 1, 5)$, Jacobi-based PCG solver requires $3,569$ iterations and 1.78 seconds. A full HQMC simulation typically requires to solve $10,000$ linear systems. By this rate, a full HQMC simulation of 8×8, would take 4.9 hours. When N is increased to 32×32, the PCG takes $10,819$ iterations and 87.80 seconds for solving one system. By this rate, a full HQMC simulation of a 32×32 lattice would need more than 20 days.

It is proposed to use an incomplete Cholesky (IC) preconditioner R, where R is lower triangular and has the same block structure of A [49]. Although the PCG convergence rate is improved considerably, it becomes impractical due to the cost of $O(N^2 L)$ in storing and applying the IC preconditioner. Furthermore, it is not robust and suffers breakdown for strongly interacting systems as we will see in section 4.4.

4.3   Cholesky factorization

We begin with a review of the Cholesky factorization of an $n \times n$ symmetric positive definite (SPD) matrix A:

$$A = RR^T, \qquad (4.3)$$

where R is lower-triangular with positive diagonal entries. R is referred to as the Cholesky factor.

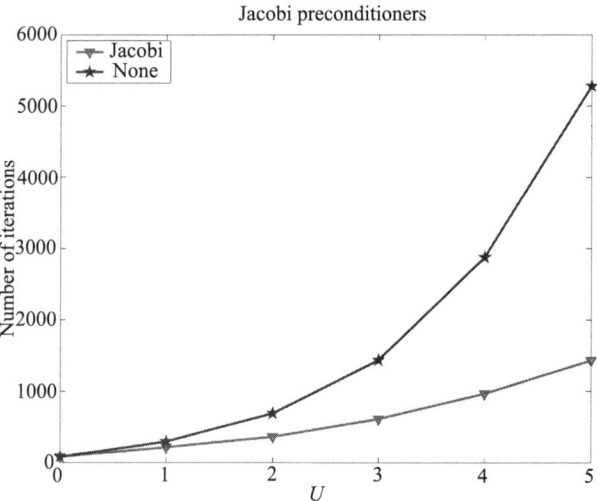

Figure 4.5. Number of PCG iterations using Jacobi preconditioner R.

We follow the presentation in [38]. The Cholesky factorization (4.3) can be computed by using the following partition and factorization:

$$A = \begin{bmatrix} a_{11} & \widehat{a}^T \\ \widehat{a} & A_1 \end{bmatrix} = \begin{bmatrix} r_{11} & 0 \\ \widehat{r} & I \end{bmatrix} \begin{bmatrix} 1 & 0 \\ 0 & A_1 - \widehat{r}\widehat{r}^T \end{bmatrix} \begin{bmatrix} r_{11} & \widehat{r}^T \\ 0 & I \end{bmatrix}. \tag{4.4}$$

By the first columns of the both sides of the factorization, we have

$$a_{11} = r_{11}^2,$$
$$\widehat{a} = \widehat{r}\, r_{11}.$$

Therefore,

$$r_{11} = \sqrt{a_{11}},$$
$$\widehat{r} = \frac{1}{r_{11}}\widehat{a}.$$

If we have the Cholesky factorization of the $(n-1) \times (n-1)$ matrix $A_1 - \widehat{r}\widehat{r}^T$:

$$A_1 - \widehat{r}\widehat{r}^T = R_1 R_1^T, \tag{4.5}$$

then the Cholesky factor R is given by

$$R = \begin{bmatrix} r_{11} & 0 \\ \widehat{r} & R_1 \end{bmatrix}.$$

Hence the Cholesky factorization can be obtained through the repeated applications of (4.4) on (4.5). The resulting algorithm is referred to as a *right-looking* Cholesky algorithm, since after the first column r_1 of R is computed, it is used to update the matrix A_1 to compute the remaining columns of R, which are on the right side of r_1.

There is a left-looking version of the Cholesky algorithm. By comparing the jth column of factorization (4.3), we have

$$a_j = \sum_{k=1}^{j} r_{jk} r_k.$$

This says that

$$r_{jj} r_j = a_j - \sum_{k=1}^{j-1} r_{jk} r_k.$$

Hence, to compute the jth column r_j of R, one first computes

$$v = a_j - \sum_{k=1}^{j-1} r_{jk} r_k,$$

and then

$$r_{ij} = \frac{v_i}{\sqrt{v_j}}, \quad \text{for } i = j, j+1, \cdots, n.$$

It is a *left-looking* algorithm since the jth column r_j of R is computed through referencing the computed columns $r_1, r_2, \cdots, r_{j-1}$ of R, which are on the left of r_j.

The Cholesky factorization could fail due to a *pivot breakdown*, namely, at the jth step, the diagonal element $a_{jj} \leqslant 0$. When A is SPD, the diagonal element a_{11} must be positive. Furthermore, $A_1 - \hat{r}\hat{r}^T$ is SPD since it is a principal submatrix of the SPD matrix $X^T A X$, where

$$X = \begin{bmatrix} 1 & -\hat{r}^T \\ 0 & I \end{bmatrix}.$$

Therefore, there is no pivot breakdown for an SPD matrix.

HQMC application. When the Cholesky factor R of A is used as a preconditioner, the HQMC linear system (4.1) is solved exactly. Table 5 records the CPU time in seconds with respect to different lattice sizes N by using CHOLMOD developed by Timothy A. Davis.[1] CHOLMOD is one of the state-of-art implementations of the sparse Cholesky factorization and is used in MATLAB version 7.2.

Table 5. Performance of CHOLMOD.

N	8×8	16×16	24×24	32×32	40×40	48×48
P-time	0.08	1.99	17.56	90.68	318.04	offmem
S-time	0.00	0.04	0.29	0.52	1.22	offmem
T-time	0.08	2.03	17.85	91.20	319.26	offmem

The other parameters are set as $(L, U, t, \beta, \mu) = (80, 4, 1, 10, 0)$. We note that, when $N = 48 \times 48$, it runs out of memory ("offmem") before completing the Cholesky factorization of $A = M^T M$. In the table, "P-time" stands for the CPU time for computing the preconditioner, "S-time" for the CPU time for PCG iterations, and "T-time" for the total CPU time. With an approximate minimum degree (AMD) recording of the matrix A, the CPU time was reduced slightly as shown in Table 6.

Table 6. Performance of CHOLMOD with AMD recording.

N	8×8	16×16	24×24	32×32	40×40	48×48
P-time	0.11	1.63	12.40	57.85	297.27	offmem
S-time	0.00	0.04	0.13	0.34	0.95	offmem
T-time	0.11	1.67	12.53	58.19	298.22	offmem

---

[1] http://www.cise.ufl.edu/research/sparse/cholmod/

The Cholesky factorization is not affected by the potential energy parameter U. Therefore, the performance of CHOLMOD is expected to be the same for different U. By an interpolation of the performance data of CHOLMOD with AMD reordering, the computational complexity of the Cholesky factorization is $O(N^2)$. By this rate, if we were able to solve the Hubbard system with $NL = 48 \times 48 \times 80 = 184,320$, the CPU elapsed time would take about 700 seconds.

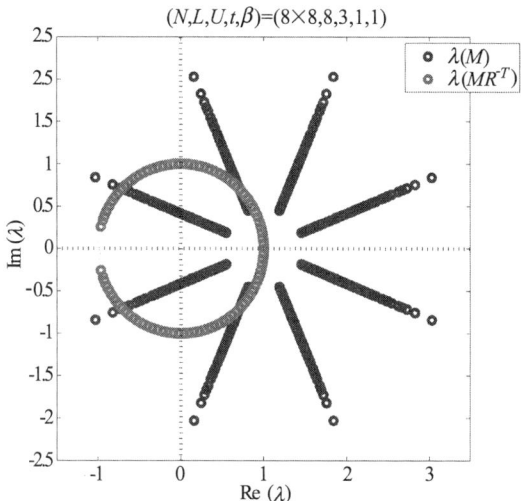

Figure 4.6. Eigenvalues of M and MR^{-1}, where R is a Cholesky factor.

To end this section, we note that with the Cholesky factor R, the preconditioned matrix MR^{-T} becomes orthogonal. The eigenvalues of MR^{-T} are on the unit circle, as shown in Figure 4.6. Therefore, to assess the quality of a preconditioner R, one can examine how close the eigenvalues of the preconditioned matrix MR^{-T} are to the unit circle. Of course, this can be checked only for small matrices.

4.4 Incomplete Cholesky factorizations

4.4.1 IC

To reduce the computational and storage costs of the Cholesky factorization (4.3), a preconditioner R can be constructed based on an incomplete Cholesky (IC) factorization:

$$A = RR^T + S + S^T, \qquad (4.6)$$

where R is lower-triangular and is referred to as an IC factor, and S is a strictly lower-triangular matrix (therefore, the diagonal elements of

A and RR^T are the same). $E = S + S^T$ is the error matrix of the incomplete factorization. The sparsity of R is controlled by a set $\mathcal{Z}$, which is a set of ordered pairs (i, j) of integers from $\{1, 2, \cdots, n\}$ such that if $(i, j) \notin \mathcal{Z}$, then $r_{ij} = 0$.

The IC factor R can be computed based on the following partition and factorization:

$$A = \begin{bmatrix} a_{11} & \widehat{a}^T \\ \widehat{a} & A_1 \end{bmatrix} = \begin{bmatrix} r_{11} & 0 \\ \widehat{r} & I \end{bmatrix} \begin{bmatrix} 1 & 0 \\ 0 & A_1 - \widehat{r}\widehat{r}^T \end{bmatrix} \begin{bmatrix} r_{11} & \widehat{r}^T \\ 0 & I \end{bmatrix} + \begin{bmatrix} 0 & \widehat{s}^T \\ \widehat{s} & 0 \end{bmatrix}. \quad (4.7)$$

By multiplying out the first column of the both sides of the factorization, we have

$$a_{11} = r_{11}^2,$$
$$\widehat{a} = \widehat{r} r_{11} + \widehat{s}.$$

Therefore, if the pivot $a_{11} > 0$, we have

$$r_{11} = \sqrt{a_{11}}.$$

The vector $\widehat{r}$ and $\widehat{s}$ are computed based on the sparsity set $\mathcal{Z}$:

for $i \leqslant n - 1$,
$$\begin{cases} \widehat{r}_i = \widehat{a}_i/r_{11}, \ \widehat{s}_i = 0, & \text{if } (i+1, 1) \in \mathcal{Z}, \\ \widehat{r}_i = 0, & \widehat{s}_i = \widehat{a}_i, \text{ otherwise.} \end{cases} \quad (4.8)$$

Therefore, if we have an IC factorization of the $(n-1) \times (n-1)$ matrix $A_1 - \widehat{r}\widehat{r}^T$:

$$A_1 - \widehat{r}\widehat{r}^T = R_1 R_1^T + S_1 + S_1^T, \quad (4.9)$$

then the IC factorization (4.6) of A is given by

$$R = \begin{bmatrix} r_{11} & 0 \\ \widehat{r} & R_1 \end{bmatrix} \quad \text{and} \quad S = \begin{bmatrix} 0 & 0 \\ \widehat{s} & S_1 \end{bmatrix}.$$

Thus, the IC factorization can be computed by the repeated application of (4.7) on (4.9) as long as $A - \widehat{r}\widehat{r}^T$ is SPD. Note that when non-zero element $\widehat{a}_i$ is discarded, i.e., $\widehat{r}_i = 0$, in (4.8), the operations to update A_1 with respect to the element $\widehat{r}_i$ in (4.9) are skipped.

The following algorithm computes an IC factor R in the right-looking fashion, where in the first line, R is initialized by the lower-triangular part of A, and the update of A_1 is performed directly on R.

IC (RIGHT-LOOKING VERSION)
1. $R = \text{lower}(A)$
2. For $j = 1, 2, \cdots, n$

3. $r(j,j) := \sqrt{r(j,j)}$ if $r(j,j) > 0$
4. for $i = j+1, j+2, \cdots, n$
5.   if $(i,j) \in \mathcal{Z}$
6. $r(i,j) := r(i,j)/r(j,j)$
7.   else
8. $r(i,j) = 0$
9.   end if
10. end for
11. for $k = j+1, j+2, \cdots, n$
12. $r(k:n,k) := r(k:n,k) - r(k,j)r(k:n,j)$
13. end for
14. end for

Note that all computational steps of the previous algorithm and the rest of algorithms presented in this section are performed with regard to the sparsity of matrices and vectors involved.

There is a left-looking algorithm. By comparing the jth column of factorization (4.6), we have

$$a_{jj} = \sum_{k=1}^{j} r_{jk}^2, \tag{4.10}$$

$$a_{ij} = \sum_{k=1}^{j} r_{jk} r_{ik} + s_{ij}, \quad \text{for } i = j+1, \cdots, n. \tag{4.11}$$

This says that

$$r_{jj}^2 = a_{jj} - \sum_{k=1}^{j-1} r_{jk}^2,$$

$$r_{jj} r_{ij} + s_{ij} = a_{ij} - \sum_{k=1}^{j-1} r_{jk} r_{ik}, \quad \text{for } i = j+1, \cdots, n.$$

Thus, to compute the jth column in the IC factorization, one first computes

$$v = a_j - \sum_{k=1}^{j-1} r_{jk} r_k. \tag{4.12}$$

Then, the pivot $v_j > 0$, the jth diagonal entry of R is then given by

$$r_{jj} = \sqrt{v_j},$$

and the rest of non-zero elements of r_j and the discarded elements of s_j

are computed based on the sparsity constraint $\mathcal{Z}$:

$$\text{for } i \geqslant j + 1,$$
$$\begin{cases} r_{ij} = v_i/r_{jj}, \ s_{ij} = 0, \ (i,j) \in \mathcal{Z}, \\ r_{ij} = 0, \qquad s_{ij} = v_i, \text{ otherwise.} \end{cases}$$

A pseudo-code of the left-looking algorithm is as the following:

IC (LEFT-LOOKING VERSION)
1. for $j = 1, 2, \cdots, n$
2. $v(j : n) = a(j : n, j)$
3.   for $k = 1, 2, \cdots, j - 1$
4. $v(j : n) := v(j : n) - r(j, k)r(j : n, k)$
5.   end for
6. $r(j, j) = \sqrt{a(j, j)}$ if $a(j, j) > 0$
7.   for $i = j + 1, j + 2, \cdots, n$
8.   if $(i, j) \in \mathcal{Z}$
9. $r(i, j) = v(i)/r(j, j)$
10.   else
11. $r(i, j) = 0$
12.   end if
13.   end for
14. end for

The following rules are often used to define the sparsity set $\mathcal{Z}$:

1. *Fixed sparsity pattern (FSP)*: the IC factor R has a prescribed sparsity pattern $\mathcal{Z}$. A popular sparsity pattern is that of the original matrix A, i.e., $\mathcal{Z} = \{(i, j) : i \geqslant j \text{ and } a_{ij} \neq 0\}$.

2. *Drop small elements (DSE)*: the small elements of R are dropped. It is controlled by a drop threshold σ. In this case, the sparsity pattern $\mathcal{Z}$ of R and the number of fill-ins in R are unknown in advance.

3. *Fixed number of non-zero elements per column*. It is similar to the DSE rule except that the maximum number of non-zero elements in each column of R is fixed.

The existence of IC factorization (4.6), for an arbitrary sparsity set $\mathcal{Z}$, is proven only for special classes of matrices [43, 44, 53]. For a general SPD matrix A, the non-zero elements introduced into the error matrix E could result in the loss of positive-definiteness of the matrix $A - E$, and the IC factorization does not exist.

HQMC application. Table 7 shows numerical results of the IC pre-conditioner R with sparsity $\mathcal{Z}$ defined by the DSE rule. The reported data is an average of successful solutions over 10 trials with the left-looking IC implementation. The table records the CPU time with different drop tolerance values σ. The first number in each cell is the time for constructing the preconditioner R and the second number is for the PCG iteration. In the table, "pbd" stands for the pivot breakdown to indicate that all of 10 trials failed due to the pivot breakdowns.

Table 7. IC with DSE, $(N, L, t, \beta) = (16 \times 16, 80, 1, 10)$.

U	0	2	4	6
$\sigma = 10^{-6}$	18.97/0.07	19.17/0.14	19.09/0.30	19.07/0.48
10^{-5}	13.20/0.11	16.92/0.20	16.41/0.80	pbd
10^{-4}	2.95/0.17	5.72/0.58	pbd	pbd
10^{-3}	pbd	pbd	pbd	pbd
10^{-2}	0.01/0.16	pbd	pbd	pbd
10^{-1}	pbd	pbd	pbd	pbd

We see that the IC factorization encounters the pivot breakdown frequently, except with an extremely small drop tolerance σ. Small drop tolerance leads to high memory and CPU time costs, and becomes impractical for large systems.

Note that the right-looking IC algorithm is mathematically equivalent to its left-looking algorithm, it also encounters the same pivot breakdown. It clearly indicates that the IC preconditioner R is neither robust nor efficient for the HQMC simulation.

4.4.2 Modified IC

To avoid the pivot breakdown, one attempt is to first make a small perturbation of A, say by simple diagonal perturbation, and then compute its IC factorization:

$$A + \alpha D_A = RR^T + S + S^T, \tag{4.13}$$

where $D_A = \text{diag}(A)$ and the scalar α is prescribed [43]. R is referred to as an IC_p preconditioner.

If the shift α is chosen such that $A + \alpha D_A$ is diagonally dominant, then it is provable that the IC factorization (4.13) exists [43]. Table 8 records the performance of the left-looking IC_p implementation, where the shift α is chosen such that $A + \alpha D_A$ is diagonally dominant. In the table, "$\text{nnz}_r(R)$" is the average number of nonzero elements per row of R, "Mem(R)" is the memory in MB for storing R in the CSC (Compressed Sparse Column) format, "Wksp." is the required workspace in MB, and "Itrs." is the total number of PCG iterations.

Table 8. IC_p/diag.dom., $\sigma = 0.001$, $(N, L, t, \beta) = (48 \times 48, 80, 1, 10)$.

U	0	1	2	3	4	5	6
α	1.29	4.15	7.58	12.37	19.70	26.14	38.80
$nnz_r(R)$	25.29	23.35	21.99	20.90	20.01	19.30	25.01
Mem(R)	57	52	49	47	45	43	56
Wksp.	22	20	19	18	18	17	20
Itrs.	94	357	882	2620	16645	68938	102500
P-time	1.40	1.22	1.13	1.06	1.01	0.97	0.93
S-time	5.03	18.21	43.48	125.58	782.96	3178.58	4621.06
T-time	6.44	19.43	44.60	126.64	783.96	3179.54	4621.98

By Table 8, we see that with the choice of the shift α such that $A + \alpha D_A$ is diagonally dominant, the pivot breakdown is avoided. However, the quality of the resulting preconditioner R is poor. In practice, we observed that good performance can often be achieved with a much smaller shift α. Although $A + \alpha D_A$ is not diagonally dominant, the IC factorization still exists. Table 9 records significant improvements of the IC_p preconditioner R computed with the fixed shift $\alpha = 0.005$.

Table 9. IC_p, $\sigma = 0.001$, $(N, L, t, \beta) = (48 \times 48, 80, 1, 10)$.

U	0	1	2	3	4	5	6
α (fixed)	0.005	0.005	0.005	0.005	0.005	0.005	0.005
$nnz_r(R)$	22.31	24.43	24.74	24.89	24.96	24.99	25.01
Mem(R)	50	55	55	56	56	56	56
Wksp.	18	19	20	20	20	20	20
Itrs.	14	32	72	190	1087	3795	5400
P-time	1.23	1.29	1.30	1.31	1.31	1.31	1.32
S-time	0.65	1.57	3.47	9.17	52.49	183.15	286.15
T-time	1.87	2.86	4.77	10.48	53.49	184.76	287.47

There is no general strategy for an optimal choice of the shift α. It is computed by a trial-and-error approach in PETSc [47].

4.5   Robust incomplete Cholesky preconditioners

In the IC factorizations (4.6) and (4.13), the discarded elements of R are simply moved to the error matrix E. As we have seen, this may result in the loss of the positive definiteness of the matrix $A - E$ and subsequently lead to the pivot breakdown. To avoid this, the error matrix E needs to be taken into account. It should be updated dynamically during the construction of an IC factorization such that the matrix $A - E$ is preserved to be SPD. Specifically, we want to have an IC factorization algorithm that computes a nonsingular lower triangular matrix R of an arbitrarily prescribed sparsity pattern $\mathcal{Z}$ satisfying

$$\begin{cases} A = RR^T + E, \\ \text{s.t. } A - E > 0. \end{cases} \tag{4.14}$$

In the rest of this section, we will discuss several approaches to construct such an IC factor R satisfying (4.14). The resulting preconditioner R is referred to as a *robust incomplete Cholesky (RIC)* preconditioner.

4.5.1 RIC1

A sufficient condition for the existence of factorization (4.14) is to ensure that the error matrix $-E$ is symmetric semi-positive definite, $-E = -E^T \geqslant 0$. For doing so, let us write

$$E = S - D + S^T,$$

where S is strictly lower-triangular and D is diagonal. Then a robust IC preconditioner should satisfy

$$\begin{cases} A = RR^T + S - D + S^T, \\ \text{s.t. } -(S - D + S^T) \geqslant 0. \end{cases} \tag{4.15}$$

The factorization is referred to as version 1 of RIC, or RIC1 in short. RIC1 was first studied in [31,39].

Note that in the IC factorization (4.6), $D = 0$ and $E = S + S^T$. In the modified IC factorization (4.13), the diagonal matrix D is prescribed $D = -\alpha D_A$ and the error matrix $E = S - \alpha D_A + S^T$. Now, in the RIC1 factorization, D will be dynamically assigned and updated during the process to satisfy the condition $-(S - D + S^T) \geqslant 0$.

The RIC1 factorization can be computed by using the following partition and factorization:

$$\begin{bmatrix} a_{11} & \widehat{a}^T \\ \widehat{a} & A_1 \end{bmatrix} = \begin{bmatrix} r_{11} & 0 \\ \widehat{r} & I \end{bmatrix} \begin{bmatrix} 1 & 0 \\ 0 & C_1 \end{bmatrix} \begin{bmatrix} r_{11} & \widehat{r}^T \\ 0 & I \end{bmatrix} + \begin{bmatrix} -d_1 & \widehat{s}^T \\ \widehat{s} & -D_1 \end{bmatrix}, \tag{4.16}$$

where $C_1 = A_1 + D_1 - \widehat{r}\widehat{r}^T$.

By the first column of the both sides of the factorization, we have

$$a_{11} = r_{11}^2 - d_1,$$
$$\widehat{a} = \widehat{r}\,r_{11} + \widehat{s}.$$

It suggests that we first compute the vector $v = \widehat{r}\,r_{11} + \widehat{s}$ based on the sparsity set $\mathcal{Z}$:

for $i \leqslant n - 1$,

$$\begin{cases} v_i = \widehat{a}_i, \ \widehat{s}_i = 0, \ \text{if } (i+1,1) \in \mathcal{Z}, \\ v_i = 0, \ \widehat{s}_i = \widehat{a}_i, \ \text{otherwise.} \end{cases} \tag{4.17}$$

To ensure $-E = -(S - D + S^T) \geqslant 0$, if there is a discarded element $\widehat{a}_i$ assigned to $\widehat{s}_i$, the diagonal element d_1 and the ith diagonal element $d_i^{(1)}$ of D_1 are updated:

$$d_1 := d_1 + \delta_1, \quad d_i^{(1)} := d_i^{(1)} + \delta_i, \tag{4.18}$$

where δ_1 and δ_i are chosen such that $\delta_1, \delta_i > 0$ and $\delta_1 \delta_i = \widehat{s}_i^2$. Subsequently, the element r_{11} is determined by

$$r_{11} = \sqrt{a_{11} + d_1},$$

and the vector $\widehat{r}$ is set by

$$\widehat{r} = \frac{1}{r_{11}} v.$$

If we have an RIC1 factorization of the $(n-1) \times (n-1)$ matrix C_1:

$$C_1 = R_1 R_1^T + S_1 - \widehat{D}_1 + S_1^T, \tag{4.19}$$

then the RIC1 factorization (4.15) of A is given by

$$R = \begin{bmatrix} r_{11} & 0 \\ \widehat{r} & R_1 \end{bmatrix}, \quad D = \begin{bmatrix} d_1 & 0 \\ 0 & D_1 + \widehat{D}_1 \end{bmatrix}, \quad S = \begin{bmatrix} 0 & 0 \\ \widehat{s} & S_1 \end{bmatrix}.$$

In other words, the RIC1 factorization can be obtained through the repeated applications of (4.16) on (4.19).

The following algorithm computes the RIC1 factor R in the right-looking fashion, where the sparsity $\mathcal{Z}$ is controlled by a prescribed drop tolerance σ. In the algorithm, δ_i and δ_j are chosen so that they result in a same factor of increase in the corresponding diagonal elements.

RIC1 (RIGHT-LOOKING VERSION)

1. $R = \text{lower}(A)$
2. $d(1:n) = 0$
3. for $j = 1, 2, \cdots, n$
4.   for $i = j+1, j+2, \cdots, n$
5. $\tau = |r(i,j)|/[(r(i,i) + d(i))(r(j,j) + d(j))]^{1/2}$
6.   if $\tau \leqslant \sigma$
7. $r(i,j) = 0$
8. $d(i) := d(i) + \tau(r(i,i) + d(i))$
9. $d(j) := d(j) + \tau(r(j,j) + d(j))$
10.   end if
11.   end for
12. $r(j,j) := \sqrt{r(j,j) + d(j)}$
13. $r(j+1:n,j) := r(j+1:n,j)/r(j,j)$
14.   for $k = j+1, j+2, \cdots, n$

15. $r(k:n,k) := r(k:n,k) - r(k,j)r(k:n,j)$
16.   end for
17. end for

The RIC1 factorization can also be computed by a left-looking algorithm. By comparing the jth column of factorization (4.15), we have

$$a_{jj} = \sum_{k=1}^{j} r_{jk}^2 - d_j,$$

$$a_{ij} = \sum_{k=1}^{j} r_{jk} r_{ik} + s_{ij}, \quad i = j+1, \cdots, n.$$

This says that

$$r_{jj}^2 - d_j = a_{jj} - \sum_{k=1}^{j-1} r_{jk}^2,$$

$$r_{jj} r_{ij} + s_{ij} = a_{ij} - \sum_{k=1}^{j-1} r_{jk} r_{ik}, \quad i = j+1, \cdots, n.$$

Thus, to compute the jth column of R, one first computes

$$v = a_j - \sum_{k=1}^{j-1} r_{jk} \, r_k,$$

and then imposes the sparsity:

for $i \geqslant j + 1$,
$$\begin{cases} s_{ij} = 0, & \text{if } (i,j) \in \mathcal{Z}, \\ s_{ij} = v_i, \, v_i = 0, \, \text{otherwise.} \end{cases} \tag{4.20}$$

To ensure $-E = -(S - D + S^T) \geqslant 0$, if there is a discarded element assigned to s_{ij}, the corresponding diagonal elements d_i and d_j are updated

$$d_i := d_i + \delta_i, \, d_j := d_j + \delta_j, \tag{4.21}$$

where δ_i and δ_j are chosen such that $\delta_i, \delta_j > 0$ and $\delta_i \delta_j = s_{ij}^2$. Initially, all d_i are set to be zero.

Subsequently, the jth column r_j of R is given by

$$r_{jj} = \sqrt{a_{jj} + d_j},$$
$$r_{ij} = v_i / r_{jj}, \quad i = j+1, \cdots, n.$$

The following algorithm computes the RIC1 factor R in the left-looking fashion, where the sparsity is controlled by a drop tolerance σ.

RIC1 (LEFT-LOOKING VERSION)
1. $d(1 : n) = 0$
2. for $j = 1, 2, \cdots, n$
3. $v(j : n) = a(j : n, j)$
4.   for $k = 1, 2, \cdots, j - 1$
5. $v(j : n) := v(j : n) - r(j, k)r(j : n, k)$
6.   end for
7.   for $i = j + 1, j + 2, \cdots, n$
8. $\tau = |v(i)|/[(a(i, i) + d(i))(a(j, j) + d(j))]^{1/2}$
9.   if $\tau \leqslant \sigma$
10. $v(i) = 0$
11. $d(i) := d(i) + \tau(a(i, i) + d(i))$
12. $d(j) := d(j) + \tau(a(j, j) + d(j))$
13.   end if
14.   end for
15. $r(j, j) = \sqrt{a(j, j) + d(j)}$
16. $r(j + 1 : n, j) = v(j + 1 : n)/r(j, j)$
17. end for

The computational cost of RIC1 is only slightly higher than the IC preconditioner (4.6). To assess the quality of the RIC1 preconditioner R, we note that the norm of the residue

$$R^{-1}AR^{-T} - I = R^{-1}(S - D + S^T)R^{-T} = R^{-1}ER^{-T} \qquad (4.22)$$

could be amplified by a factor of $\|R^{-1}\|^2$ of the error matrix E. When a large number of diagonal updates are necessary, some elements of D could be large. Consequently, the residue norm is large and R is a poor preconditioner.

HQMC application. Table 10 records the performance of the RIC1 preconditioner computed by the left-looking implementation. The drop threshold is set to be $\sigma = 0.003$. With this drop threshold, the resulting RIC1 preconditioners R is are about the same sparsity as the IC_p preconditioners reported in Table 9 of section 4.4.

Table 10. RIC1/left-looking, $\sigma = 0.003$, $(N, L, t, \beta) = (48 \times 48, 80, 1, 10)$.

U	0	1	2	3	4	5	6
$nnz_r(R)$	22.01	24.28	24.45	24.43	24.33	24.21	24.07
Mem(R)	49	54	55	55	55	54	54
Wksp.	18	19	20	19	19	19	19
Itrs.	20	57	127	342	1990	7530	11460
P-time	1.83	1.93	1.93	1.93	1.92	1.90	1.89
S-time	1.00	3.02	6.69	17.95	104.12	393.60	596.00
T-time	2.82	4.96	8.62	19.88	106.03	395.51	597.90

We note that the quality of the RIC1 preconditioner in terms of the number of PCG iterations is worse than the IC_p preconditioner. This is due to the fact that it is necessary to have a large diagonal matrix D to guarantee the semi-positive definiteness of the error matrix $-E = -(S - D + S^T)$. On the other hand, the RIC1 factorization is provably robust and does not breakdown. In the following sections, we will discuss how to improve the quality of the RIC1 preconditioner.

The right-looking implementation requires the updating of the unfactorized block, i.e., forming the matrix C_1 in (4.16). It causes significant computational overhead. It is less efficient than the left-looking implementation. Therefore, we only present the performance data for the left-looking implementation.

4.5.2   RIC2

One way to improve the quality of the RIC1 preconditioner is by setting the error matrix E as $E = RF^T + FR^T$, where F is strictly lower-triangular. This was proposed in [51]. In this scheme, we compute an IC factorization of the form

$$A = RR^T + RF^T + FR^T. \qquad (4.23)$$

Note that the factorization can be equivalently written as

$$A + FF^T = (R + F)(R + F)^T.$$

Hence, the existence of R is guaranteed. With factorization (4.23), the residue becomes

$$R^{-1}AR^{-T} - I = FR^{-T} + R^{-1}F^T.$$

The residue norm could be amplified at most by the factor of $\|R^{-1}\|$ of the error matrix F, instead of $\|R^{-1}\|^2$ in the RIC1 factorization. We refer (4.23) as version 2 of RIC factorization, or RIC2 in short.

The RIC2 factorization (4.23) can be constructed by using the following partition and factorization:

$$A = \begin{bmatrix} a_{11} & \widehat{a}^T \\ \widehat{a} & A_1 \end{bmatrix} = \begin{bmatrix} r_{11} & 0 \\ \widehat{r} & I \end{bmatrix} \begin{bmatrix} 1 & 0 \\ 0 & C_1 \end{bmatrix} \begin{bmatrix} r_{11} & \widehat{r}^T \\ 0 & I \end{bmatrix} +$$

$$\begin{bmatrix} r_{11} & 0 \\ \widehat{r} & 0 \end{bmatrix} \begin{bmatrix} 0 & \widehat{f}^T \\ 0 & 0 \end{bmatrix} + \begin{bmatrix} 0 & 0 \\ \widehat{f} & 0 \end{bmatrix} \begin{bmatrix} r_{11} & \widehat{r}^T \\ 0 & 0 \end{bmatrix}, \qquad (4.24)$$

where $C_1 = A_1 - \widehat{r}\widehat{r}^T - \widehat{r}\widehat{f}^T - \widehat{f}\widehat{r}^T$.

By the first column of the both sides of the factorization, we have

$$a_{11} = r_{11}^2,$$
$$\widehat{a} = \widehat{r}\, r_{11} + \widehat{f}\, r_{11}.$$

Hence, we have

$$r_{11} = \sqrt{a_{11}}.$$

The vectors $\widehat{r}$ and $\widehat{f}$ are computed based on the sparsity set $\mathcal{Z}$:

for $i \leqslant n - 1$,

$$\begin{cases} \widehat{r}_i = \widehat{a}_i/r_{11}, \, \widehat{f}_i = 0, & \text{if } (i, 1) \in \mathcal{Z}, \\ \widehat{f}_i = 0, & \widehat{f}_i = \widehat{a}_i/r_{11}, \text{ otherwise.} \end{cases}$$

If an RIC2 factorization of the $(n - 1) \times (n - 1)$ matrix C_1 is given by

$$C_1 = R_1 R_1^T + R_1 F_1^T + F_1 R_1^T, \tag{4.25}$$

then the RIC2 factorization of A is given by

$$R = \begin{bmatrix} r_{11} & 0 \\ \widehat{r} & R_1 \end{bmatrix}, \quad F = \begin{bmatrix} 0 & 0 \\ \widehat{f} & F_1 \end{bmatrix}.$$

Hence the RIC2 factorization can be computed by the repeated applications of (4.24) on (4.25).

The following algorithm computes the RIC2 factor R in the right-looking implementation. The sparsity $\mathcal{Z}$ is controlled by dropping small elements of R with the drop tolerance σ.

         RIC2 (RIGHT-LOOKING VERSION)
1. $R = \text{lower}(A)$
2. for $j = 1, 2, \cdots, n$
3. $r(j, j) := \sqrt{r(j, j)}$
4.   for $i = j + 1, j + 2, \cdots, n$
5.   if $|a(i, j)|/r(j, j) > \sigma$
6. $r(i, j) := r(i, j)/r(j, j)$
7. $f(i, j) = 0$
8.   else
9. $r(i, j) = 0$
10. $f(i, j) = r(i, j)/r(j, j)$
11.   end if
12.   end for
13.   for $k = j + 1, j + 2, \cdots, n$
14. $r(k : n, k) := r(k : n, k) - r(k, j)\, r(k : n, j)$
15. $r(k : n, k) := r(k : n, k) - r(k, j)\, f(k : n, j)$

16. $r(k : n, k) := r(k : n, k) - f(k, j)\, r(k : n, j)$
17.   end for
18. end for

The RIC2 factorization can also be computed by a left-looking algorithm. By comparing the jth column in factorization (4.23), we have

$$a_j = \sum_{k=1}^{j} (r_{jk} r_k + r_{jk} f_k + f_{jk} r_k). \tag{4.26}$$

This says that

$$r_{jj}(r_j + f_j) = a_j - \sum_{k=1}^{j-1} (r_{jk} r_k + r_{jk} f_k + f_{jk} r_k).$$

Thus, to compute the jth column of R, one first computes

$$v = a_j - \sum_{k=1}^{j-1} (r_{jk} r_k + r_{jk} f_k + f_{jk} r_k). \tag{4.27}$$

Then, the jth diagonal entry of R is given by

$$r_{jj} = \sqrt{v_j},$$

and the rest of the non-zero elements in r_j and f_j are computed based on the sparsity set $\mathcal{Z}$:

for $i \geqslant j + 1$,
$$\begin{cases} r_{ij} = v_i/r_{jj}, \ f_{ij} = 0, & \text{if } (i, j) \in \mathcal{Z}, \\ r_{ij} = 0, & f_{ij} = v(i)/r_{ii}, \text{ otherwise.} \end{cases}$$

The following algorithm computes the RIC2 factor R in the left-looking fashion, where the sparsity of R is controlled by a drop tolerance σ.

RIC2 (LEFT-LOOKING VERSION)
1. for $j = 1, 2, \cdots, n$
2. $v(j : n) = a(j : n, j)$
3.   for $k = 1, 2, \cdots, j - 1$
4. $v(j : n) := v(j : n) - r(j, k)r(j : n, k)$
5. $v(j : n) := v(j : n) - r(j, k)f(j : n, k)$
6. $v(j : n) := v(j : n) - f(j, k)r(j : n, k)$
7.   end for
8. $r(j, j) = \sqrt{v(j)}$
9.   for $i = j + 1, j + 2, \cdots, n$

10.   if $|v(i)|/r(j,j) > \sigma$

11. $r(i,j) = v(i)/r(j,j)$

12. $f(i,j) = 0$

13.   else

14. $r(i,j) = 0$

15. $f(i,j) = v(i)/r(j,j)$

16.   end if

17.   end for

18. end for

Note that in the above algorithm, the columns $f_1, f_2, \cdots, f_{j-1}$ of the matrix F are required to compute r_j.

HQMC application. Since the left-looking implementation of RIC2 needs to store the entire error matrix F, it requires a large amount of workspace. For example, Table 11 shows that the workspace is about 128MB. It runs out of core memory for $N = 48 \times 48$ and $L = 80$. The right-looking implementation reduces the workspace by a factor of more than 10 but with a significant increase of the CPU time as shown in Table 12. Note that the left-looking and right-looking implementations of RIC2 produce the same preconditioner R. Therefore, the storage requirement for R and the number of the PCG iterations are the same.

Table 11. RIC2/left, $\sigma = 0.012$, $(N, L, t, \beta) = (16 \times 16, 80, 1, 10)$.

U	0	1	2	3	4	5	6
Mem(R)	5	5	5	5	5	5	5
Wksp.	129	129	129	129	127	127	127
Iters.	16	36	73	194	344	453	539
P-time	1.79	1.86	1.88	1.90	1.90	1.90	1.88
S-time	0.12	0.27	0.57	1.52	2.70	3.57	4.24
T-time	1.91	2.13	2.45	3.42	4.60	5.47	6.11

Table 12. RIC2/right, $\sigma = 0.012$, $(N, L, t, \beta) = (16 \times 16, 80, 1, 10)$.

U	0	1	2	3	4	5	6
Mem(R)	5	5	5	5	5	5	5
Wksp.	11	11	11	11	11	11	10
Iters.	16	36	73	194	344	453	539
P-time	19.29	19.29	19.31	19.31	19.34	19.28	19.19
S-time	0.12	0.27	0.57	1.52	2.70	3.57	4.24
T-time	19.43	19.58	19.90	20.84	22.06	22.87	23.45

4.5.3   RIC3

One way to reduce the large workspace requirement of the RIC2 factorization (4.23) is to impose additional sparsity of the error matrix F with a secondary drop threshold. This was proposed in [40]. It begins with setting the error matrix E as

$$E = RF^T + FR^T + S - D + S^T,$$

where S is a strictly lower-triangular and represents the discarded elements from F, and D is diagonal. It means that we compute a preconditioner R satisfying

$$\begin{cases} A = RR^T + RF^T + FR^T + S - D + S^T, \\ \text{s.t. } -(S - D + S^T) > 0. \end{cases} \tag{4.28}$$

The sparsity of R and F are controlled by the primary and secondary drop thresholds σ_1 and σ_2, respectively. Similar to the RIC1 factorization (4.15), the diagonal elements D is dynamically updated such that $-(S - D + S^T) \geqslant 0$. As a result, the robustness of the factorization is guaranteed. We called this as version 3 of the RIC factorization, or RIC3 in short.

With factorization (4.28), the residue becomes

$$R^{-1}AR^{-T} - I = FR^{-T} + R^{-1}F^T + R^{-1}(S - D - S^T)R^{-T}.$$

Therefore, we see that the residue norm is amplified by a factor of $\|R^{-1}\|$ on the primary error $\|F\| = O(\sigma_1)$, and a factor of $\|R^{-1}\|^2$ on the secondary error $\|S - D - S^T\| = O(\sigma_2)$. The RIC3 factorization will be able to preserve at least the same quality of the RIC2 preconditioner R as long as σ_2 is small enough.

The RIC3 factorization (4.28) can be constructed by using the following partition and factorization:

$$A = \begin{bmatrix} a_{11} & \hat{a}^T \\ \hat{a} & A_1 \end{bmatrix} = \begin{bmatrix} r_{11} & 0 \\ \hat{r} & I \end{bmatrix} \begin{bmatrix} 1 & 0 \\ 0 & C_1 \end{bmatrix} \begin{bmatrix} r_{11} & \hat{r}^T \\ 0 & I \end{bmatrix} + \begin{bmatrix} r_{11} & 0 \\ \hat{r} & 0 \end{bmatrix} \begin{bmatrix} 0 & \hat{f}^T \\ 0 & 0 \end{bmatrix}$$

$$+ \begin{bmatrix} 0 & 0 \\ \hat{f} & 0 \end{bmatrix} \begin{bmatrix} r_{11} & \hat{r}^T \\ 0 & 0 \end{bmatrix} + \begin{bmatrix} -d_1 & \hat{s}^T \\ \hat{s} & -D_1 \end{bmatrix}, \tag{4.29}$$

where $C_1 = A_1 + D_1 - \hat{r}\hat{r}^T - \hat{r}\hat{f} - \hat{f}\hat{r}^T$.

By the first column of the both sides of the factorization, we have

$$a_{11} = r_{11}^2 - d_1,$$
$$\hat{a} = \hat{r}r_{11} + \hat{f}r_{11} + \hat{s}.$$

It suggests that we first compute the vector $v = (\widehat{r} + \widehat{f})r_{11}$ by imposing the sparsity constraint with the secondary drop tolerance σ_2, i.e.,

$$\text{for } i \leqslant n - 1,$$

$$\begin{cases} v_i = \widehat{a}_i, \ \widehat{s}_i = 0, & \text{if } \tau > \sigma_2, \\ v_i = 0, \ \widehat{s}_i = \widehat{a}_i, & \text{otherwise,} \end{cases}$$

where

$$\tau = \left[\frac{\widehat{a}_i^2}{(a_{11} + d_1)(a_{ii}^{(1)} + d_i^{(1)})} \right]^{1/2},$$

and $a_{ii}^{(1)}$ and $d_i^{(1)}$ denote the ith diagonal elements of A_1 and D_1, respectively.

To ensure $-(S - D + S^T) \geqslant 0$, if a discarded element $\widehat{a}_i$ is assigned to the position $\widehat{s}_i$, the corresponding diagonal element d_1 and $d_i^{(1)}$ are updated,

$$d_1 := d_1 + \delta_1, \quad d_i^{(1)} := d_i^{(1)} + \delta_i,$$

where δ_1 and δ_i are chosen such that $\delta_1, \delta_i > 0$ and $\delta_1 \delta_i = \widehat{s}_i^2$. Initially, all d_i are set to be zero.

Subsequently, the entry r_{11} of R is given by

$$r_{11} = \sqrt{a_{11} + d_1}.$$

Finally, vectors $\widehat{r}$ and $\widehat{f}$ are computed by imposing the primary sparsity constraint on v with the drop threshold σ_1, i.e.,

$$\text{for } i < n - 1,$$

$$\begin{cases} \widehat{r}_i = v_i/r_{11}, \ \widehat{f}_i = 0, & \text{if } |v_i|/r_{11} > \sigma_1, \\ \widehat{r}_i = 0, \qquad \widehat{f}_i = v_i/r_{11}, & \text{otherwise.} \end{cases}$$

Note that the vectors $\widehat{r}$ and $\widehat{f}$ are structurally "orthogonal", i.e., $\widehat{r}_i \widehat{f}_i = 0$ for all i.

If we have an RIC3 factorization of the $(n - 1) \times (n - 1)$ matrix C_1,

$$C_1 = R_1 R_1^T + R_1 F_1^T + F_1 R_1^T + S_1 - \widehat{D}_1 + S_1^T, \tag{4.30}$$

then the RIC3 factorization is given by

$$R = \begin{bmatrix} r_{11} & 0 \\ \widehat{r} & R_1 \end{bmatrix}, \quad F = \begin{bmatrix} 0 & 0 \\ \widehat{f} & F_1 \end{bmatrix}, \quad S = \begin{bmatrix} 0 & 0 \\ \widehat{s} & S_1 \end{bmatrix}, \quad D = \begin{bmatrix} d_1 & 0 \\ 0 & D_1 + \widehat{D}_1 \end{bmatrix}.$$

Thus, the RIC3 factorization can be computed by the repeated application of (4.29) on (4.30).

The following algorithm computes the RIC3 factor R in the right-looking fashion, where δ_i and δ_j are chosen in the same way as in the RIC1 algorithms for the same increasing of the diagonal elements of D.

RIC3 (RIGHT-LOOKING VERSION)
1. $R = \text{lower}(A)$
2. $d(1:n) = 0$
3. for $j = 1, 2, \cdots, n$
4.   for $i = j + 1, j + 2, \cdots, n$
5. $\tau = |r(i,j)|/[(a(i,i) + d_i)(a(j,j) + d(j))]^{1/2}$
6.   if $\tau \leqslant \sigma_2$
7. $r(i,j) = 0$
8. $d(i) := d(i) + \tau(a(i,i) + d(i))$
9. $d(j) := d(j) + \tau(a(j,j) + d(j))$
10.   end if
11.   end for
12. $r(j,j) = \sqrt{a(j,j) + d(j)}$
13.   for $i = j + 1, j + 2, \cdots, n$
14.   if $|r(i,j)|/r(j,j) > \sigma_1$
15. $r(i,j) := r(i,j)/r(j,j)$
16. $f(i,j) = 0$
17.   else
18. $r(i,j) = 0$
19. $f(i,j) = r(i,j)/r(j,j)$
20.   end if
21.   end for
22.   for $k = j + 1, j + 2, \cdots, n$
23. $r(k:n,k) := r(k:n,k) - r(k,j)r(k:n,j)$
24. $r(k:n,k) := r(k:n,k) - r(k,j)f(k:n,j)$
25. $r(k:n,k) := r(k:n,k) - f(k,j)r(k:n,j)$
26.   end for
27. end for

We remark that in the previous right-looking algorithm, the jth column f_j of F can be discarded after it is used to update the remaining column $r_{j+1}, \cdots, r_n$.

The RIC3 factorization can also be computed by a left-looking algorithm. By comparing the jth column in factorization (4.28), we have

$$a_{jj} = \sum_{k=1}^{j} r_{jk}^2 - d_{jj},$$

$$a_{ij} = s_{ij} + \sum_{k=1}^{i}(r_{jk}r_{ik} + r_{jk}f_{ik} + f_{jk}r_{ik}), \quad i = j + 1, j + 2, \cdots, n.$$

This says that

$$r_{jj}^2 + d_{jj} = a_{jj} - \sum_{k=1}^{j-1} r_{jk}^2,$$

and

$$r_{jj}(r_{ij} + f_{ik}) + s_{ij} = a_{ij} - \sum_{k=1}^{j-1}(r_{jk}(r_{ik} + f_{ik}) + f_{jk}r_{ik}), \quad i = j+1, \cdots, n.$$

Therefore, to compute the jth column of R, one first computes the vector

$$v = a_j - \sum_{k=1}^{j-1}(r_{jk}(r_k + f_k) + f_{jk}r_k).$$

Then the sparsity of v is imposed with the secondary drop threshold σ_2, i.e.,

$$\text{for } i \geqslant j+1,$$

$$\begin{cases} s_{ij} = 0, & \text{if } \tau > \sigma_2, \\ s_{ij} = v_i, \ v_i = 0, & \text{otherwise,} \end{cases}$$

where

$$\tau = \left[\frac{v_i^2}{(a_{ii} + d_i)(a_{jj} + d_j)} \right]^{1/2}.$$

To ensure $-(S - D + S^T) \geqslant 0$, if a discarded element $\hat{a}_i$ is entered into the position $\hat{s}_i$, the diagonal elements d_i and d_j are updated,

$$d_i = d_i + \delta_i, \quad d_j = d_j + \delta_j,$$

where δ_i and δ_j are chosen such that $\delta_i, \delta_j > 0$ and $\delta_i \delta_j = s_{ij}^2$. Initially, all d_i are set to be zero.

Subsequently, the jth diagonal entry of R is given by

$$r_{jj} = \sqrt{v_j + d_j},$$

and the rest of non-zero elements in r_j and f_j are computed by imposing the primary sparsity constraint on v with the primary drop threshold σ_1, i.e.,

$$\text{for } i \geqslant j+1,$$

$$\begin{cases} r_{ij} = v_i/r_{jj}, \ f_{ij} = 0, & \text{if } |v_i|/r_{jj} > \sigma_1 \\ r_{ij} = 0, & f_{ij} = v_i/r_{jj}, \text{ otherwise.} \end{cases}$$

The following algorithm computes the RIC3 factor R in the left-looking fashion.

RIC3 (LEFT-LOOKING VERSION)
1. $d(1:n) = 0$
2. for $j = 1, 2, \cdots, n$
3. $\quad v(j:n) = a(j:n, j)$

4. for $k = 1, 2, \cdots, k - 1$
5. $\quad v(j : n) := v(j : n) - r(j, k)r(j : n, k)$
6. $\quad v(j : n) := v(j : n) - r(j, k)f(j : n, k)$
7. $\quad v(j : n) := v(j : n) - f(j, k)r(j : n, k)$
8. end for
9. for $i = j + 1, j + 2, \cdots, n$
10. $\quad \tau = |v(i)| / [(a(i, i) + d(i))(a(j, j) + d(j))]^{1/2}$
11. $\quad$ if $\tau \leqslant \sigma_2$
12. $\quad\quad v(i) = 0$
13. $\quad\quad d(i) := d(i) + \tau(a(i, i) + d(i))$
14. $\quad\quad d(j) := d(j) + \tau(a(j, j) + d(j))$
15. $\quad$ end if
16. end for
17. $r(j, j) = \sqrt{v(j) + d(j)}$
18. for $i = j + 1, j + 2, \cdots, n$
19. $\quad$ if $|v(i)| / r(j, j) > \sigma_1$
20. $\quad\quad r(i, j) = v(i) / r(j, j)$
21. $\quad\quad f(i, j) = 0$
22. $\quad$ else
23. $\quad\quad r(i, j) = 0$
24. $\quad\quad f(i, j) = v(i) / r(j, j)$
25. $\quad$ end if
26. end for
27. end for

Note that in the above algorithm, the columns $f_1, f_2, \cdots, f_{j-1}$ are needed to compute the jth column r_j.

HQMC application. Table 13 shows the numerical results of the RIC3 preconditioner computed by the left-looking implementation. The drop thresholds are set to be $\sigma_1 = 0.005$ and $\sigma_2 = 0.00025$. With these drop thresholds, the RIC3 preconditioners R are of about the same sparsity as the IC_p and RIC1 preconditioners presented in Tables 9 and 10, respectively.

Table 13. RIC3/left, $\sigma_1 = 0.005$, $\sigma_2 = 0.00025$, $(N, L, t, \beta) = (48 \times 48, 80, 1, 10)$.

U	0	1	2	3	4	5	6
$\mathrm{nnz}_r(R)$	25.84	27.24	26.98	26.73	26.53	26.35	26.20
$\mathrm{nnz}_r(F)$	42.65	51.84	49.04	46.90	45.21	43.85	42.69
$\mathrm{Mem}(R)$	58	61	60	60	59	59	59
Wksp.	200	236	226	217	211	206	201
Itrs.	12	29	66	106	1026	3683	5412
P-time	5.54	6.36	6.07	5.84	5.65	5.51	5.46
S-time	0.60	1.49	3.33	9.22	51.12	182.24	296.24
T-time	6.13	7.86	9.40	15.05	56.77	188.05	301.71

The RIC3 factorization introduces smaller diagonal updates D and results a preconditioner of better quality than the RIC1 preconditioner. Even though the quality of the IC_p preconditioner for the particular choice of the shift reported in Table 9 is as good as the RIC3 preconditioner, the robustness of the IC_p factorization is not guaranteed, and the quality strongly depends on the choice of the shift α.

The right-looking algorithm is not competitive. Similar to the RIC2 implementations, although the right-looking implementation reduces the workspace requirement, it significantly increases the CPU time.

4.6   Performance evaluation

The numerical results presented so far in this section indicate that the IC_p and RIC3 preconditioners are the most competitive ones for solving the HQMC linear system (4.1). In this section, we focus on these two preconditioners and evaluate their performance for solving HQMC linear systems (4.1) with respect to the length-scale parameter N and energy-scale parameter U. The rest of parameters of the linear systems are $(L, t, \beta, \mu) = (80, 1, 10, 0)$. The IC_p preconditioners are computed with the diagonal shift $\alpha = 10^{-3}$ and the drop tolerance $\sigma = 10^{-3}$. On the other hand, the RIC3 preconditioners are computed with the drop tolerances $\sigma_1 = 10^{-2}$ and $\sigma_2 = 10^{-3}$.

4.6.1   Moderately interacting systems

Figures 4.7 and 4.8 show the performance of the PCG solvers using IC_p and RIC3 preconditioners for moderate interacting systems, namely $U = 0, 1, 2, 3$. These plots show that as lattice size N increases, the numbers of PCG iterations are essentially constants for $U = 0, 1, 2$ and only increases slightly for $U = 3$.

The number of PCG iterations indicates the linear-scaling of PCG solver with respect to the lattice size N. Figures 4.9 and 4.10 show the CPU elapsed time. The black dashed lines indicate the linear-scaling for $U = 1$ and 3. The CPU time at $N = 40 \times 40$ is used as the reference point.

To summarize, the quality of the IC_p and RIC3 preconditioners are comparable. The IC_p preconditioner is slightly more efficient than the RIC3 preconditioner in terms of the total CPU elapsed time. We should note that even though the pivot breakdown did not occur with the shift $\alpha = \sigma$, the IC_p factorization is not provable robust.

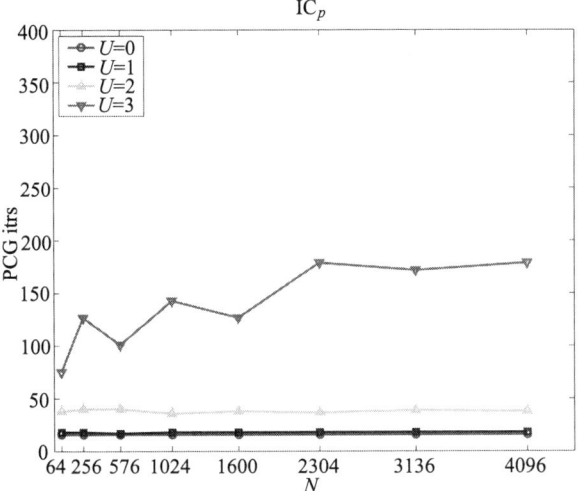

Figure 4.7. Number of PCG iterations using IC_p preconditioner, $U = 0, 1, 2, 3$.

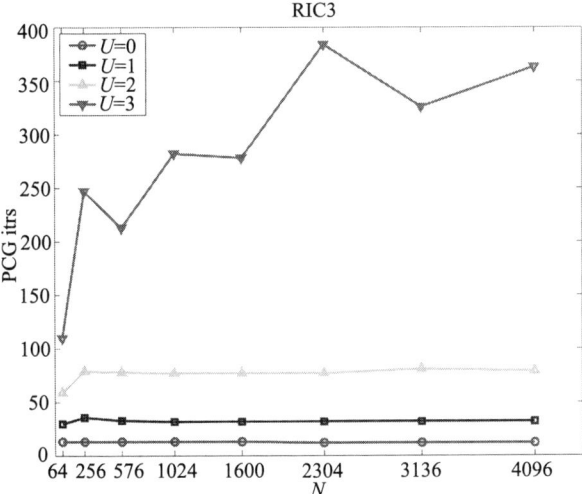

Figure 4.8. Number of PCG iterations using RIC3 preconditioner, $U = 0, 1, 2, 3$.

4.6.2 Strongly interacting systems

For strongly interacting systems, namely $U \geqslant 4$, the number of PCG iterations grows rapidly as the lattice sizes N increasing as shown in Figures 4.11 and 4.12. The CPU elapsed time are shown in Figures 4.13 and 4.14. As we see that the RIC3 preconditioner slightly outperforms the

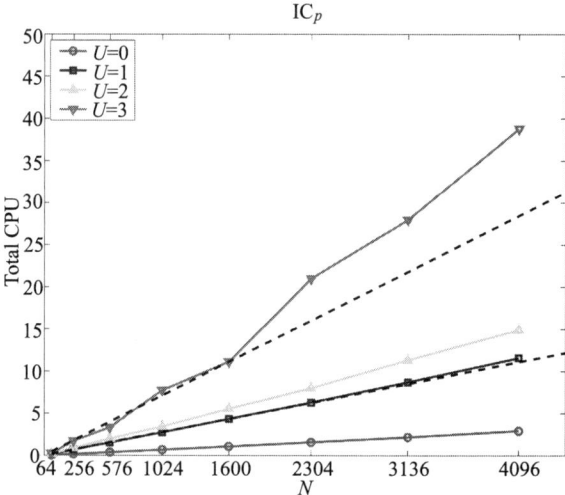

Figure 4.9. CPU time of PCG using IC_p preconditioner, $U = 0, 1, 2, 3$.

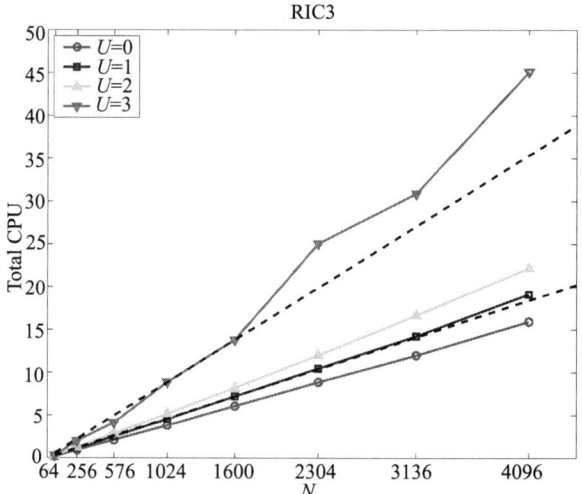

Figure 4.10. CPU time of PCG using RIC3 preconditioner, $U = 0, 1, 2, 3$.

IC_p preconditioner. However, for both preconditioners, the total CPU time of the PCG solver scales at the order of N^2. The dashed line indicates the desired linear-scaling for $U = 4$. The CPU time at $N = 40 \times 40$ is used as the reference point.

To summarize, for strongly-interacting systems, the linear equations (4.1) are ill-conditioned. It remains an open problem whether there is a

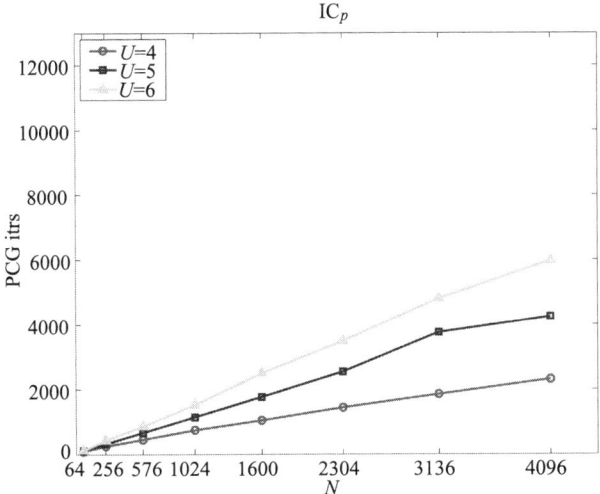

Figure 4.11. Number of PCG iterations using IC_p preconditioner, $U = 4, 5, 6$.

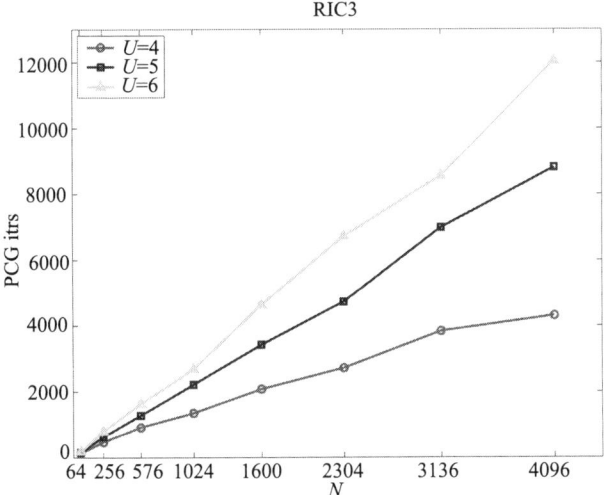

Figure 4.12. Number of PCG iterations using RIC3 preconditioner, $U = 4, 5, 6$.

preconditioning technique to achieve a linear-scaling PCG solver for the strongly-interacting systems.

Remark 4.1. We have observed that for strongly-interacting systems, the residual norm stagnates after initial rapid decline. Figure 4.15 shows the relative residual norm of the PCG iteration for $(N, L, U, t, \beta, \mu) = (32 \times 32, 80, 6, 1, 10, 0)$.

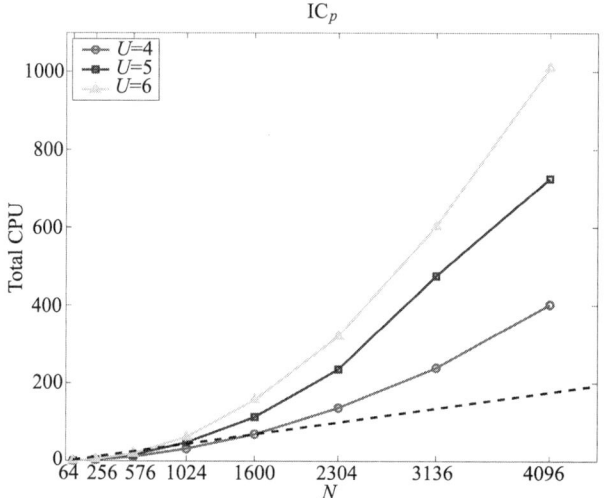

Figure 4.13. CPU time of PCG using IC_p preconditioner, $U = 4, 5, 6$.

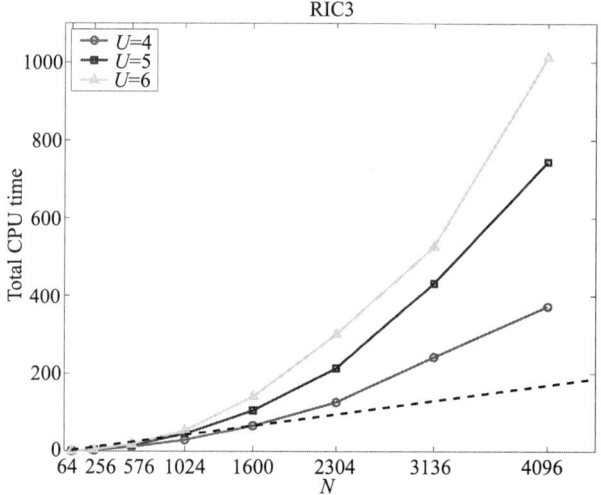

Figure 4.14. CPU time of PCG using RIC3 preconditioner, $U = 4, 5, 6$.

The plateau is largely due to the slow decay of the components of the residual vector associated with the small eigenvalues of the preconditioned matrix $R^{-1}AR^{-T}$. Several techniques have been proposed to deflate these components from the residual vector as a way to avoid the plateau of the convergence, see [32, 35, 36, 45, 46] and references within. It remains to be studied about the applicability of these techniques to our HQMC applications.

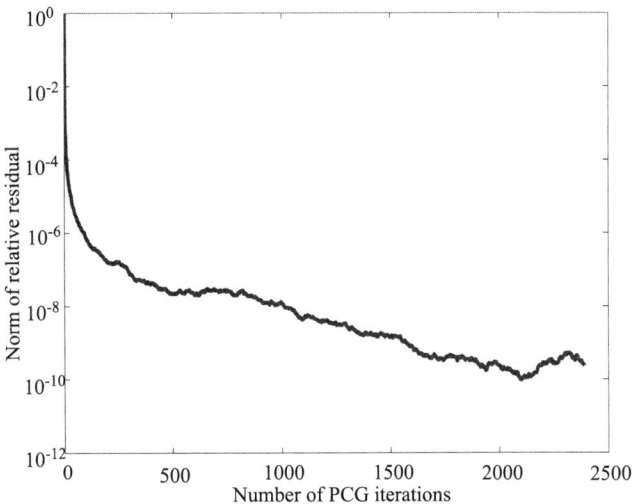

Figure 4.15. Relative residual norms of PCG iterations using RIC3 preconditioner.

Appendix A. Updating algorithm in DQMC

In this appendix, we discuss the single-state MC updating algorithm to provide a fast means to compute the Metropolis ratio in DQMC described in section 1.2.2.

A.1   Rank-one updates

Consider matrices M_1 and M_2 of the forms

$$M_1 = I + FV_1 \quad \text{and} \quad M_2 = I + FV_2,$$

where F is a given matrix. V_1 and V_2 are diagonal and nonsingular, and moreover, they differ only at the $(1,1)$-element, i.e.,

$$V_1^{-1}V_2 = I + \alpha_1 e_1 e_1^T,$$

where

$$\alpha_1 = \frac{V_2(1,1)}{V_1(1,1)} - 1,$$

and e_1 is the first column of the identity matrix I.

It is easy to see that M_2 is a rank-one update of M_1:

$$\begin{aligned}
M_2 &= I + FV_1 + FV_1(V_1^{-1}V_2 - I) \\
&= M_1 + \alpha_1(M_1 - I)e_1 e_1^T \\
&= M_1 \left[I + \alpha_1(I - M_1^{-1})e_1 e_1^T \right].
\end{aligned}$$

The ratio of the determinants of the matrices M_1 and M_2 is immediately given by[①]

$$r_1 = \frac{\det[M_2]}{\det[M_1]} = 1 + \alpha_1(1 - e_1^T M_1^{-1} e_1). \qquad (A.1)$$

Therefore, computing the ratio r_1 is essentially about computing the $(1,1)$-element of the inverse of the matrix M_1.

By Sherman-Morrison-Woodbury formula,[②] the inverse of the matrix M_2 is a rank-one update of M_1^{-1}:

$$M_2^{-1} = \left[I - \frac{\alpha_1}{r_1}(I - M_1^{-1})e_1 e_1^T\right] M_1^{-1}$$

$$= M_1^{-1} - \left(\frac{\alpha_1}{r_1}\right) u_1 w_1^T, \qquad (A.2)$$

where

$$u_1 = (I - M_1^{-1})e_1, \quad w_1 = M_1^{-T} e_1.$$

Now, let us consider a sequence of matrices M_{i+1} generated by rank-one updates

$$M_{i+1} = I + F V_{i+1}$$

for $i = 1, 2, \cdots, n-1$, where

$$V_i^{-1} V_{i+1} = I + \alpha_i e_i e_i^T, \quad \alpha_i = \frac{V_{i+1}(i,i)}{V_i(i,i)} - 1.$$

Then by equation (A.1), we immediately have

$$r_i = \frac{\det[M_{i+1}]}{\det[M_i]} = 1 + \alpha_i(1 - e_i^T M_i^{-1} e_i),$$

and

$$M_{i+1}^{-1} = M_i^{-1} - \left(\frac{\alpha_i}{r_i}\right) u_i w_i^T,$$

where $u_i = (I - M_i^{-1})e_i$ and $w_i = M_i^{-T} e_i$.

Denote

$$U_k = [u_1, u_2, \cdots, u_{k-1}] \quad \text{and} \quad W = [w_1, w_2, \cdots, w_{k-1}].$$

Then it is easy to see that the inverse of M_k can be written as a rank-$(k-1)$ update of M_1^{-1}:

$$M_k^{-1} = M_1^{-1} - U_{k-1} D_k W_{k-1}^T,$$

where $D_k = \text{diag}(\frac{\alpha_1}{r_1}, \frac{\alpha_2}{r_2}, \cdots, \frac{\alpha_{k-1}}{r_{k-1}})$.

Numerical stability of the rank updating procedure have been examined in [54] and [55].

[①]Here we use the fact that $\det[I + xy^T] = 1 + y^T x$ for any two column vectors x and y.

[②]$(A + UV^T)^{-1} = A^{-1} - A^{-1}(I + V^T A^{-1} U)^{-1} U^T A^{-1}$.

A.2   Metropolis ratio and Green's function computations

As we discussed in section 1.2.2 of the DQMC simulation, it is necessary to repeatedly compute the Metropolis ratio

$$r = \frac{\det[M_+(h')]\det[M_-(h')]}{\det[M_+(h)]\det[M_-(h)]},$$

for configurations $h = (h_1, h_2, \cdots, h_L)$ and $h' = (h'_1, h'_2, \cdots, h'_L)$, where $M_\sigma(h)$ is defined in (1.18), namely

$$M_\sigma(h) = I + B_{L,\sigma}(h_L)B_{L-1,\sigma}(h_{L-1})\cdots B_{2,\sigma}(h_2)B_{1,\sigma}(h_1).$$

The Green's function associated with the configuration h is defined as

$$G^\sigma(h) = M_\sigma^{-1}(h).$$

In the DQMC simulation, the elements of configurations h' and h are the same except at a specific imaginary time slice ℓ and spatial site i,

$$h'_{\ell,i} = -h_{\ell,i}.$$

It says that the configuration h' is obtained by a simple flipping at the site (ℓ, i).

The Monte Carlo updates run in the double-loop for $\ell = 1, 2, \cdots, L$ and $i = 1, 2, \cdots, N$. Let us start with the imaginary time slice $\ell = 1$:

- At the spatial site $i = 1$:

$$h'_{1,1} = -h_{1,1}.$$

By the relationship between $M_\sigma(h')$ and $M_\sigma(h)$ and equation (A.1), one can derive that the Metropolis ratio r_{11} is given by

$$r_{11} = d_+d_-, \tag{A.3}$$

where for $\sigma = \pm$,

$$d_\sigma = 1 + \alpha_{1,\sigma}\left(1 - e_1^T M_\sigma^{-1}(h)e_1\right)$$
$$= 1 + \alpha_{1,\sigma}\left(1 - G_{11}^\sigma(h)\right),$$

and

$$\alpha_{1,\sigma} = e^{-2\sigma\nu h_{1,1}} - 1.$$

Therefore, the gist of computing the Metropolis ratio r_{11} is to compute the $(1,1)$-element of the inverse of the matrix $M_\sigma(h)$. If the

Green's function $G^\sigma(h)$ has been computed explicitly in advance, then it is essentially *free* to compute the ratio r_{11}.

In the DQMC simulation, if the proposed h' is accepted, then by equality (A.2), the Green's function $G^\sigma(h)$ is updated by a rank-one matrix:

$$G^\sigma(h) \leftarrow G^\sigma(h) - \frac{\alpha_{1,\sigma}}{r_{11}} u_\sigma w_\sigma^T,$$

where

$$u_\sigma = (I - G^\sigma(h))e_1 \quad \text{and} \quad w_\sigma = (G^\sigma(h))^T e_1.$$

- At the spatial site $i = 2$:

$$h'_{1,2} = -h_{1,2}.$$

By a similar derivation as for the previous case, we have

$$r_{12} = d_+ d_-, \tag{A.4}$$

where for $\sigma = \pm$,

$$d_\sigma = 1 + \alpha_{2,\sigma}\left(1 - G_{12}^\sigma(h)\right), \quad \alpha_{2,\sigma} = e^{-2\sigma h_{1,2}} - 1.$$

Correspondingly, if necessary, the Green's function is updated by the rank-one matrix

$$G^\sigma(h) \leftarrow G^\sigma(h) - \frac{\alpha_{2,\sigma}}{r_{12}} u_\sigma w_\sigma^T,$$

where

$$u_\sigma = (I - G^\sigma(h))e_2 \quad \text{and} \quad w_\sigma = (G^\sigma(h))^T e_2.$$

- In general, for $i = 3, 4, \cdots, N$, we can immediately see that the same procedure can be used for computing the Metropolis ratios r_{1i} and updating the Green's functions.

Remark A.1. For high performance computing, one may delay the update of the Green's functions to lead to a block high rank update instead of rank-one update. This is called a "delayed update" technique.

Now, let us consider how to do the DQMC configuration update for the time slice $\ell = 2$. We first notice that the matrices $M_\sigma(h)$ and $M_\sigma(h')$ can be rewritten as

$$M_\sigma(h) = B_{1,\sigma}^{-1}(h_1)\widehat{M}_\sigma(h)B_{1,\sigma}(h_1),$$
$$M_\sigma(h') = B_{1,\sigma}^{-1}(h'_1)\widehat{M}_\sigma(h')B_{1,\sigma}(h'_1),$$

where

$$\widehat{M}_\sigma(h) = I + B_{1,\sigma}(h_1)B_{L,\sigma}(h_L)B_{L-1,\sigma}(h_{L-1})\cdots B_{2,\sigma}(h_2),$$
$$\widehat{M}_\sigma(h') = I + B_{1,\sigma}(h'_1)B_{L,\sigma}(h'_L)B_{L-1,\sigma}(h'_{L-1})\cdots B_{2,\sigma}(h'_2).$$

The Metropolis ratios r_{2i} corresponding to the time slice $\ell = 2$ can be written as

$$r_{2i} = \frac{\det[M_+(h')]\det[M_-(h')]}{\det[M_+(h)]\det[M_-(h)]} = \frac{\det[\widehat{M}_+(h')]\det[\widehat{M}_-(h')]}{\det[\widehat{M}_+(h)]\det[\widehat{M}_-(h)]},$$

and the associated Green's functions are given by "wrapping":

$$\widehat{G}^\sigma(h) \leftarrow B_{1,\sigma}^{-1}(h_1)G^\sigma(h)B_{1,\sigma}(h_1).$$

As a result of the wrapping, the configurations h_2 and h'_2 associated with the time slice $\ell = 2$ appear at the same location of the matrices $\widehat{M}_\sigma(h)$ and $\widehat{M}_\sigma(h')$ as the configurations h_1 and h'_1 at the time slice $\ell = 1$. Therefore, we can use the same formulation as for the time slice $\ell = 1$ to compute the Metropolis ratios r_{2i} and update the associated Green's functions.

For $\ell \geqslant 3$, it is clear that we can repeat the wrapping trick to compute the Metropolis ratios $r_{\ell i}$ and updating the associated Green's functions.

Remark A.2. By the discussion, see that the main computing cost of computing the Metropolis ratios $r_{\ell i}$ is on the Green's function updating. It costs $2N^2$ flops for each update. The total cost of one sweep through all $N \times L$ Hubbard-Stratonovich variables h is $2N^3 L$. An important issue is about numerical stability and efficiency of computation, updating and wrapping of the Green's functions. A QR decomposition with partial pivoting based method is currently used in the DQMC implementation [11].

Appendix B   Particle-hole transformation

In this appendix, we present an algebraic derivation for the so-called *particle-hole transformation*.

B.1   Algebraic identities

We first present a few algebraic identities.

Lemma B.1. *For any nonsingular matrix A,*

$$(I + A^{-1})^{-1} = I - (I + A)^{-1}.$$

Proof. By straightforward verification.   □

Lemma B.2. *Let the matrices A_ℓ be symmetric and nonsingular for $\ell = 1, 2, \cdots, m$. Then*

$$(I + A_m^{-1} A_{m-1}^{-1} \cdots A_1^{-1})^{-1} = I - (I + A_m A_{m-1} \cdots A_1)^{-T}.$$

Proof. By straightforward verification.   □

Theorem B.1. *For any square matrices A and B, if there exists a nonsingular matrix Π such that*

$$\Pi A + A\Pi = 0 \quad and \quad \Pi B - B\Pi = 0,$$

namely, Π anti-commutes with A and commutes with B. Then we have

$$(I + e^{A-B})^{-1} = I - \Pi^{-1}(I + e^{A+B})^{-1}\Pi \tag{B.1}$$

and

$$\det\left[I + e^{A-B}\right] = e^{\text{Tr}(A-B)} \det\left[I + e^{A+B}\right]. \tag{B.2}$$

Proof. First, we prove the inverse identity (B.1),

$$(I + e^{A-B})^{-1} = I - (I + e^{-A+B})^{-1} = I - (I + e^{\Pi^{-1}(A+B)\Pi})^{-1}$$
$$= I - (I + \Pi^{-1}e^{A+B}\Pi)^{-1} = I - \Pi^{-1}(I + e^{A+B})^{-1}\Pi.$$

Now, let us prove the determinant identity (B.2). Note that

$$I + e^{A-B} = e^{A-B}(I + e^{-(A-B)}) = e^{A-B}(I + e^{-A+B})$$
$$= e^{A-B}(I + e^{\Pi^{-1}A\Pi + \Pi^{-1}B\Pi}) = e^{A-B}(I + \Pi^{-1}e^{A+B}\Pi)$$
$$= e^{A-B}\Pi^{-1}(I + e^{A+B})\Pi.$$

Hence, we have

$$\det\left[I + e^{A-B}\right] = \det[e^{A-B}] \cdot \det[\Pi^{-1}] \cdot \det[I + e^{A+B}] \cdot \det[\Pi]$$
$$= e^{\text{Tr}(A-B)} \det[I + e^{A+B}].$$

For the last equality, we used the identity $\det e^W = e^{\text{Tr}W}$ for any square matrix W.   □

The following theorem gives relations of the inverses and determinants of matrices $I + e^A e^{-B}$ and $I + e^A e^B$.

Theorem B.2. *For symmetric matrices A and B, if there exists a nonsingular matrix Π such that*

$$\Pi A + A\Pi = 0 \quad and \quad \Pi B - B\Pi = 0,$$

then we have

$$(I + e^A e^{-B})^{-1} = I - \Pi^{-T}(I + e^A e^B)^{-T}\Pi^T \tag{B.3}$$

and

$$\det[I + e^A e^{-B}] = e^{\mathrm{Tr}(A-B)} \det[I + e^A e^B]. \tag{B.4}$$

Proof. Similar to the proof of Theorem B.1. □

The following two theorems are the generalization of Theorem B.2.

Theorem B.3. *Let $M_\sigma = I + e^A e^{\sigma B_k} e^A e^{\sigma B_{k-1}} \cdots e^A e^{\sigma B_1}$, where A and $\{B_\ell\}$ are symmetric, $\sigma = +, -$. If there exists a nonsingular matrix Π that anti-commutes with A and commutes with B_ℓ, i.e.,*

$$\Pi A + A\Pi = 0 \quad and \quad \Pi B_\ell - B_\ell \Pi = 0 \quad for\ \ell = 1, 2, \cdots, k,$$

then we have

$$M_-^{-1} = I - \Pi^{-T} M_+^{-T}\Pi^T \tag{B.5}$$

and

$$\det[M_-] = e^{k\mathrm{Tr}(A) - \sum_{\ell=1}^k \mathrm{Tr}(B_\ell)} \det[M_+]. \tag{B.6}$$

Theorem B.4. *Let A and B be symmetric matrices and W be a non-singular matrix. If there exists a nonsingular matrix Π such that it anti-commutes with A and commutes with B, i.e.,*

$$\Pi A + A\Pi = 0 \quad and \quad \Pi B - B\Pi = 0$$

and furthermore, it satisfies the identity

$$\Pi = W\Pi W^T,$$

then

$$(I + e^A e^{-B} W)^{-1} = I - \Pi^{-T}(I + e^A e^B W)^{-T}\Pi^T \tag{B.7}$$

and

$$\det[I + e^A e^{-B} W] = e^{\mathrm{Tr}(A-B)} \cdot \det[W] \cdot \det[I + e^A e^B W]. \tag{B.8}$$

B.2   Particle-hole transformation in DQMC

For the simplest 1-D lattice of N_x sites:

$$K_x = \begin{bmatrix} 0 & 1 & & & 1 \\ 1 & 0 & 1 & & \\ & 1 & 0 & 1 & \\ & & \ddots & \ddots & \ddots \\ 1 & & & 1 & 0 \end{bmatrix}_{N_x \times N_x}$$

and $N_x \times N_x$ diagonal matrices V_ℓ for $\ell = 1, 2, \cdots, L$, if N_x is even, the matrix

$$\Pi_x = \text{diag}(1, -1, 1, -1, \cdots, 1, -1)$$

anti-commutes with K_x and commutes with V_ℓ:

$$\Pi_x K_x + K_x \Pi_x = 0$$

and

$$\Pi_x V_\ell - V_\ell \Pi_x = 0 \quad \text{for} \quad \ell = 1, 2, \cdots, L,$$

then by Theorem B.3, the determinants of the matrices M_- and M_+ satisfy the relation

$$\det[M_-] = e^{-\sum_{\ell=1}^{L} \text{Tr}(V_\ell)} \det[M_+].$$

For the Green's functions:

$$G^\sigma = M_\sigma^{-1} = \left(I + e^{\Delta \tau t K_x} e^{\sigma V_L} e^{\Delta \tau t K_x} e^{\sigma V_{L-1}} \cdots e^{\Delta \tau t K_x} e^{\sigma V_1} \right)^{-1}$$

where $\sigma = +$ or $-$, we have

$$G^- = I - \Pi_x (G^+)^T \Pi_x.$$

This is referred to as the *particle-hole transformation* in the condensed matter physics literature because it can be viewed as a change of operators $c_{i\downarrow} \rightarrow c_{i\downarrow}^\dagger$.

For a 2-D rectangle lattice with $N_x \times N_y$ sites:

$$K = K_x \otimes I + I \otimes K_y$$

and $N_x N_y \times N_x N_y$ diagonal matrices V_ℓ for $\ell = 1, 2, \cdots, L$, if N_x and N_y are even, the matrix

$$\Pi = \Pi_x \otimes \Pi_y$$

anti-commutes with K and commutes with V_ℓ:

$$\Pi K + K \Pi = 0$$

and

$$\Pi V_\ell - V_\ell \Pi = 0 \quad \text{for} \quad \ell = 1, 2, \cdots, L,$$

then by Theorem B.3, we have

$$\det[M_-] = e^{-\sum_{\ell=1}^{L} \text{Tr}(V_\ell)} \det[M_+].$$

This is the identity used for equation (1.26). For the Green's functions, we have

$$G^\sigma = M_\sigma^{-1} = \left(I + e^{\Delta \tau t K} e^{\sigma V_L} e^{\Delta \tau t K} e^{\sigma V_{L-1}} \cdots e^{\Delta \tau t K} e^{\sigma V_1} \right)^{-1},$$

where $\sigma = +$ (spin up) or $-$ (spin down), we have

$$G^- = I - \Pi (G^+)^T \Pi.$$

This is the *particle-hole transformation* for the 2D rectangle lattice.

B.3   Particle-hole transformation in the HQMC

In the HQMC, we consider the matrix M_σ of the form

$$
M_\sigma =
\begin{bmatrix}
I & & & & & e^{\Delta\tau t K}e^{\sigma V_1} \\
-e^{\Delta\tau t K}e^{\sigma V_2} & I & & & & \\
& -e^{\Delta\tau t K}e^{\sigma V_2} & I & & & \\
& & & \ddots & \ddots & \\
& & & & -e^{\Delta\tau t K}e^{\sigma V_L} & I
\end{bmatrix}
$$

$$
= I + e^A e^{\sigma D} P,
$$

where $A = \mathrm{diag}(\Delta\tau t K, \Delta\tau t K, \cdots, \Delta\tau t K)$ and $D = \mathrm{diag}(V_1, V_2, \cdots, V_L)$
and

$$
P =
\begin{bmatrix}
0 & & & & I \\
-I & 0 & & & \\
& -I & 0 & & \\
& & \ddots & \ddots & \\
& & & -I & 0
\end{bmatrix}.
$$

Note that $\det[P] = 1$. It can be verified that for the 1-D or 2-D rectangle
lattice, i.e., $K = K_x$ or $K = K_x \otimes I + I \otimes K_y$ as defined in B.2, matrix

$$
\Pi = I \otimes \Pi_x \quad \text{(1-D)}
$$

or

$$
\Pi = I \times \Pi_x \otimes P_y \quad \text{(2-D)}
$$

anti-commutes with A and commutes with D, i.e.,

$$
\Pi A + A\Pi = 0, \quad \Pi D - D\Pi = 0.
$$

Furthermore, it satisfies

$$
\Pi = P\Pi P^T.
$$

Then by Theorem B.4, the determinants of M_+ and M_- are related by

$$
\det[M_-] = e^{-\sum_{\ell=1}^{L} \mathrm{Tr}(V_\ell)} \cdot \det[M_+].
$$

and the Green's functions $G^\sigma = M_\sigma^{-1}$ satisfy the relation

$$
G^- = I - \Pi (G^+)^T \Pi.
$$

Remark B.1. Besides the 1-D and 2-D rectangle lattices, namely the
lattice structure matrices K_x and K as defined in B.2, are there other
types of lattices (and associated structure matrices K) such that we can
apply Theorems B.4 to establish the relationships between the inverses
and determinants in the DQMC? It is known that for the honeycomb
lattices, it is true, but for the triangle lattices, it is false. A similar
question is also valid for the HQMC. Indeed, it works on any "bipartite"
lattice, i.e., any geometry in which sites divides into two disjoint sets $\mathcal{A}$
and $\mathcal{B}$ and K connects sites in $\mathcal{A}$ and $\mathcal{B}$ only.

B.4　Some identities of matrix exponentials

1. In general, $e^{A+B} \neq e^A e^B$, and $e^A e^B \neq e^B e^A$.

2. If A and B commute, namely $AB = BA$, then $e^{A+B} = e^A e^B = e^B e^A$.

3. $(e^A)^{-1} = e^{-A}$.

4. $e^{P^{-1}AP} = P^{-1} e^A P$.

5. $(e^A)^H = e^{A^H}$ for every square matrix A,
   e^A is Hermitian if A is Hermitian,
   e^A is unitary if A is skew-Hermitian.

6. $\det e^A = e^{\operatorname{Tr} A}$ for every square matrix A.

7. $e^{A \otimes I + I \otimes B} = e^A \otimes e^B$.

Acknowledgments

This paper first assembled for the lecture notes used in the Shanghai Summer School of Mathematics held in 2006. We are exceedingly grateful to Professor Ta-Tsien Li (Li Daqian) of Fudan University and Professor Thomas Yizhao Hou of California Institute of Technology for inviting us to present this work at this summer school and providing us an opportunity to put the materials together in the first place.

References

[1] V. I. Arnold. Mathematical Methods of Classical Mechanics, 2nd ed. Springer-Verlag, New York, 1989.

[2] R. Blankenbecler, D. J. Scalapino and R. L. Sugar. Monte Carlo calculations of coupled Boson-fermion systems I. Phys. Rev. D, 24(1981), 2278–2286.

[3] R. P. Feynman and A. R. Hibbs. Quantum Mechanics and Path Integrals. McGraw-Hill, New York, 1965

[4] E. Hairer, C. Lubich and G. Wanner. Geometric numerical integration illustrated by the Störmer-Verlet method. Acta Numerica, 12(2003), 399–450.

[5] J. E. Hirsch. Two-dimensional Hubbard model: numerical simulation study. Phy. Rev. B, 31(1985), 4403–4419.

[6] J. E. Hirsch. Discrete Hubbard-Stratonovich transformation for fermion lattice models. Phys. Rev. B, 28(1983), 4059–4061.

[7] J. E. Hirsch. Erratum: Discrete Hubbard-Stratonovich transformation for fermion lattice models. Phy. Rev. B, 29(1984), 4159.

[8] J. Hubbard. Electron correlations in narrow energy bands. Proc. Roy. Soc. London, A, 276(1963), 238–257.

[9] J. Hubbard. Electron correlations in narrow energy bands III: an improved solution. Proc. Roy. Soc. London, A, 281(1964), 401–419.

[10] Jun S. Liu. Monte Carlo Strategies in Scientific Computing. Springer, 2001.

[11] E. Y. Loh Jr. and J. E. Gubernatis. Stable numerical simulations of models of interacting electrons. In: Electronic Phase Transition (edited by W. Hanks and Yu. V. Kopaev). Elsevier Science Publishers B.V., 1992, 177–235.

[12] R. K. Pathria. Statistical Mechanics, 2nd ed. Elsevier, 2001.

[13] R. Schumann and E. Heiner. Transformations of the Hubbard interaction to quadratic forms. Phy. Let. A, 134(1988), 202–204.

[14] D. J. Scalapino and R. L. Sugar. Monte Carlo calculations of coupled Boson-fermion systems II. Phys. Rev. B, 24(1981), 4295–4308.

[15] R. T. Scalettar, D. J. Scalapino, R. L. Sugar, and D. Toussaint. Hybrid molecular-dynamics algorithm for the numerical simulation of many-electron systems. Phy. Rev. B, 36(1987), 8632–8640.

[16] R. T. Scalettar, D. J. Scalapino and R. L. Sugar. New algorithm for the numerical simulation of fermions. Phy. Rev. B, 34(1986), 7911–7917.

[17] J. P. Wallington and J. F. Annett. Discrete symmetries and transformations of the Hubbard model. Phy. Rev. B, 58(1998), 1218–1221.

[18] T. Hanaguri, C. Lupien, Y. Kohsaka, D. -H. Lee, M. Azuma, M. Takono, H. Takagi and J. C. Davis. A 'checkerboard' electronic crystal state in lightly hole-doped $Ca_{2-x}Na_xCuO_2Cl_2$. Nature, 430(2004), 1001–1005.

[19] C. Moler and C. Van Loan. Nineteen dubious ways to compute the exponential of a matrix, twenty-five years later. SIAM Review, 45(2003), 3–49.

[20] R. S. Varga. Matrix Iterative Analysis. Prentice-Hall, Englewood Cliffs, 1962; 2nd ed, Springer, Berlin/Heidelberg, 2000.

[21] S. R. White and D. J. Scalapino. Density matrix renormalization group study of the striped phase in the 2D t-J model. Phys. Rev. Lett., 80(1998), 1272–1275.

[22] B. L. Buzbee, G. H. Golub and C. W. Nielson. On direct methods for solving Poisson's equations. SIAM J. Numer. Anal., 7(1970), 627–656.

[23] G. Fairweather and I. Gladwell. Algorithms for almost block diagonal linear systems. SIAM Review, 46(2004), 49–58.

[24] N. Higham. Accuracy and Stability of Numerical Algorithms, 2nd ed. SIAM, 2002.

[25] B. Philippe, Y. Saad and W. J. Stewart. Numerical methods in Markov chain modeling. Operations Research, 40(1992), 1156–1179.

[26] W. J. Stewart. Introduction to the Numerical Solution of Markov Chains. Princeton University Press, 1994.

[27] G. J. Tee. An application of p-cyclic matrices for solving periodic parabolic problems. Numer. Math., 6(1964), 142–159.

[28] U. M. Ascher, R. M. M. Mattheij and R. D. Russell. Numerical Solution of Boundary Value Problems for Ordinary Differential Equations. Prentice-Hall, Englewood Cliffs, 1988.

[29] S. J. Wright. Stable parallel algorithms for two-point boundary value problems. SIAM J. Sci. Statist. Comput., 13(1992), 742–764.

[30] S. J. Wright. A collection of problems for which gaussian elimination with partial pivoting is unstable. SIAM J. Sci. Statist. Comput., 14(1993), 231–238.

[31] M. Ajiz and A. Jennings. A robust incomplete choleski-conjugate gradient algorithm. Inter. J. Numer. Meth. Engrg., 20(1984), 949–966.

[32] M. Arioli and and D. Ruiz. A Chevyshev-based two-stage iterative method as an alternative to the direct solution of linear systems. Technical Report, RAL-TR-2002-021, Rutherford Appleton Laboratory, 2000

[33] O. Axelsson and L. Y. Kolotilina. Diagonally compensated reduction and related preconditioning methods. Numer. Linear Algebra Appl., 1(1994), 155–177.

[34] Z. Bai and R. T. Scalettar. private communication, 2005

[35] M. Bollhoefer and V. Mehrmann. A new approach to algebraic multilevel methods based on sparse approximate inverses. Preprint, Numerische Simulation auf Massiv Parallelen Rechnern, 1999.

[36] B. Capentieri, I. S. Duff, and L. Giraud. A class of spectral two-level preconditioners. SIAM J. Scient. Comp., 25(2003), 749–765.

[37] V. Eijkhout. On the existence problem of incomplete factorization methods. LAPACK Working Note 144, UT-CS-99-435, Computer Science Department, University of Tennessee, 1991.

[38] G. H. Golub and C. F. van Loan. Matrix Computations, 3rd ed. Johns Hopkins University Press, 1996.

[39] A. Jennings and G. M. Malik. Partial elimination. J. Inst. Math. Appl., 20(1977), 307–316.

[40] I. E. Kaporin. High quality preconditioning of a general symmetric positive definite matrix based on its $u^t u + u^t r + r^t u$-decomposition. Numer. Lin. Alg. Appl., 5(1998), 483–509.

[41] D. S. Kershaw. The incomplete Cholesky conjugate gradient method for the iterative solution of systems of linear equations. J. of Comp. Phys., 26(1978), 43–65.

[42] R. Lancaze, A. Morel, B. Petersson and J. Schroper. An investigation of the 2D attractive Hubbard model. Eur. Phys. J. B., 2(1998), 509–523.

[43] T. A. Manteuffel. An incomplete factorization technique for positive definite linear systems. Math. Comp., 150(1980), 473–497.

[44] J. Meijerinkand and H. A. van der Vorst. An iterative solution method for linear systems of which the coefficient matrix is a symmetric m-matrix. Math. Comp., 137(1977), 134–155.

[45] R. A. Nicolaides. Deflation of conjugate gradients with application to boundary value problems. SIAM J. Numer. Anal., 24(1987), 355–365.

[46] A. Padiy, O. Axelsson and B. Polman. Generalized augmented matrix preconditioning approach and its application to iterative solution of ill-conditioned algebraic systems. SIAM J. Matrix Anal. Appl., 22 (2000), 793–818.

[47] PETSc: Portable, Extensible toolkit for scientific computation. http://www-unix.mcs.anl.gov/petsc/petsc-2/.

[48] R. T. Scalettar, D. J. Scalapino and R. L. Sugar. A new algorithm for numerical simulation of fermions. Phys. Rev. B, 34(1986), 7911–7917.

[49] R. T. Scalettar, D. J. Scalapino, R. L. Sugar and D. Tousaint. A hybrid-molecular dynamics algorithm for the numerical simulation of many electron systems. Phys. Rev. B, 36(1987), 8632–8641.

[50] R. B. Schnabel and E. Eskow. A new modified Cholesky factorization. SIAM J. Sci. Comput., 11(1990), 1136–1158.

[51] M. Tismenetsky. A new preconditioning technique for solving large sparse linear systems. Lin. Alg. Appl., 154–156(1991), 331–353.

[52] H. A. van der Vorst. Iterative solution methods for certain sparse linear systems with a non-symmetric matrix arising from PDE-problems. J. of Comp. Phys., 44(1981), 1–19.

[53] R. S. Varga, E. B. Saff and V. Mehrmann. Incomplete factorizations of matrices and connections with H-matrices. SIAM J. Numer. Anal., 17(1980), 787–793.

[54] G. W. Stewart. Modifying pivot elements in Gaussian elimination. Math. Comp., 28(1974), 537–542.

[55] E. L. Yip. A note on the stability of solving a rank-p modification of a linear system by the Sherman-Morrison-Woodbury formula. SIAM J. Sci. Stat. Comput., 7(1986), 507–513.

Introduction to Propagation, Time Reversal and Imaging in Random Media

Albert C. Fannjiang

Department of Mathematics, University of California
Davis, CA 95616-8633, USA
E-mail: fannjiang@math.ucdavis.edu

1 Scalar diffraction theory

1.1 Introduction

One of the most useful partial differential equations in applied mathematics is the (scalar or vector) wave equation. It describes propagation of linear waves and has a great variety of applications some of which will be discussed below. For instance, the acoustic wave is governed by the scalar wave equation while the electromagnetic wave in dielectric media is governed by the vector wave equation for its vector potential. In media such as the earth's turbulent atmosphere there is negligible depolarization. Thus, by restricting the source to be linearly polarized or by considering two orthogonal polarization components independently, the scalar wave equation is suitable. In this review, we will focus on the scalar waves. Main references for this topic include [3], [43].

Monochromatic waves correspond to the time-harmonic solution $u(\mathbf{r})e^{-i\omega t}$ where ω is the frequency. The spatial component u then satisfies reduced wave equation is

$$[\nabla^2 + k^2(1 + \tilde{\epsilon}(\mathbf{r}))]u = f, \quad \mathbf{r} \in \mathbb{R}^3, \tag{1}$$

where f represents source, $\tilde{\epsilon}$ the deviation from the constant background and $k = 2\pi/\lambda$ the wavenumber. Suitable boundary conditions are required to solve eq. (1). $\tilde{\epsilon}$ is related to the relative fluctuation of index of refraction $\tilde{n}$ as $\tilde{\epsilon} \approx 2\tilde{n}$ for $\tilde{n} \ll 1$.

The free, undisturbed propagation is described by the free-space Green function

$$G_0(\mathbf{r}, \mathbf{r}') = -\frac{1}{4\pi|\mathbf{r} - \mathbf{r}'|}e^{ik|\mathbf{r}-\mathbf{r}'|}$$

which solves

$$[\nabla^2 + k^2]G_0 = \delta(\mathbf{r})$$

and satisfies the outward radiation condition

$$\left| \frac{\partial G_0}{\partial r} - ikG_0 \right| = O(r^{-2}), \quad r = |\mathbf{r}|.$$

The wave field u should satisfy the same radiation condition at far field.

Two main phenomena are present when there are multiple sources or fluctuations in the medium: diffraction and interference.

It is important to realize that there is no physical difference between interference and diffraction. However, it is traditional to consider a phenomenon as interference when it involves the superposition of only a few waves, and as diffraction when a large number of waves are involved.

Another aspect that is important to understand is that every optical instrument only uses a portion of the full incident wavefront. Because of this, diffraction plays a significant role in the detailed understanding of the light train through the device. Even in all of the potential defects in the lens system were eliminated, the ultimate sharpness of the image would be limited by diffraction.

In the modern treatment, diffraction effects are not connected with light transmission through apertures and obstacles only. Diffraction is examined as a natural property of wavefield with the nonhomogeneous transverse intensity distribution. It commonly appears even if the beam is transversally unbounded. The Gaussian beam is the best known example. In optics, nondiffracting propagation of the beam-like fields can be obtained in convenient media such as waveguides or nonlinear materials. The beams then propagate as waveguide modes and spatial solitons, respectively. In the *free* space, one can easily verify that the field

$$u(\rho, z) = e^{i\beta z} \int_0^{2\pi} A(\phi) e^{i\alpha(x \cos \phi + y \sin \phi)} d\phi, \quad \rho = \sqrt{x^2 + y^2}, \quad (2)$$

with $\alpha^2 + \beta^2 = k^2$ and an arbitrary, complex-valued A satisfies the Helmholtz equation. The choice of $A = 1/(2\pi)$ leads to the zeroth order Bessel beam

$$u(\rho, z) = e^{i\beta z} J_0(\alpha \rho),$$

where J_0 is the zeroth order Bessel function, Figure 1. Such a beam is non-diffracting because the intensity distribution on the transverse plane does not change with the distance of propagation. More generally, we have the higher-order non-diffracting beams

$$u(\rho, \phi, z) = J_m(\alpha \rho) e^{im\phi} e^{i\beta z}, \quad m \in \mathbb{N}$$

[13]. The most useful case is $m = 0$ which gives rise to a cental bright spot.

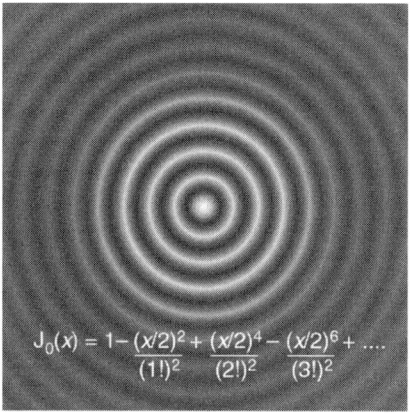

Figure 1   Non-diffracting Bessel beam with $m = 0$ (adapted from [36]).

Formula (2) suggests that a Bessel beam can be formed by a plane wave passing through an axicon, Figure 2. The summation, or interference of all these waves leads to a bright spot in the centre of the beam. However, since $J_0^2(\alpha\rho)$ decays like $1/\rho$, a Bessel beam requires infinite energy and cannot be physically realized exactly. Still, experimental approximations to Bessel beams have extremely low divergence, and are therefore very useful in many applications, specifically in optical tweezers, where an extremely tightly focused hollow beam is optimal.

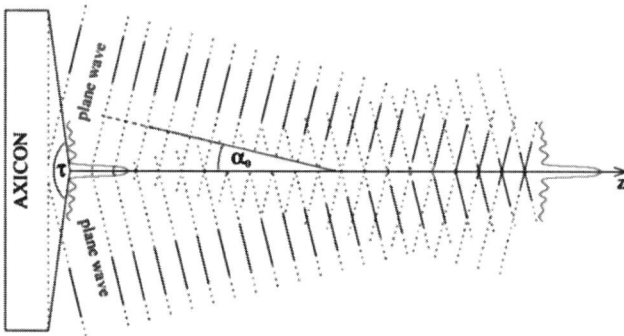

Figure 2   A Bessel beam is formed after an incident plane wave passes through a conical lens.

1.2 Kirchhoff's theory of diffraction

As said, diffraction corresponds to boundary value problem for the reduced wave equation. For simplicity we consider first the free space with $\tilde{\epsilon} = 0$.

Let a point source be situated at the origin. Green's second identity says that for any $u, v \in C^2$,

$$\int_D \left[v\nabla^2 u - u\nabla^2 v \right] d\mathbf{r}_0 = \int_{\partial D} \left[v\frac{\partial u}{\partial n} - u\frac{\partial v}{\partial n} \right] d\sigma(\mathbf{r}_0), \qquad (3)$$

where D is the domain of interest, say the space behind an aperture or obstacle. For a fixed probing point $\mathbf{r}$, consider the test function

$$v = \frac{e^{ikr'}}{r'}, \quad r' = |\mathbf{r} - \mathbf{r}_0|$$

on the punctured-at-$\mathbf{r}$ domain D where r' the distance between the point of probing $\mathbf{r}$ and the point of integration $\mathbf{r}_0$. Due to the radiation condition, the far-field boundary on the right-hand side of (3) would not contribute at all. And we can use (3) for an unbounded domain D.

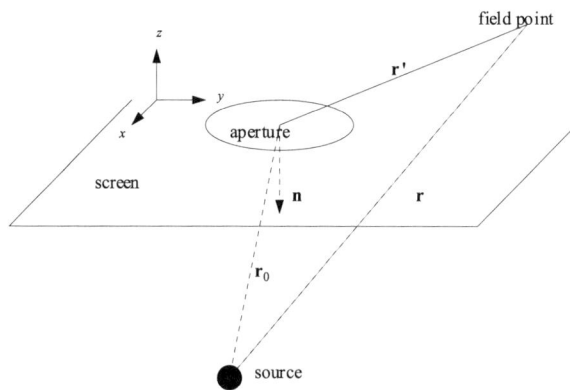

Figure 3   Geometry of a diffracting aperture

Consider the finite boundary $S \cup A$ where S is the (black) screen and A the aperture (pupil), see Figure 3. St. Venant's hypothesis states that the optical field in an aperture is the same as if the aperture were not present and $u = 0$ on the screen. Kirchhoff's hypothesis is even stronger:

$$A : u = u_i, \quad \frac{\partial u}{\partial n} = \frac{\partial u_i}{\partial n} \text{ (free, undisturbed propagation)},$$

$$S : u = 0, \quad \frac{\partial u}{\partial n} = 0, \quad \text{(vanishing excitation, black screen)},$$

where $u_i = 4\pi a G_0$ is the incident wave assuming a point source of strength a. With it, we obtain

$$u(\mathbf{r}) = -\int_A \left[G_0(\mathbf{r}, \mathbf{r}_0) \frac{\partial u_i}{\partial n} - u_i \frac{\partial G_0(\mathbf{r}, \mathbf{r}_0)}{\partial n} \right] d\sigma(\mathbf{r}_0). \tag{4}$$

Since the problem is linear, eq. (4) holds for an arbitrary incident wave u_i.

Kirchhoff's diffraction formula is an exact solution of the Helmholtz equation but does *not* satisfy Kirchhoff's boundary conditions exactly since one can specify only the Dirichlet or the Neumann condition, but not both in general. In a sense, Kirchhoff's formula is not self-consistent but it is a good approximation when the aperture is sufficiently large and the field point is sufficiently away from the aperture/screen as compared with the wavelength and when the diffraction angle is small. As such, Kirchhoff theory can be viewed as attempt to turn the boundary value problem into an *initial value* problem with data posed on the screen and aperture.

Self-consistency can be obtained so that only one boundary condition is specified but nothing essential is gained. For example one can consider the following test functions for a *planar* diffracting surface

$$v = \frac{e^{ikr_1'}}{r_1'} \mp \frac{e^{ikr_2'}}{r_2'}, \quad r_1' = |\mathbf{r} - \mathbf{r}_0|, \quad r_2' = |\mathbf{r}_i - \mathbf{r}_0|,$$

where r_1' is the distance between the field point and the point on the diffracting surface, r_2' the distance between the *image* point of the field point and the diffracting surface, $\mathbf{r}_i$ is the image point of $\mathbf{r}$ with respect to the planar screen. For such test functions we have $v = 0$ and $\partial v/\partial n = 0$, respectively, on ∂D and therefore

$$u(\mathbf{r}) = -\frac{1}{4\pi} \int_{\partial D} u \frac{\partial v}{\partial n} d\sigma(\mathbf{r}_0) = \frac{1}{2\pi} \int_{\partial D} u \frac{\partial}{\partial n} \frac{e^{ikr_1'}}{r_1'} d\sigma(\mathbf{r}_0),$$

$$u(\mathbf{r}) = \frac{1}{4\pi} \int_{\partial D} v \frac{\partial u}{\partial n} d\sigma(\mathbf{r}_0) = -\frac{1}{2\pi} \int_{\partial D} \frac{e^{ikr_1'}}{r_1'} \frac{\partial u}{\partial n} d\sigma(\mathbf{r}_0)$$

respectively. They are called Rayleigh's diffraction formulae of the first and the second kind, respectively, and solves the Dirichlet and the Neumann boundary value problem, respectively. The Rayleigh diffraction formulae are mathematically consistent but they are limited to planar diffraction surfaces and are not necessarily in closer agreement with observation than the Kirchhoff diffraction formula as the exact boundary values of a black screen is not known [3].

From Green's third identity we know

$$u_i(\mathbf{r}) = -\int_{\partial D} \left[G_0 \frac{\partial u_i}{\partial n} - u_i \frac{\partial G_0}{\partial n} \right] d\sigma$$

and hence

$$u(\mathbf{r}) = u_i(\mathbf{r}) - \int_S \left[G_0 \frac{\partial u_i}{\partial n} - u_i \frac{\partial G_0}{\partial n} \right] d\sigma. \tag{5}$$

This implies Babinet's theorem: Let u_1 be the solution for one setting and u_2 the solution if the aperture and the screens are interchanged. Then, u_1 is the left-hand side of (5) and u_2 is the second term on the right-hand side of (5). Therefore we have

$$u_1(\mathbf{r}) + u_2(\mathbf{r}) = u_i(\mathbf{r}).$$

Babinet's theorem connects diffraction by an finite obstacle to diffraction through a finite aperture.

Now we apply Babinet's theorem to show the self-repair property of a Bessel beam. Let the incident field u_i be a Bessel beam propagating through an obstacle of a finite extent. The wave field behind the obstacle is u_2 which equals $u_i - u_1$ by Babinet's theorem. The scattered field u_1, as given by (4) with A being the obstacle, decays like $1/z$ pointwise where z is the distance behind the obstacle. Thus, for a sufficiently large distance u_2 is approximately u_i. In other words, a Bessel beam reforms after passing through an obstacle [5].

Physically, this can be understood as follows, see Figure 4. If the bright centre of the beam is distorted it creates a shadow after the distortion. Parts of the light waves far removed from the centre are able to move past the obstacle unhindered and recreate the beam centre at some distance beyond the obstacle. Of course, the same argument applies to an incident, plane wave which is also nondiffracting except the higher intensity at the central spot of a Bessel beam makes this self-repair property more apparent.

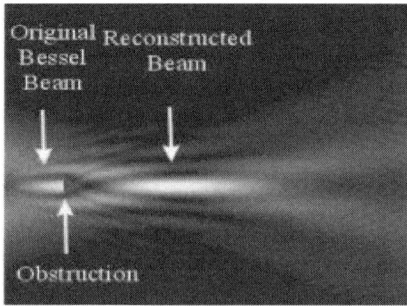

Figure 4   The beam propagates from left to right. The beam is blocked at the marked position and is seen to reform a short distance later.

1.3   Huygens-Fresnel principle

Equation (4) is a manifestation of the Huygens-Fresnel principle of the $3 - d$ wave equation, namely every unobstructed point of a wavefront, at a given instant in time, serves as a source of spherical secondary wavelets, with the same frequency as that of the primary wave. The amplitude of the optical field at any point beyond is the superposition of all these wavelets, taking into consideration their amplitudes and relative phases.

First we note that

$$\frac{\partial v}{\partial n} = \frac{\partial}{\partial n} \frac{e^{ikr'}}{r'} = |\hat{\mathbf{n}} \cdot \hat{\mathbf{r}}'| \frac{\partial}{\partial r'} \frac{e^{ikr'}}{r'} = |\hat{\mathbf{n}} \cdot \hat{\mathbf{r}}'| ik \frac{e^{ikr'}}{r'} \left(1 - \frac{1}{ikr'}\right), \tag{6}$$

$$\frac{\partial u}{\partial n} = \frac{\partial}{\partial n} \frac{e^{ikr_0}}{r_0} = -|\hat{\mathbf{n}} \cdot \hat{\mathbf{r}}_0| \frac{\partial}{\partial r_0} \frac{e^{ikr_0}}{r_0} = -|\hat{\mathbf{n}} \cdot \hat{\mathbf{r}}_0| ik \frac{e^{ikr_0}}{r_0} \left(1 - \frac{1}{ikr_0}\right). \tag{7}$$

It follows that

$$u(\mathbf{r}) = -\frac{a}{4\pi} \int_A \frac{e^{ik(r'+r_0)}}{r_0 r'} ik \left[|\hat{\mathbf{n}} \cdot \hat{\mathbf{r}}_0| \left(1 - \frac{1}{ikr_0}\right) + |\hat{\mathbf{n}} \cdot \hat{\mathbf{r}}'| \left(1 - \frac{1}{ikr'}\right)\right] d\sigma, \tag{8}$$

where $r' = |\mathbf{r} - \mathbf{r}_0|$ and $r_0 = |\mathbf{r}_0|$.

We would like to work out a simplification of (8) under the assumption

$$kr_0 \gg 1, \quad kr' \gg 1,$$

that is, the Kirchhoff diffraction formula

$$u(\mathbf{r}) = -\frac{ika}{4\pi} \int_A \frac{e^{ik(r'+r_0)}}{r_0 r'} (|\cos\theta_0| + |\cos\theta'|) d\sigma, \tag{9}$$

which means that the wave field is the repropagation of the wave front ae^{ikr_0}/r_0 by the Green function G_0 times the inclination factor. We note that this is not the first term of any iterative scheme of the boundary value problem for the reduced wave equation.

For a planar screen we can write $\mathbf{r}' = (z, \mathbf{x} - \mathbf{y})$ where z is the longitudinal coordinate and $\mathbf{x}$ (or $\mathbf{y}$) are the transverse coordinates of the aperture. In the case of *planar* incident wave $u_i = ae^{ikz}$ or a faraway source, then $\theta_0 = 0$ and (9) becomes

$$u(\mathbf{r}) = -\frac{ik}{4\pi} ae^{ikz_0} \int_A \frac{e^{ikr'}}{r'} [1 + \cos\theta'] d\sigma \tag{10}$$

$$= -\frac{ik}{2\pi} ae^{ikz_0} \int_A \frac{e^{ikr'}}{r'} \cos^2 \frac{\theta'}{2} d\sigma. \tag{11}$$

Note that the Kirchhoff diffraction formula does *not* predict $u = 0$ on the screen (i.e., $\theta' = \pi/2$) as does the Rayleigh diffraction formula of the first kind. Instead, the Kirchhoff diffraction formula predicts $u = 0$ in the back-propagation direction (i.e., $\theta' = \pi$).

Using the Huygens-Fresnel principle we can generalize Kirchhoff's formula to the case with arbitrary source and aperture (transmission) function $\tau(\mathbf{r}_0)$:

$$u(\mathbf{r}) = -ik \int_A \tau(\mathbf{r}_0) u_i(\mathbf{r}_0) \frac{e^{ikr'}}{r'} \big(|\cos\theta_0| + |\cos\theta'|\big) d\sigma(\mathbf{r}_0), \qquad (12)$$

where the contribution from θ_0 represents a "history" term.

For example for a thin parabolic lens the aperture (transmission) function is given by

$$\tau(\mathbf{x}) \approx |\tau(\mathbf{x})| e^{-ik|\mathbf{x}|^2/(2f)}.$$

In the case of plano-convex lens we have $f = R/(n-1)$; in the case of bi-convex lens with radii of curvature R_1, R_2 we have

$$\frac{1}{f} = (n-1)\Big(\frac{1}{R_1} + \frac{1}{R_2}\Big).$$

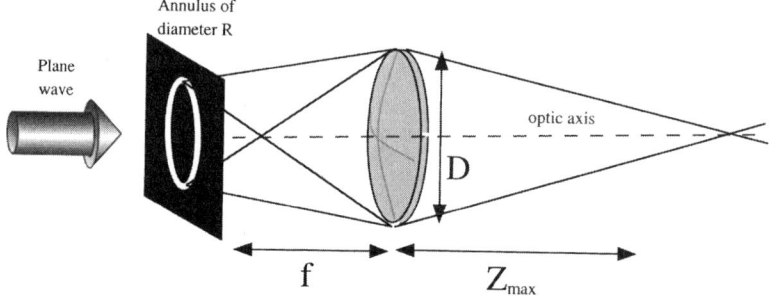

Figure 5   A He-Ne laser is used to illuminate the annular slit to produce the Bessel beam (adapted from [44]).

The axicon (Figure 2) can be replaced with an annular slit (the aperture) at the focal plane of a thin parabolic lens as depicted in Figure 5. After the incident plane wave passing through the annular slit each point of the slit becomes a point source of a spherical wave which is converted back into a plane wave again with a transverse wavevector given by minus the transverse coordinates of the point on the aperture plane. An annular slit produces the entire set of the wavevectors of an axicon. An incident plane wave corresponds to a constant A in (2) and thus the zeroth order Bessel beam.

1.4   Fresnel and Fraunhofer diffraction

We derive the Fresnel diffraction formula which is the paraxial approximation of (11) under the small-diffraction-angle condition $A \ll z$. In the Fresnel diffraction one approximates the radius function by the quadratic approximation

$$|\mathbf{r}| \approx z + \frac{|\mathbf{x} - \mathbf{y}|^2}{2z}$$

and set $\theta' = 0$ (hence called "paraxial approximation"). Let $\mathbb{I}_A$ be the indicator function of the aperture and

$$G(z, \mathbf{x}) = e^{ikz} \frac{k}{i2\pi z} \exp\left[i\frac{k}{2z}|\mathbf{x}|^2\right].$$

Then (11) can be written as

$$u(z, \mathbf{x}) = u_0(\cdot) * G(z, \cdot), \tag{13}$$

where $u_0 = u_i \tau \mathbb{I}_A$. Note that the Fresnel formula (13) satisfies the equation

$$\left(\frac{\partial}{\partial z} - ik\right)u = \frac{i}{2k}\Delta_x u \tag{14}$$

with u_0 as initial data.

For far field, the Fraunhofer diffraction formula is a good approximation under the condition $L_f = \sqrt{\lambda z} \gg A$. This means Fresnel number

$$\gamma = \frac{z}{kA^2} \gg 1.$$

This leads to negligible quadratic phase factor and the curvature of the wave front and the approximation of spherical waves by planar waves of different angles (or spatial frequencies):

$$u(z, \mathbf{x}) = e^{ikz} e^{i\frac{k|\mathbf{x}|^2}{2z}} \frac{k}{i2\pi z} \int_A u_0(\mathbf{y}) \exp\left(-i\mathbf{p} \cdot \mathbf{y}\right) d\mathbf{y} \tag{15}$$

with the Fourier variables, called *spatial frequencies*

$$\mathbf{p} = 2\pi \frac{\mathbf{x}}{\lambda z} = \frac{k\mathbf{x}}{z}.$$

1.5   Focal spot size and resolution

Let us consider the Fraunhofer diffraction by a slit and a circular aperture.

Consider the planar incident wave $u_0 = ae^{ikz}$. Let A be the x-width of the y-infinitely long slit. Then the Fraunhofer formula amounts to the Fourier transformation of a finite interval of length A modulo a phase factor independent of the transverse variable $\mathbf{x}$. We obtain

$$|u|^2(x) = |a|^2 \left| \frac{\sin(pA/2)}{p/2} \right|^2, \quad p = kx/z,$$

which has the minima for $pA/2 = \pm\pi, \pm2\pi, \cdots$.

The size of the main lobe determines the resolution of the system. We set $pA/2 = \pi$ and obtain

$$\rho = \frac{\lambda z}{A},$$

which is called the Rayleigh (or Abbe) resolution formula. Note that as the derivation relies on (15) the Rayleigh formula is not valid for $z/A \to 0$.

In the case of a circular aperture of diameter A, the Fraunhofer diffraction formula amounts to the Fourier transform of a circular disk of diameter A modulo a phase factor, i.e.

$$|u|^2(\rho) = |a|^2 \left| \frac{J_1(kA\tan\theta/2)}{kA\tan\theta/2} \right|^2,$$

where θ is the diffraction angle and

$$J_1(\xi) = \frac{1}{\pi} \int_{-\pi/2}^{\pi/2} \sin(\xi\cos\phi)\cos\phi\, d\phi$$

is the Bessel function of order one. The first zero of $J_1(\xi)$ is at $\xi = 1.22\pi$. Consequently, the first dark ring occurs at the diffraction angle

$$\tan\theta_1 = 1.22\frac{\lambda}{A}. \tag{16}$$

For the resolving power of a microscope we need to consider the optical geometry of the instrument as shown in Figure 6. We use the unprimed notion for the quantities on the object plane and the primed notion for those on the image plane. Let us observe the relation

$$\overline{PO}\tan\alpha = \overline{P'O'}\tan\alpha',$$

which follows from the following string of identities

$$\frac{\overline{P'O'}}{\overline{PO}} = \frac{\overline{O'Q}}{\overline{OQ}} = \frac{\overline{O'Q}/\overline{QM}}{\overline{OQ}/\overline{QM}} = \frac{\tan\alpha}{\tan\alpha'},$$

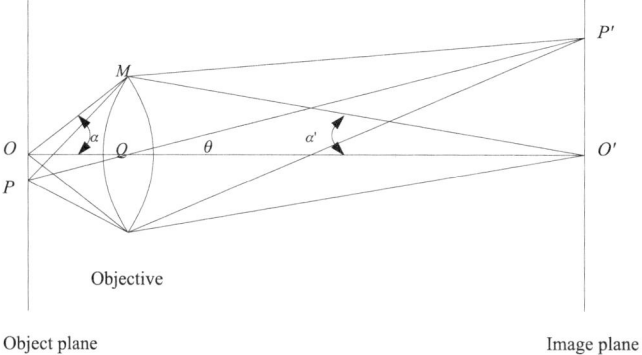

Figure 6   Ray geometry of a microscope.

cf. Figure 6. A more accurate relation, called the sine condition,

$$\overline{OP} \sin \alpha = \overline{O'P'} \sin \alpha', \tag{17}$$

can be derived by using the Fermat principle in geometrical optics [3].

According to Rayleigh's criterion (16) θ needs to be large than θ_1 for P' and O' to be distinguishable, see Figure 7. This implies by the sine condition that

$$\overline{OP} > 1.22 \frac{\lambda}{n \tan \alpha}.$$

Therefore, the ultimate resolution of a microscope with a circular lens is limited by $1.22\lambda/(2n)$.

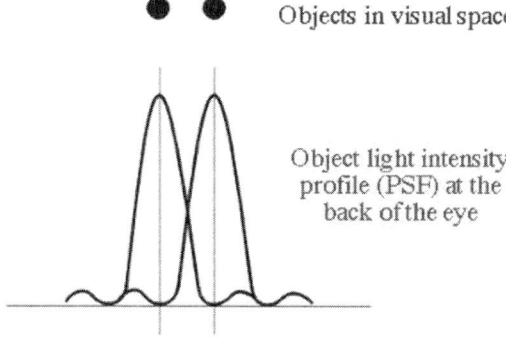

Figure 7   Rayleigh resolution criterion.

2 Approximations: weak fluctuation

First we consider several widely used approximation for propagation in the weak fluctuation regime where $\tilde{\epsilon}$ is small.

2.1 Born approximation

We rewriting (1) in the form

$$(\nabla^2 + k^2)u = -k^2\tilde{\epsilon}u$$

and convert it into the Lippmann-Schwinger integral equation

$$u(\mathbf{r}) = u_0(\mathbf{r}) - k^2 \int G_0(\mathbf{r} - \mathbf{r}')\tilde{\epsilon}(\mathbf{r}')u(\mathbf{r}')d\mathbf{r}',$$

where u_0 is the wave field in the absence of the heterogeneity. This formulation suggests the iteration scheme for solving for u which produces the Born series for u. Substituting u_0 in the right-hand side of the above integral we obtain the first term u_1 in the Born series. Repeating this process we can develop the entire Born series.

2.2 Rytov approximation

The Rytov approximation employs the exponential transformation

$$u(\mathbf{r}) = u_0 e^{\phi}, \tag{18}$$

where u_0 is the solution of the Helmholtz equation in the absence of medium fluctuation, and develop a series solution for ϕ in a way analogous to the Born approximation. The equation for ϕ is

$$(\nabla^2 + k^2)(u_0\phi) + u_0|\nabla\phi|^2 + k^2\tilde{\epsilon}u_0 = 0.$$

Note that the equation is nonlinear but the multiplicative heterogeneity becomes additive.

Once again we can formulate the equation as the integral equation

$$\phi(\mathbf{r}) = -\frac{1}{u_0(\mathbf{r})} \int G_0(\mathbf{r} - \mathbf{r}')u_0(\mathbf{r}')\left[|\phi(\mathbf{r}')|^2 + k^2\tilde{\epsilon}(\mathbf{r}')\right]d\mathbf{r}'$$

and develop a series expansion for ϕ. The Rytov approximation amounts to neglecting the quadratic term in ϕ:

$$\phi_1(\mathbf{r}) = -\frac{k^2}{u_0(\mathbf{r})} \int G_0(\mathbf{r} - \mathbf{r}')u_0(\mathbf{r}')\tilde{\epsilon}(\mathbf{r}')d\mathbf{r}'.$$

The Rytov approximation is consistent with the Born approximation in that the latter is the two-term approximation of $u_0 e^{\phi_1} = u_0(1 + \phi_1 + \cdots)$ in view of the fact $u_0 \phi_1 = u_1$. However, the Rytov approximation is generally believed to be superior to the Born approximation for propagation in turbulent media. The Rytov approximation is the basic propagation model in diffractive tomography of which the computerized tomography is a limit case as $k \to \infty$ [46]. One can find a comprehensive treatment of the Rytov method in [53].

More generally, let us divide a heterogeneous medium as a series of thin layers for each of which the Rytov approximation applies. Let u_0 be the incident wave. Then after the 1-st layer the wave field is $u_0 u_1$ where $u_1 = e^{\phi_1}$ is the Rytov solution. For the second layer, $u_0 u_1$ is the incident wave and $u_0 u_1 u_2$ is the transmitted wave where $u_2 = e^{\phi_2}$. In this way, the total output field becomes $u_0 u_1 u_2 u_3 \cdots = u_0 \exp(\phi_1 + \phi_2 + \phi_3 + \cdots)$. Namely the exponential representation (18) turns the product of many contributions into a summation which is more convenient to analyze. That is why the Rytov method is more suitable for the so-called line-of-sight propagation. On the other hand, for the single scattering problem where the output field is a sum of contributions from different parts of the medium fluctuation, the sum form $u = u_0 + u_1$ is more suitable [8], [37]. In reality, however, the true picture is the combination of the two.

2.3   The extended Huygens-Fresnel principle

The extended Huygens-Fresnel principle is the extension of (12) as

$$u(\mathbf{r}) = -2ik \int u_i(\mathbf{r}_0) G_0(\mathbf{r} - \mathbf{r}_0) e^{\phi_1(\mathbf{r}, \mathbf{r}_0)} \cos^2 \frac{\theta'}{2} d\sigma \qquad (19)$$

where ϕ_1 is the first Rytov approximation.

In the paraxial regime (19) can be approximated by

$$u(\mathbf{r}) = -\frac{ik}{2\pi z} \int e^{ikz} e^{i\frac{k|\mathbf{x} - \mathbf{x}_0|^2}{2z}} e^{\phi_1} u_i(\mathbf{r}_0) d\sigma. \qquad (20)$$

A useful application of the Huygens-Fresnel principle is in treating the case where the extended medium can be approximated by a series of phase screens. A phase screen is defined as having the transmission function of the form $\tau(\mathbf{x}) = e^{\phi(\mathbf{x})}$ where ϕ may be complex-valued and zero reflection coefficient. Mathematically this amounts to the forward-scattering (or paraxial) approximation [38], [14].

Consider a series of phase screens with the transmission functions $\tau_j = e^{i\phi_j}$ as in Figure 8. By iterating the extended Huygens-Fresnel

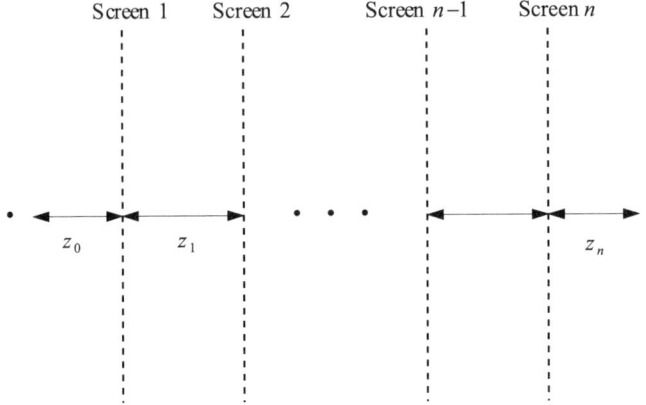

Figure 8   Multiple phase screens.

principle we obtain the wave field at the end of the screens [51]

$$u(z, \mathbf{x}) = \int d\mathbf{x}_0 \cdots \int d\mathbf{x}_n u_i(\mathbf{x}_0) e^{\phi_1 + \cdots + \phi_n} G_0(z_0, \mathbf{x}_1 - \mathbf{x}_0)$$

$$G_0(z_1, \mathbf{x}_2 - \mathbf{x}_1) \cdots G_0(z_n, \mathbf{x} - \mathbf{x}_n).$$

The concept of phase screen is not restricted to electromagnetic wave propagation. For example, the basis of image formation in the transmission electron microscope is the interaction of the electron with the object. In most applications, the elastic scattering interaction can be described as phase shift of the incident wave traveling in the z direction

$$\Phi(\mathbf{x}) = \int C(z, \mathbf{x}) dz, \tag{21}$$

where $C(z, \mathbf{x})$ is the Coulomb potential within the object and the incident wave u_0 is modified according to

$$u(\mathbf{x}) = u_0 e^{i\Phi(\mathbf{x})}.$$

In contrast, tomography concerns mostly the inelastic scattering (absorption). The Born approximation would then lead to $u = u_0(1 + i\Phi)$ [32]. In diffraction theory, a *field*, rather than a material object, can often be modeled as a *phase* object, cf. Section 7.1.

2.4 Paraxial approximation

The forward-scattering (or paraxial) approximation concerns the propagation of modulated high-frequency carrier wave written as

$$E(\mathbf{x}) = \Psi(z, \mathbf{x}) e^{ikz}$$

with $k \gg 1$. The equation for Ψ is

$$\frac{\partial^2}{\partial z^2}\Psi + \nabla^2_\perp \Psi + 2ik\frac{\partial}{\partial z}\Psi + 2k^2\tilde{n}\Psi = 0, \tag{22}$$

where $\tilde{n}$ is the relative fluctuation of the index of refraction. Then under the assumption

$$|\Psi_{zz}| \ll 2k|\Psi_z|,$$

we obtain

$$i\frac{\partial}{\partial z}\Psi + \frac{1}{2k}\Delta\Psi + k\tilde{n}\Psi = 0, \tag{23}$$

which is analogous to the Fresnel approximation (14).

A more formal approach is to factorize eq. (22) as

$$\left(\frac{\partial}{\partial z} + ik + ikQ\right)\left(\frac{\partial}{\partial z} + ik - ikQ\right)\Psi + ik\left[Q, \frac{\partial}{\partial z}\right]\Psi = 0, \tag{24}$$

where

$$Q = \left(1 + \frac{1}{k^2}\nabla^2_\perp + 2\tilde{n}\right)^{1/2}$$

and $\left[Q, \frac{\partial}{\partial z}\right] = Q\frac{\partial}{\partial z} - \frac{\partial}{\partial z}Q$ is the commutator. For Q to be well-defined, it is necessary that it acts on a wave field whose maximum transverse wavenumber is smaller than k (i.e. the evanescent waves are negligible) and $\tilde{n}$ is sufficiently small. For weak fluctuation $\tilde{n} \ll 1$ or z-independent $\tilde{n}$ the commutator can be dropped and the remaining equation is the product of two nearly commutative operators describing forward and backward propagating waves. If only the forward-propagating term of eq. (24) is retained, we have the generalized paraxial wave equation

$$\left(\frac{\partial}{\partial z} + ik - ikQ\right)\Psi = 0. \tag{25}$$

The paraxial approximation turns the two-sided boundary value problem for the reduced wave equation into the one-sided initial value problem for the Schrödinger equation [51], [50].

The standard paraxial wave equation is the simplest approximation of (25) by Taylor expanding Q around the identity and using

$$Q' = 1 + \frac{1}{2k^2}\nabla^2_\perp + \tilde{n}$$

to approximate Q. The resulting equation is (23). Formally this approximation requires that $\tilde{n}$ is small and that the transverse wavenumber of Ψ is much smaller than k (i.e. small diffraction angle).

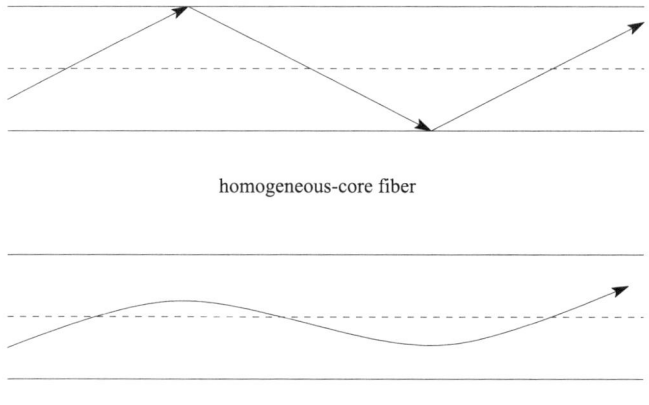

homogeneous-core fiber

parabolic index fiber

Figure 9   Homogeneneous vs. graded index wave guide.

In the graded-index optical fibers the index of refraction has the profile

$$n^2 = \begin{cases} n_0^2[1 - 2\sigma|\mathbf{x}|^2/a^2], & 0 < |\mathbf{x}| < a \\ n_0^2[1 - 2\sigma], & |\mathbf{x}| > a. \end{cases}$$

Note that in this case $\left[Q, \frac{\partial}{\partial z}\right] = 0$. The parabolic index fiber can reduce pulse dispersion in optical communication because the rays making larger angles with the axis also traverse a larger path length in a region of lower refractive index [33], see Figure 9. In the wave-guide with infinitely extended parabolic profile

$$n^2 = n_0^2\left[1 - \frac{|\mathbf{x}|^2}{a^2}\right],$$

the eigenfunctions of the paraxial wave equation can be solved for exactly. The resulting eigenfunctions form a complete and orthogonal set of solutions for the square-integrable functions on the transverse plane. When the transverse coordinate is one-dimensional, the normalized eigenfunctions are the Hermite functions

$$\Psi_m(x) = \sqrt{\frac{1}{\sqrt{\pi}2^m m!}} H_m(x)e^{-\frac{x^2}{2}}, \quad m = 0, 1, 2, \cdots, \qquad (26)$$

where $H_j(x)$ stands for the Hermite polynomial of order j

$$H_j(x) = (-1)^j e^{x^2}\frac{d^j}{dx^j}e^{-x^2}$$

or equivalently defined recursively as $H_0(x) = 1, H_1(x) = 2x, H_{j+1}(x) = 2xH_j(x) - 2jH_{j-1}(x)$. A laser beam profile can be decomposed into the so called Hermite-Gaussian modes, each mode being characterized by two integer numbers n and m

$$\text{TEM}_{mn}(x, y) = w(z)^{-1} \Psi_m \left(\frac{x}{w(z)} \right)$$
$$\times \Psi_n \left(\frac{y}{w(z)} \right) e^{-\frac{ik(x^2+y^2)}{2R(z)} - i(kz - \phi_m(z) - \phi_n(z))}, \quad (27)$$

where Φ_{mn} is the Guoy phase defined as

$$\phi_m(z) = (m + 1/2) \arctan \left(\frac{\lambda z}{\pi w_0^2} \right).$$

TEM is the acronym for *Transverse Electro-Magnetic* mode, where m refers to the number of intensity minima in the direction of the electric field oscillation, and n refers to the number of minima in the direction of the magnetic field oscillation, Figure 10.

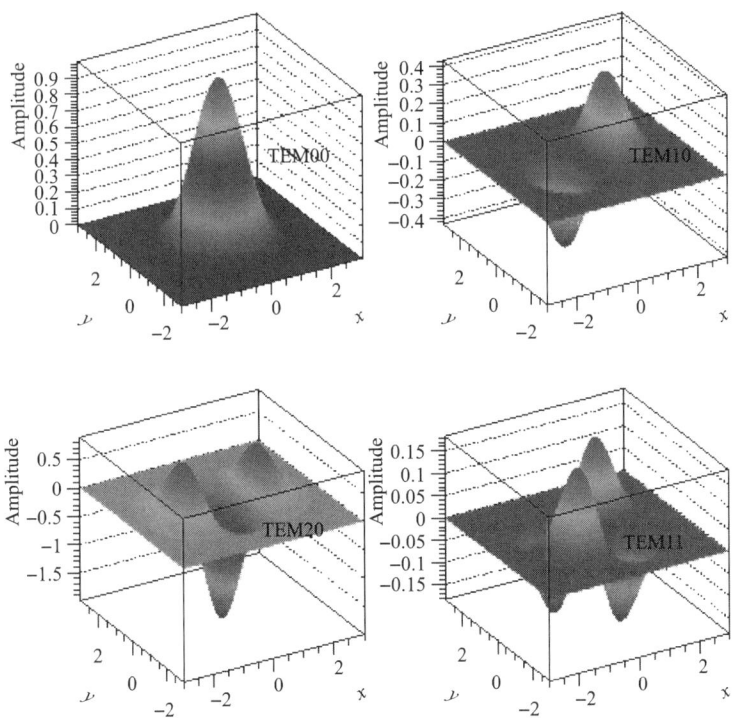

Figure 10   Graphs of $\text{TEM}_{mn}, m, n = 0, 1$. The fundamental mode TEM_{00} has a Gaussian profile in the transverse plane while TEM_{10} exhibits a left-right asymmetry.

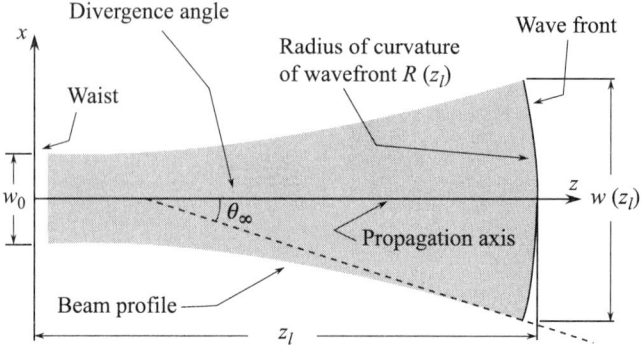

Figure 11   Longitudinal profile of a laser beam.

The set of modes is characterized at every point along the propagation axis by two functions: $R(z)$ and $w(z)$ as shown on Figure 11. The first describes the radius of curvature of the wavefront that intersects the propagation axis, while the second parameter, with respect to the fundamental mode TEM_{00}, gives the radius in the transverse plane for which the amplitude of the field has decreased by a factor e^{-1} with respect to the amplitude value along the propagation axis. The transversal intensity distribution of the TEM_{00} has a Gaussian dependence and its radius $w(z)$ contracts to a minimum w_0 known as the waist of the beam. The two parameters $R(z)$ and $w(z)$ are determined by the waist size w_0 and by the distance z from the waist position:

$$w^2(z) = w_0^2 \left[1 + \left(\frac{\lambda z}{\pi w_0^2} \right)^2 \right], \tag{28}$$

$$R(z) = z \left[1 + \left(\frac{\pi w_0^2}{\lambda z} \right)^2 \right]. \tag{29}$$

The longitudinal profile of a laser beam is a hyperbola with asymptotes forming an angle

$$\theta_\infty = \frac{\lambda}{\pi w_0}$$

with the propagation axis, which defines the divergence of the beam.

Nevertheless, the paraxial wave equation is not exactly solvable in general. A general way of understanding and numerical solution of the standard paraxial equation is through the *step-splitting* method. By Trotter's product formula we can write the solution to (23) for a small step as

$$\Psi(z + dz, \mathbf{x}) \approx e^{i \frac{k}{2} \Delta dz} e^{ik \int_z^{z+dz} \tilde{n}(z', \mathbf{x}) dz'} \Psi(z, \mathbf{x}).$$

The evolution with only $\tilde{n}$ present is the effect of medium fluctuation represented as a simple phase screen in the geometrical optics while the evolution with only Δ present is the diffraction effect in the free space. Thus the entire propagation consists of a series of phase screens separated by uniform medium in between cf. Figure 8. When dz is greater than the correlation length of medium fluctuation the phase screens can be considered statistically independent. The phase screens cause the fluctuation in phase and the diffraction causes the fluctuation in amplitude (and phase). This algorithm is the discrete analog of path-integral method of the paraxial wave equation in the limit of $dz \to 0$ [51]. Moreover, the multiple phase screen model is not restricted to the paraxial wave equation and can be extended to deal with point source and spherical wave by writing the equation in spherical coordinates.

We will be mostly interested in a randomly heterogeneous medium such as the turbulent atmosphere for which $\tilde{n}(\mathbf{r})$ is a random function. We shall use $\langle \cdot \rangle$ and $\mathbb{E}$ to denote the averaging with respect to the ensemble of noise and media, respectively. There are two regimes of interest: the weak fluctuation regime and the strong fluctuation regime and they require different treatments. The weak fluctuation regime can be defined as

$$\frac{\mathbb{E}(I - \mathbb{E}I)^2}{|\mathbb{E}I|^2} < 1, \quad I = |\Psi|^2,$$

whose left-hand side is also known as the scintillation index. The strong fluctuation regime is when the scintillation index is much larger than one.

3   The Wigner distribution

In this section, we discuss a useful phase-space tool for analyzing imaging properties of optical elements [12, 31]. This is a quadratic transform of the wave field.

The standard Wigner distribution (or transform) for a wave field Ψ is defined as

$$W[\Psi](\mathbf{x}, \mathbf{p}) = \frac{1}{(2\pi)^d} \int e^{-i\mathbf{p}\cdot\mathbf{y}} \Psi\left(\mathbf{x} + \frac{\mathbf{y}}{2}\right) \Psi^*\left(\mathbf{x} - \frac{\mathbf{y}}{2}\right) d\mathbf{y} \qquad (30)$$

from which the wave amplitude Ψ can be recovered up to a constant phase factor by using

$$\Psi(\mathbf{x}_1)\Psi^*(\mathbf{x}_2) = \int W[\Psi]\left(\frac{1}{2}(\mathbf{x}_1 + \mathbf{x}_2), \mathbf{q}\right) \exp\left[i\mathbf{q} \cdot (\mathbf{x}_1 - \mathbf{x}_2)\right] d\mathbf{q}.$$

For example, the Gaussian beam

$$\Psi(x) = C e^{-\frac{(x-x_0)^2}{2w_0^2} + ip_0 x}$$

gives rise to a Gaussian Wigner distribution as shown in Figure 12.

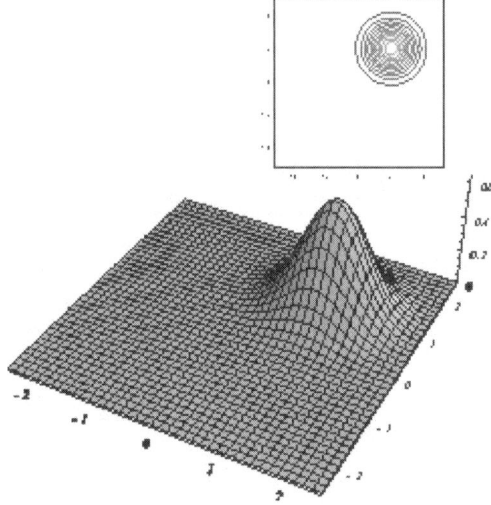

Figure 12   The Wigner distribution associated with the Gaussian beam.

The Wigner distribution is the Fourier transform in the phase space of the ambiguity function

$$A(\mathbf{y}, \mathbf{q}) = \frac{1}{(2\pi)^d} \int e^{-i\mathbf{q}\cdot\mathbf{x}} \Psi\left(\mathbf{x} + \frac{\mathbf{y}}{2}\right) \Psi^*\left(\mathbf{x} - \frac{\mathbf{y}}{2}\right) d\mathbf{x}$$

widely used in radar signal processing, also called the Fourier-Wigner transform of Ψ [35]. While the ambiguity function is an expression for correlative structure, the Wigner distribution describes the energy distribution in the phase space. This is manifest in the following properties. For instance, partial integration of W gives rise to the marginal distributions

$$\int W[\Psi](\mathbf{x}, \mathbf{p})d\mathbf{p} = |\Psi(\mathbf{x})|^2,$$

$$\int W[\Psi](\mathbf{x}, \mathbf{p})d\mathbf{x} = (2\pi)^d |\hat{\Psi}(\mathbf{p})|^2.$$

Also, the energy flux is given by

$$\frac{1}{2i}(\Psi\nabla\Psi^* - \Psi^*\nabla\Psi) = \int_{\mathbb{R}^d} \mathbf{p}W(z, \mathbf{x}, \mathbf{p})d\mathbf{p}. \tag{31}$$

We consider the following operators and their action on the corresponding Wigner distribution.

- Fourier transform $\mathfrak{F}\Psi(\mathbf{k}) = \int e^{-i\mathbf{k}\cdot\mathbf{x}}\Psi(\mathbf{x})d\mathbf{x}$ (Fraunhofer diffraction).
- Dilation $D_a, a > 0$: $D_a\Psi(\mathbf{x}) = a^{-1}\Psi(\frac{\mathbf{x}}{a})$.
- Symmetry: $S\Psi(\mathbf{x}) = \Psi(-\mathbf{x})$.
- Translation $T_{\mathbf{y}}, \mathbf{y} \in \mathbb{R}^2$: $T_{\mathbf{y}}\Psi(\mathbf{x}) = \Psi(\mathbf{x} - \mathbf{y})$.
- Modulation $M_{\mathbf{k}}, \mathbf{k} \in \mathbb{R}^2$: $M_{\mathbf{k}}\Psi(\mathbf{x}) = e^{i\mathbf{k}\cdot\mathbf{x}}\Psi(\mathbf{x})$.
- Chirp multiplication $P_{\mathbf{k}}, \mathbf{k} \in \mathbb{R}^2$: $P_{\mathbf{k}}\Psi(\mathbf{x}) = e^{i|\mathbf{k}||\mathbf{x}|^2}\Psi(\mathbf{x})$ (Lens effect).
- Chirp convolution $Q_{\mathbf{k}}, \mathbf{k} \in \mathbb{R}^2$: $Q_{\mathbf{k}}\Psi(\mathbf{x}) = e^{i|\mathbf{k}||\mathbf{x}|^2} \star \Psi(\mathbf{x})$ (free space propagation).

We have

- $W[\mathfrak{F}\Psi](\mathbf{x}, \mathbf{p}) = W[\Psi](-\mathbf{p}, \mathbf{x})$,
- $W[D_a\Psi](\mathbf{x}, \mathbf{p}) = W[\Psi](\frac{\mathbf{x}}{a}, a\mathbf{p})$,
- $W[S\Psi](\mathbf{x}, \mathbf{p}) = W[\Psi](-\mathbf{x}, -\mathbf{p})$,
- $W[T_{\mathbf{y}}\Psi](\mathbf{x}, \mathbf{p}) = W[\Psi](\mathbf{x} - \mathbf{y}, \mathbf{p})$,
- $W[M_{\mathbf{q}}\Psi](\mathbf{x}, \mathbf{p}) = W[\Psi](\mathbf{x}, \mathbf{p} - \mathbf{q})$,
- $W[P_{\mathbf{k}}\Psi](\mathbf{x}, \mathbf{p}) = W[\Psi](\mathbf{x}, \mathbf{p} - 2|\mathbf{k}|\mathbf{x})$,
- $W[Q_{\mathbf{k}}\Psi](\mathbf{x}, \mathbf{p}) = W[\Psi](\mathbf{x} - \frac{\mathbf{p}}{2|\mathbf{k}|}, \mathbf{p})$.

Moreover, $W[\Psi^*](\mathbf{x}, \mathbf{p}) = W[\Psi](\mathbf{x}, -\mathbf{p})$.

We note that all the above operations result in *volume-preserving* affine transformations of the phase space coordinates. Another non-trivial operator leading to a linear transformation in the phase space coordinates is the Schrödinger semigroup $e^{i\alpha\pi H/2}$ for the harmonic oscillator Hamiltonian $H = -\Delta + |\mathbf{x}|^2 - 1$. We have

$$W[e^{i\alpha\pi H/2}\Psi](\mathbf{x}, \mathbf{p}) = W[\Psi]\left(\mathbf{x}\cos\frac{\alpha\pi}{2} + \mathbf{p}\sin\frac{\alpha\pi}{2}, \mathbf{p}\cos\frac{\alpha\pi}{2} - \mathbf{x}\sin\frac{\alpha\pi}{2}\right)$$

corresponding to $\alpha\pi/2$ rotation [39]. Note that $\mathfrak{F} = \exp(i\pi H/2)$. In view of the fact the Fourier transform corresponds to $\frac{\pi}{2}$-rotation in the phase plane one can define $e^{i\alpha\pi H/2}\Psi$ to be the *fractional Fourier transform* $\mathfrak{F}^\alpha$ *of order* α. As a consequence of the above, the integration of W on any hyperplane in the phase space is proportional to the square modulus of some fractional Fourier transform of Ψ and hence is non-negative pointwise. It is noteworthy that all the above transformation can be realized by simple optical systems [12].

Let us state a few more properties of the Wigner transform. If $\Psi = \Psi_1 \star \Psi_2$ where $\star$ stands for the spatial convolution, then

$$W[\Psi] = \int W[\Psi_1](\mathbf{x} - \mathbf{y}, \mathbf{p})W[\Psi_2](\mathbf{y}, \mathbf{p})d\mathbf{y},$$

which is not obvious since the Wigner transform is quadratic. Likewise the pointwise product leads to the momentum convolution

$$W[\Psi_1\Psi_2] = \int W[\Psi_1](\mathbf{x}, \mathbf{p} - \mathbf{q})W[\Psi_2](\mathbf{x}, \mathbf{q})d\mathbf{q}.$$

The next property is called the Moyal identity:

$$\int W[\Psi_1]W[\Psi_2]d\mathbf{x}d\mathbf{p} = \frac{1}{(2\pi)^d}\left|\int \Psi_1\Psi_2^* d\mathbf{x}\right|^2, \quad \forall \Psi_1, \Psi_2 \in L^2(\mathbb{R}^d).$$

The fundamental property of the Wigner distribution in application to signal analysis is this theorem [39]:

Theorem 1. *Let $\Psi_j, j \in \mathbb{N}$, be the sequence of L^2-functions and let W_j be the Wigner distribution of Ψ_j. Then the following two properties are equivalent:*

$$\Psi_j, j \in \mathbb{N}, \text{ is an orthonormal basis for } L^2, \tag{32}$$

$$\begin{cases} \sum_j W_j(\mathbf{x}, \mathbf{p}) = 1, \forall \mathbf{x}, \mathbf{p}, \\ \int W_i W_j d\mathbf{x}d\mathbf{p} = \frac{\delta_{ij}}{(2\pi)^d}. \end{cases} \tag{33}$$

The first property in (33) is the partition of unity in the phase-space coordinates. The second property (33) is the consequence of the Moyal identity. As a result of the theorem, the set of Wigner distributions associated with the Hermite functions $\{\Psi_m\}$ satisfies (33), i.e. partition of unity in the phase space and the orthogonality.

The most "troublesome" feature of the Wigner distribution is its possible negative value (Figures 13 and 14) and the resulting *lack* of uniform L^1-estimate like

$$\int |W|d\mathbf{x}d\mathbf{p} < C \tag{34}$$

for some constant C and all $\|\Psi\|_2 = 1$. As a result, the first property holds only in the sense of distribution. On the other hand, the Wigner distribution satisfies uniform bound in L^∞ and L^2.

Before ending this section, let us note that if $u(z, \mathbf{x})$ is governed by (14) then $W[u]$ satisfies

$$\frac{\partial}{\partial z}W + \frac{\mathbf{p}}{k} \cdot \nabla_{\mathbf{x}}W = 0,$$

which can be solved by method of characteristics.

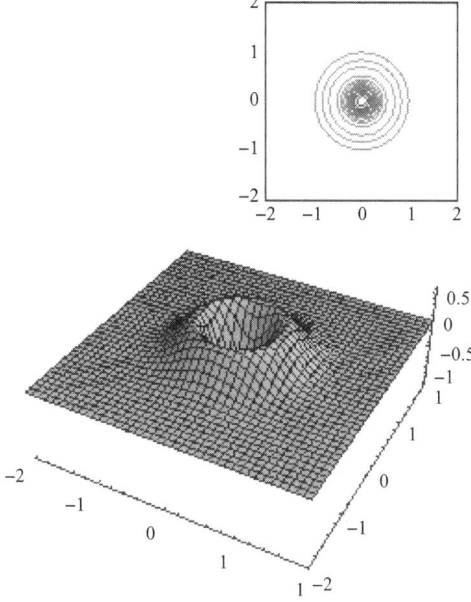

Figure 13   The Wigner distribution associated with Ψ_1.

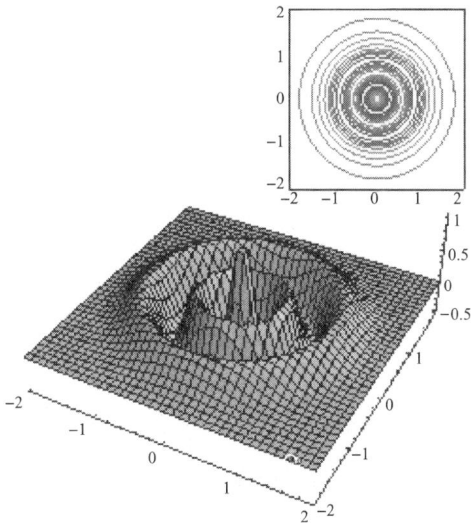

Figure 14   The Wigner distribution associated with Ψ_4.

4   Markovian approximation

For the weak fluctuation regime the Rytov method is suitable. The strong fluctuation regime is harder. For the statistically homogeneous random media, the Markovian model is fundamental and widely used [52].

4.1   White-noise scaling

We will take a somewhat different formulation in terms of the Wigner distribution. First let us non-dimensionalize the paraxial wave equation by setting

$$z \to z/L_z, \quad \mathbf{x} \to \mathbf{x}/L_x, \tag{35}$$

where L_z is roughly the distance of propagation and L_x is some reference length. For example, it is natural to choose L_x as the correlation length of the index fluctuation. We obtain

$$i\frac{\partial \Psi}{\partial z} + \frac{\gamma}{2}\Delta\Psi + kL_z\tilde{n}(zL_z, \mathbf{x}L_x)\Psi = 0, \tag{36}$$

where γ is the dimensionless Fresnel number

$$\gamma = \frac{L_z}{kL_x^2}.$$

The Fraunhofer diffraction corresponds to $\gamma \to \infty$ when, e.g. $L_z \to \infty$ with k_0, L_x fixed; the Fresnel diffraction corresponds to $\gamma = O(1)$; the geometrical optics limit corresponds to $\gamma \to 0$ when, e.g. $k_0 \to \infty$ with L_z, L_x fixed.

Note that when $k_0 \gg 1$ or $L_z \gg 1$ the fluctuation can be large even when $\tilde{n} \ll 1$. Below we shall write

$$kL_z\tilde{n}(zL_z, \mathbf{x}L_x) = \frac{\sqrt{L_z}}{\gamma}V(zL_z, \mathbf{x}).$$

The purpose of this is to introduce the normalized potential V which has $O(1)$ magnitude and transverse correlation length so that the right-hand side manifests the central-limit scaling.

To study the long distance propagation $L_z \equiv \varepsilon^{-2} \to \infty$ limit while μ, γ are fixed.

For arbitrary Fresnel number we redefine the Wigner distribution as

$$W(\mathbf{x}, \mathbf{k}) = \frac{1}{(2\pi)^d}\int e^{-i\mathbf{k}\cdot\mathbf{y}}\Psi(\mathbf{x} + \frac{\gamma\mathbf{y}}{2})\Psi^*(\mathbf{x} - \frac{\gamma\mathbf{y}}{2})d\mathbf{y}. \tag{37}$$

All the nice properties of the Wigner distribution for $\gamma = 1$ survive with suitable rescaling.

The Wigner distribution satisfies an evolution equation, called the Wigner-Moyal equation,

$$\frac{\partial W}{\partial z} + \mathbf{p} \cdot \nabla_{\mathbf{x}} W + \mathcal{V}W = 0 \tag{38}$$

with the initial data

$$W_0(\mathbf{x}, \mathbf{k}) = \frac{1}{(2\pi)^d} \int e^{i\mathbf{k}\cdot\mathbf{y}} \Psi_0(\mathbf{x} - \frac{\gamma \mathbf{y}}{2}) \Psi_0^*(\mathbf{x} + \frac{\gamma \mathbf{y}}{2}) d\mathbf{y}, \tag{39}$$

where the operator $\mathcal{V}$ is formally given as

$$\mathcal{V}W = i \int e^{i\mathbf{q}\cdot\mathbf{x}} \gamma^{-1} \left[W(\mathbf{x}, \mathbf{p} + \gamma \mathbf{q}/2) - W(\mathbf{x}, \mathbf{p} - \gamma \mathbf{q}/2) \right] \widehat{V}(\frac{z}{\varepsilon^2}, d\mathbf{q}).$$

Before taking the limit $\varepsilon \to 0$, let us pause to comment on the geometrical optics limit $\gamma \to 0$. It is not hard to show that the Wigner-Moyal equation converges formally to

$$\frac{\partial}{\partial z} W + \mathbf{p} \cdot \nabla_{\mathbf{x}} W + \nabla_{\mathbf{x}} V \cdot \nabla_{\mathbf{p}} W = 0,$$

which is known as the Liouville equation in classical Hamiltonian mechanics. The Liouville equation is equivalent to the Hamiltonian system with the Hamiltonian $|\mathbf{p}|^2/2 + V$. Conventionally, the geometrical optics limit is approached by using the WKB method.

4.2   Markovian limit

Let us return to the limit with $\varepsilon \to 0$. It can be proved under very general condition that the weak solution of the Wigner-Moyal equation converges in law to the Markov process governed by the Itô equation [16]

$$dW_z = (-\mathbf{p} \cdot \nabla_{\mathbf{x}} + \mathcal{Q}_0) W_z \, dz + d\mathcal{B}_z W_z, \quad W_0(\mathbf{x}) \in L^2(\mathbb{R}^{2d}) \tag{40}$$

or as the Stratonovich's equation

$$dW_z = -\mathbf{p} \cdot \nabla_{\mathbf{x}} + d\mathcal{B}_z \circ W_z, \quad W_0(\mathbf{x}) \in L^2(\mathbb{R}^{2d}),$$

where $\mathcal{B}_z$ is the operator-valued Brownian motion with the covariance operator $\mathcal{Q}$, i.e.

$$\mathbb{E}\left[d\mathcal{B}_z \theta(\mathbf{x}, \mathbf{p}) d\mathcal{B}_{z'} \theta(\mathbf{y}, \mathbf{q}) \right] = \delta(z - z') \mathcal{Q}(\theta \otimes \theta)(\mathbf{x}, \mathbf{p}, \mathbf{y}, \mathbf{q}) dz dz'.$$

Here the covariance operators $\mathcal{Q}, \mathcal{Q}_0$ are defined as

$$\mathcal{Q}_0 \theta = \int \Phi(\mathbf{q}) \gamma^{-2} \left[-2\theta(\mathbf{x}, \mathbf{p}) + \theta(\mathbf{x}, \mathbf{p} - \gamma \mathbf{q}) + \theta(\mathbf{x}, \mathbf{p} + \gamma \mathbf{q}) \right] d\mathbf{q}.$$

$$\mathcal{Q}(\theta \otimes \theta)(\mathbf{x}, \mathbf{p}, \mathbf{y}, \mathbf{q})$$
$$= \int e^{i\mathbf{q}' \cdot (\mathbf{x} - \mathbf{y})} \Phi(\mathbf{q}') \gamma^{-2} \left[\theta(\mathbf{x}, \mathbf{p} - \gamma \mathbf{q}'/2) - \theta(\mathbf{x}, \mathbf{p} + \gamma \mathbf{q}'/2) \right]$$
$$\times \left[\theta(\mathbf{y}, \mathbf{q} - \gamma \mathbf{q}'/2) - \theta(\mathbf{y}, \mathbf{q} + \gamma \mathbf{q}'/2) \right] d\mathbf{q}'.$$

If we take the simultaneous limit $\gamma, \epsilon \to 0$ then the covariance operators become

$$\mathcal{Q}_0 \theta(\mathbf{x}, \mathbf{p}) = \nabla_{\mathbf{p}} \cdot \int \Phi(\mathbf{q}) \mathbf{q} \otimes \mathbf{q} \, d\mathbf{q} \cdot \nabla_{\mathbf{p}} \theta(\mathbf{x}, \mathbf{p})$$

$$\mathcal{Q}(\theta \otimes \theta)(\mathbf{x}, \mathbf{p}, \mathbf{y}, \mathbf{q})$$
$$= \nabla_{\mathbf{p}} \theta(\mathbf{x}, \mathbf{p}) \cdot \left[\int e^{i\mathbf{q}' \cdot (\mathbf{x} - \mathbf{y})} \Phi(\mathbf{q}') \mathbf{q}' \otimes \mathbf{q}' d\mathbf{q}' \right] \cdot \nabla_{\mathbf{q}} \theta(\mathbf{y}, \mathbf{q}).$$

The most useful feature of the Markovian model is that all the moments satisfy closed form equations we arrive after some algebra the following equation

$$\frac{\partial F^{(n)}}{\partial z} = -\sum_{j=1}^{n} \mathbf{p}_j \cdot \nabla_{\mathbf{x}_j} F^{(n)} + \sum_{j=1}^{n} \mathcal{Q}_0(\mathbf{x}_j, \mathbf{p}_j) F^{(n)}$$
$$+ \sum_{\substack{j,k=1 \\ j \neq k}}^{n} \mathcal{Q}(\mathbf{x}_j, \mathbf{p}_j, \mathbf{x}_k, \mathbf{p}_k) F^{(n)} \tag{41}$$

for the n-point correlation function

$$F^{(n)}(z, \mathbf{x}_1, \mathbf{p}_1, \ldots, \mathbf{x}_n, \mathbf{p}_n) \equiv \mathbb{E}\left[W_z(\mathbf{x}_1, \mathbf{p}_1) \cdots W_z(\mathbf{x}_n, \mathbf{p}_n) \right],$$

where $\mathcal{Q}_0(\mathbf{x}_j, \mathbf{p}_j)$ is the operator $\mathcal{Q}_0$ acting on the variables $(\mathbf{x}_j, \mathbf{p}_j)$ and $\mathcal{Q}(\mathbf{x}_j, \mathbf{p}_j, \mathbf{x}_k, \mathbf{p}_k)$ is the operator $\mathcal{Q}$ acting on the variables $(\mathbf{x}_j, \mathbf{p}_j, \mathbf{x}_k, \mathbf{p}_k)$.
    Equation (41) can be compactly written as

$$\frac{\partial F^{(n)}}{\partial z} = -\sum_{j=1}^{n} \mathbf{p}_j \cdot \nabla_{\mathbf{x}_j} F^{(n)} + \sum_{j,k=1}^{n} \mathcal{Q}(\mathbf{x}_j, \mathbf{p}_j, \mathbf{x}_k, \mathbf{p}_k) F^{(n)} \tag{42}$$

with the identification $\mathcal{Q}(\mathbf{x}_j, \mathbf{p}_j, \mathbf{x}_j, \mathbf{p}_j) = \mathcal{Q}_0(\mathbf{x}_j, \mathbf{p}_j)$. The operator

$$\mathcal{Q}_{\text{sum}} = \sum_{j,k=1}^{n} \mathcal{Q}(\mathbf{x}_j, \mathbf{p}_j, \mathbf{x}_k, \mathbf{p}_k) \tag{43}$$

is a non-positive symmetric operator.

In the case of the Liouville equation, eq. (42) can be more explicitly written as the Fokker-Planck equation on the phase space

$$\frac{\partial F^{(n)}}{\partial z} = \sum_{j=1}^{n} \mathbf{p}_j \cdot \nabla_{\mathbf{x}_j} F^{(n)} + \sum_{j,k=1}^{n} \mathbf{D}(\mathbf{x}_j - \mathbf{x}_k) : \nabla_{\mathbf{p}_j} \nabla_{\mathbf{p}_k} F^{(n)} \quad (44)$$

with

$$\mathbf{D}(\mathbf{x}_j - \mathbf{x}_k) = \int e^{i\mathbf{q} \cdot (\mathbf{x}_j - \mathbf{x}_k)} \Phi(\mathbf{q}) \mathbf{q} \otimes \mathbf{q} d\mathbf{q}.$$

Equation (42) for $n = 1$ takes the following form

$$\frac{\partial}{\partial z} \bar{W} + \mathbf{p} \cdot \nabla_{\mathbf{x}} \bar{W} = \mathcal{Q}_0 \bar{W}$$

which is exactly solvable since $\mathcal{Q}_0$ is a convolution operator. The Green function is

$$G_w(z, \mathbf{x}, \mathbf{p}, \bar{\mathbf{x}}, \bar{\mathbf{p}})$$
$$= \frac{1}{(2\pi)^2} \int \exp\left[i(\mathbf{q} \cdot (\mathbf{x} - \bar{\mathbf{x}}) - \mathbf{y} \cdot (\mathbf{p} - \bar{\mathbf{p}}) - z\mathbf{q} \cdot \bar{\mathbf{p}})\right] \quad (45)$$
$$\times \exp\left[-\frac{1}{\gamma^2} \int_0^z D_*(\gamma \mathbf{y} + \mathbf{q}\gamma s) ds\right] d\mathbf{y} d\mathbf{q},$$

where the (medium) structure function D_* is given by

$$D_*(\mathbf{x}) = \int \Phi(0, \mathbf{q}) \left[1 - e^{i\mathbf{x} \cdot \mathbf{q}}\right] d\mathbf{q}. \quad (46)$$

We shall refer to $\exp\left[-\gamma^{-2} \int_0^z D_*(\gamma \mathbf{y} + \mathbf{q}\gamma s) ds\right]$ as the *wave* structure function. The case $n = 2$ can be approximately solved in certain circumstances. In the next section, we discuss the application of these equations to the time reversal of waves in random media.

The most important quantity for us is the mutual coherence function $\Gamma_2(z, \mathbf{x}_1, \mathbf{x}_2; \bar{\mathbf{x}}_1, \bar{\mathbf{x}}_2) = \mathbb{E}\left[\Psi_1(z, \mathbf{x}_1)\Psi_2(z, \mathbf{x}_2)\right]$ with $\Psi_1(0, \mathbf{x}_1) = \delta(\mathbf{x}_1 - \bar{\mathbf{x}}_1)$ and $\Psi_2(0, \mathbf{x}_1) = \delta(\mathbf{x}_1 - \bar{\mathbf{x}}_2)$. From (37) it follows that

$$\bar{W}(0, \mathbf{x}, \mathbf{p}) = \frac{1}{(2\pi)^2} e^{\frac{i}{\gamma}(\bar{\mathbf{x}}_2 - \bar{\mathbf{x}}_1) \cdot \mathbf{P}} \delta(2\mathbf{x} - \bar{\mathbf{x}}_1 - \bar{\mathbf{x}}_2),$$

which then yields by (45)
$$\bar{W}(z, \mathbf{x}, \mathbf{p}) =$$
$$\frac{1}{(2\pi)^2} \int e^{i\mathbf{q} \cdot \left(\mathbf{x} - \frac{\bar{\mathbf{x}}_1 + \bar{\mathbf{x}}_2}{2}\right)} e^{i\left(\frac{\bar{\mathbf{x}}_2 - \bar{\mathbf{x}}_1}{\gamma} - z\mathbf{q}\right) \cdot \mathbf{P}} e^{-\frac{1}{\gamma^2} \int_0^z D_*(\bar{\mathbf{x}}_2 - \bar{\mathbf{x}}_1 - \mathbf{q}\gamma s) ds} d\mathbf{q}.$$

By (31) we have

$$\Gamma_2(z, \mathbf{x}_1, \mathbf{x}_2; \bar{\mathbf{x}}_1, \bar{\mathbf{x}}_2) = \int \bar{W}(z, \frac{1}{2}(\mathbf{x}_1 + \mathbf{x}_2), \mathbf{p}) \exp[i\mathbf{p} \cdot (\mathbf{x}_1 - \mathbf{x}_2)/\gamma]d\mathbf{p},$$

and hence

$$\Gamma_2(z, \mathbf{x}_1, \mathbf{x}_2; \bar{\mathbf{x}}_1, \bar{\mathbf{x}}_2) = \frac{1}{z^2} e^{\frac{i}{2z\gamma}(\mathbf{x}_1 - \mathbf{x}_2 - \bar{\mathbf{x}}_1 + \bar{\mathbf{x}}_2) \cdot (\mathbf{x}_1 + \mathbf{x}_2 - \bar{\mathbf{x}}_1 - \bar{\mathbf{x}}_2)}$$
$$\times e^{-\frac{z}{\gamma^2} \int_0^1 D_*((1-s)(\bar{\mathbf{x}}_2 - \bar{\mathbf{x}}_1) + s(\mathbf{x}_2 - \mathbf{x}_1))ds}. \quad (47)$$

5   Two-frequency transport theory

When the wavelength is comparable to the spatial scale of medium fluctuations then a different scaling and approximation, called radiative transfer, is valid.

Instead of the standard one-frequency transport theory, we will present the two-frequency formulation and deduce the one-frequency theory as a special case.

5.1   Paraxial waves

Analysis of pulsed signal propagation in random media often requires spectral decomposition of the time-dependent signal and the correlation information of two different frequency components. In the conventional approach, the analysis is in terms of the two-frequency mutual coherence function

$$\Gamma_{12}(z, \mathbf{x}, \mathbf{y}) = \mathbb{E}[\Psi_1(z, \mathbf{x} + \frac{\gamma \mathbf{y}}{2})\Psi_2(z, \mathbf{x} - \frac{\gamma \mathbf{y}}{2})]$$

and uses various ad hoc approximations [38].

Let k_1, k_2 be two (relative) wavenumbers nondimensionlized by the central wavenumber k_0. We write the paraxial wave equation in the dimensionless form

$$i\frac{\partial}{\partial z}\Psi_j(z, \mathbf{x}) + \frac{\gamma}{2k_j}\nabla^2\Psi_j(z, \mathbf{x}) + \frac{\mu k_j}{\gamma}V(\frac{z}{\varepsilon^2}, \frac{\mathbf{x}}{\varepsilon^{2\alpha}})\Psi_j(z, \mathbf{x}) = 0, \quad (48)$$
$$j = 1, 2,$$

where γ is the Fresnel number with respect to the central wavenumber.

An important regime for classical wave propagation takes place when the transverse correlation length is much smaller than the propagation distance but is comparable or much larger than the central wavelength which is proportional to the Fresnel number. This is the radiative transfer regime for monochromatic waves described by the following scaling limit

$$\gamma = \theta\varepsilon^{2\alpha}, \quad \mu = \varepsilon^{2\alpha-1}, \quad \theta > 0, \quad \text{such that} \quad \lim_{\varepsilon \to 0} \theta < \infty, \quad (49)$$

(see [15], [22] and references therein). With two different frequencies, the most interesting scaling limit requires another simultaneous limit

$$\lim_{\varepsilon \to 0} k_1 = \lim_{\varepsilon \to 0} k_2 = k, \quad \lim_{\varepsilon \to 0} \gamma^{-1} k^{-1} (k_2 - k_1) = \beta > 0. \qquad (50)$$

We shall refer to the conditions (49) and (50) as the two-frequency radiative transfer scaling limit.

But in the radiative transfer regime the two-frequency mutual coherence function is not as convenient as the two-frequency Wigner distribution, introduced in [17], which is a natural extension of the standard Wigner distribution and is self-averaging in the radiative transfer regime.

The two-frequency Wigner distribution is defined as

$$W_z(\mathbf{x}, \mathbf{p})$$
$$= \frac{1}{(2\pi)^d} \int e^{-i\mathbf{p}\cdot\mathbf{y}} \Psi_1(z, \frac{\mathbf{x}}{\sqrt{k_1}} + \frac{\gamma \mathbf{y}}{2\sqrt{k_1}}) \Psi_2^*(z, \frac{\mathbf{x}}{\sqrt{k_2}} - \frac{\gamma \mathbf{y}}{2\sqrt{k_2}}) d\mathbf{y}, \quad (51)$$

where the scaling factor $\sqrt{k_j}$ is introduced so that W_z satisfies a closed-form equation (see below).

The following property can be derived easily from the definition

$$\|W_z\|_2 = \left(\frac{\sqrt{k_1 k_2}}{2\gamma\pi} \right)^{d/2} \|\Psi_1(z, \cdot)\|_2 \|\Psi_2(z, \cdot)\|_2.$$

Hence the L^2-norm is conserved $\|W_z\|_2 = \|W_0\|_2$. The Wigner distribution has the following obvious properties:

$$\int W_z(\mathbf{x}, \mathbf{p}) e^{i\mathbf{p}\cdot\mathbf{y}} d\mathbf{p}$$
$$= \Psi_1(z, \frac{\mathbf{x}}{\sqrt{k_1}} + \frac{\gamma \mathbf{y}}{2\sqrt{k_1}}) \Psi_2^*(z, \frac{\mathbf{x}}{\sqrt{k_2}} - \frac{\gamma \mathbf{y}}{2\sqrt{k_2}}), \qquad (52)$$

$$\int_{\mathbb{R}^d} W_z(\mathbf{x}, \mathbf{p}) e^{-i\mathbf{x}\cdot\mathbf{q}} d\mathbf{x}$$
$$= \left(\frac{\pi^2 \sqrt{k_1 k_2}}{\gamma} \right)^d \widehat{\Psi}_1(z, \frac{\mathbf{p}\sqrt{k_1}}{4\gamma} + \frac{\sqrt{k_1}\mathbf{q}}{2}) \widehat{\Psi}_2^*(z, \frac{\mathbf{p}\sqrt{k_2}}{4\gamma} - \frac{\sqrt{k_2}\mathbf{q}}{2}) \quad (53)$$

and so contains essentially all the information in the two-point two-frequency function.

The Wigner distribution W_z satisfies the Wigner-Moyal equation exactly [17]

$$\frac{\partial W_z^\varepsilon}{\partial z} + \mathbf{p} \cdot \nabla_{\mathbf{x}} W_z^\varepsilon + \frac{1}{\varepsilon} \mathcal{L}_z W_z^\varepsilon = 0 \qquad (54)$$

where the operator $\mathcal{L}_z$ is formally given as

$$\mathcal{L}_z W_z = i \int \theta^{-1} \left[e^{i\mathbf{q}\cdot\tilde{\mathbf{x}}/\sqrt{k_1}} k_1 W_z^\varepsilon(\mathbf{x}, \mathbf{p} + \frac{\theta\mathbf{q}}{2\sqrt{k_1}}) \right. \tag{55}$$

$$\left. - e^{i\mathbf{q}\cdot\tilde{\mathbf{x}}/\sqrt{k_2}} k_2 W_z^\varepsilon(\mathbf{x}, \mathbf{p} - \frac{\theta\mathbf{q}}{2\sqrt{k_2}}) \right] \hat{V}(\frac{z}{\varepsilon^2}, d\mathbf{q})$$

with $\tilde{\mathbf{x}} = \mathbf{x}/\varepsilon^{2\alpha}$ being the 'fast' transverse variable.

5.1.1 Two-frequency radiative transfer equations

Under the same assumptions as in the one-frequency theory [15,22], one can show the convergence to one of the two types of transport equations as ε tends to zero.

In the first case, let $\theta > 0$ be fixed. The limit equation is

$$\frac{\partial}{\partial z}\bar{W} + \mathbf{p}\cdot\nabla\bar{W} =$$

$$\frac{2\pi k^2}{\theta^2} \int K(\mathbf{p}, \mathbf{q}) \left[e^{-i\beta\theta\mathbf{q}\cdot\mathbf{x}/(2\sqrt{k})} \bar{W}(\mathbf{x}, \mathbf{p} + \frac{\theta\mathbf{q}}{\sqrt{k}}) - \bar{W}(\mathbf{x}, \mathbf{p}) \right] d\mathbf{q}, \tag{56}$$

where the kernel K is given by

$$K(\mathbf{p}, \mathbf{q}) = \Phi(0, \mathbf{q}), \quad \text{for} \quad \alpha \in (0, 1),$$

and

$$K(\mathbf{p}, \mathbf{q}) = \Phi\big((\mathbf{p} + \frac{\theta\mathbf{q}}{2\sqrt{k}}) \cdot \mathbf{q}, \mathbf{q}\big), \quad \text{for} \quad \alpha = 1.$$

For $\alpha > 1$, then with the choice of $\mu = \varepsilon^\alpha$ the limit kernel becomes

$$K(\mathbf{p}, \mathbf{q}) = \delta\big((\mathbf{p} + \frac{\theta\mathbf{q}}{2\sqrt{k}}) \cdot \mathbf{q}\big) \int \Phi(w, \mathbf{q}) dw.$$

In the second case, let $\lim_{\varepsilon\to 0} \theta = 0$. The limit equation becomes

$$\frac{\partial}{\partial z}W_z + \mathbf{p}\cdot\nabla W_z = k\left(\nabla_\mathbf{p} - \frac{i}{2}\beta\mathbf{x}\right) \cdot \mathbf{D} \cdot \left(\nabla_\mathbf{p} - \frac{i}{2}\beta\mathbf{x}\right) W_z, \tag{57}$$

where the (momentum) diffusion coefficient $\mathbf{D}$ is given by

$$\mathbf{D} = \pi \int \Phi(0, \mathbf{q})\mathbf{q} \otimes \mathbf{q} d\mathbf{q}, \quad \text{for} \quad \alpha \in (0, 1), \tag{58}$$

$$\mathbf{D}(\mathbf{p}) = \pi \int \Phi(\mathbf{p}\cdot\mathbf{q}, \mathbf{q})\mathbf{q} \otimes \mathbf{q} d\mathbf{q}, \quad \text{for} \quad \alpha = 1. \tag{59}$$

For $\alpha > 1$, then with the choice of $\mu = \varepsilon^{\alpha}$ the limit coefficients become

$$\mathbf{D}(\mathbf{p}) = \pi|\mathbf{p}|^{-1} \int_{\mathbf{p}\cdot\mathbf{p}_{\perp}=0} \int \Phi(w, \mathbf{p}_{\perp})dw \; \mathbf{p}_{\perp} \otimes \mathbf{p}_{\perp} d\mathbf{p}_{\perp}. \qquad (60)$$

When $k_1 = k_2$ or $\beta = 0$, eqs. (56) and (57) reduce to the standard radiative transfer equations derived in [15].

5.1.2   The longitudinal and transverse cases

To illustrate the utility of these equations, we proceed to discuss the two special cases for the transverse dimension $d = 2$. For simplicity, we will assume the isotropy of the medium in the transverse coordinates such that $\Phi(w, \mathbf{p}) = \Phi(w, |\mathbf{p}|)$. As a consequence the momentum diffusion coefficient is a scalar. In the longitudinal case $\mathbf{D} = D\mathbf{I}$ with a constant scalar D whereas in the transverse case $\mathbf{D}(\mathbf{p}) = C|\mathbf{p}|^{-1}\hat{\mathbf{p}}_{\perp} \otimes \hat{\mathbf{p}}_{\perp}$ with the constant C given by

$$C = \frac{\pi}{2} \int \int \Phi(w, \mathbf{p}_{\perp})dw|\mathbf{p}_{\perp}|^2 d\mathbf{p}_{\perp}.$$

Here $\hat{\mathbf{p}}_{\perp} \in \mathbb{R}^2$ is a unit vector normal to $\mathbf{p} \in \mathbb{R}^2$.

First of all, the equation (57) by itself gives qualitative information about three important parameters of the stochastic channel: the spatial spread σ_*, the coherence length ℓ_c and the coherence bandwidth β_c, through the following scaling argument. One seeks the change of variables

$$\tilde{\mathbf{x}} = \frac{\mathbf{x}}{\sigma_*\sqrt{k}}, \quad \tilde{\mathbf{p}} = \mathbf{p}\ell_c\sqrt{k}, \quad \tilde{z} = \frac{z}{L}, \quad \tilde{\beta} = \frac{\beta}{\beta_c}, \qquad (61)$$

where L is the propagation distance to remove all the physical parameters from (57) and to aim for the form

$$\frac{\partial}{\partial \tilde{z}}W + \tilde{\mathbf{p}} \cdot \nabla_{\tilde{\mathbf{x}}}W = \left(\nabla_{\tilde{\mathbf{p}}} + \frac{i\tilde{\beta}}{2}\tilde{\mathbf{x}}\right) \cdot \left(\nabla_{\tilde{\mathbf{p}}} + \frac{i\tilde{\beta}}{2}\tilde{\mathbf{x}}\right)W \qquad (62)$$

in the longitudinal case and the form

$$\frac{\partial}{\partial \tilde{z}}W + \tilde{\mathbf{p}} \cdot \nabla_{\tilde{\mathbf{x}}}W = \left(\nabla_{\tilde{\mathbf{p}}} + \frac{i\tilde{\beta}}{2}\tilde{\mathbf{x}}\right) \cdot \frac{\hat{\mathbf{p}}_{\perp} \otimes \hat{\mathbf{p}}_{\perp}}{|\tilde{\mathbf{p}}|} \cdot \left(\nabla_{\tilde{\mathbf{p}}} + \frac{i\tilde{\beta}}{2}\tilde{\mathbf{x}}\right)W \quad (63)$$

in the transverse case. From the left side of (57) it immediately follows the first duality relation $\ell_c\sigma_* \sim L/k$. The balance of terms inside each pair of parentheses leads to the second duality relation $\beta_c \sim \ell_c/\sigma_*$. Finally the removal of D or C determines the spatial spread σ_* which has a different expression in the longitudinal and transverse case. In the longitudinal case,

$$\sigma_* \sim D^{1/2}L^{3/2}, \quad \ell_c \sim k^{-1}D^{-1/2}L^{-1/2}, \quad \beta_c \sim k^{-1}D^{-1}L^{-2}$$

whereas in the transverse case

$$\sigma_* \sim k^{-1/6}C^{1/3}L^{4/3}, \quad \ell_c \sim k^{-5/6}C^{-1}L^{-1}, \quad \ell_c \sim \beta_c \sim k^{-2/3}C^{-2/3}L^{-5/3}.$$

In the longitudinal case, the inverse Fourier transform in $\tilde{\mathbf{p}}$ renders eq. (62) to the form

$$\frac{\partial \tilde{W}}{\partial \tilde{z}} - i\nabla_{\tilde{\mathbf{y}}} \cdot \nabla_{\tilde{\mathbf{x}}}\tilde{W} = -\left|\tilde{\mathbf{y}} - \frac{\tilde{\beta}}{2}\tilde{\mathbf{x}}\right|^2\tilde{W} \tag{64}$$

which can be solved exactly and whose Green function at $\tilde{z} = 1$ is

$$\frac{(1+i)^{d/2}\tilde{\beta}^{d/4}}{(2\pi)^d \sin^{d/2}\left[\tilde{\beta}^{1/2}(1+i)\right]} \exp\left[i\frac{|\tilde{\mathbf{y}} - \mathbf{y}'|^2}{2\tilde{\beta}}\right]$$

$$\times \exp\left[i\frac{(\tilde{\mathbf{y}} - \mathbf{y}') \cdot (\tilde{\mathbf{x}} - \mathbf{x}')}{2}\right] \exp\left[i\frac{\tilde{\beta}|\tilde{\mathbf{x}} - \mathbf{x}'|^2}{8}\right]$$

$$\times \exp\left[\frac{1-i}{2\tilde{\beta}^{1/2}}\cot\left(\tilde{\beta}^{1/2}(1+i)\right)\left|\tilde{\mathbf{y}} - \tilde{\beta}\tilde{\mathbf{x}}/2 - \frac{\mathbf{y}' - \tilde{\beta}\mathbf{x}'/2}{\cos\left(\tilde{\beta}^{1/2}(1+i)\right)}\right|^2\right]$$

$$\times \exp\left[-\frac{1-i}{2\tilde{\beta}^{1/2}}\left|\mathbf{y}' - \tilde{\beta}\mathbf{x}'/2\right|^2 \tan\left(\tilde{\beta}^{1/2}(1+i)\right)\right] \tag{65}$$

[19]. This solution gives asymptotically precise information about the cross-frequency correlation, important for analyzing the information transfer and time reversal with broadband signals in the channel described by the random Schrödinger equation [19]. It is unclear if the transverse case is exactly solvable or not.

5.2   Spherical waves

The two-frequency radiative transfer theory can be extended to the spherical scalar wave as follows [21].

Let $U_j, j = 1, 2$ be governed by the reduced wave equation

$$\Delta U_j(\mathbf{r}) + k_j^2\left(\nu_j + V_j(\mathbf{r})\right)U_j(\mathbf{r}) = f_j(\mathbf{r}), \quad \mathbf{r} \in \mathbb{R}^3, \quad j = 1, 2, \tag{66}$$

where ν_j and V_j are respectively the mean and fluctuation of the refractive index associated with the wavenumber k_j and are in general complex-valued. The source terms f_j may result from the initial data or the external sources. Here and below the vacuum phase speed is set to be unity. To solve (66) one needs also some boundary condition which is assumed to be vanishing at the far field.

Radiative transfer regime is characterized by the scaling limit which replaces $\nu_j + V_j$ in eq. (66) with

$$\frac{1}{\theta^2\varepsilon^2}\left(\nu_j + \sqrt{\varepsilon}V_j\left(\frac{\mathbf{r}}{\varepsilon}\right)\right), \quad \theta > 0, \quad \varepsilon \ll 1, \tag{67}$$

where ε is the ratio of the scale of medium fluctuation to the $O(1)$ propagation distance and θ the ratio of the wavelength to the scale of medium fluctuation.

Anticipating small-scale fluctuation due to (67) we define the two-frequency Wigner distribution in the following way

$$W(\mathbf{x},\mathbf{p}) = \frac{1}{(2\pi)^3}\int e^{-i\mathbf{p}\cdot\mathbf{y}} U_1\left(\frac{\mathbf{x}}{k_1} + \frac{\theta\varepsilon\mathbf{y}}{2k_1}\right) U_2^*\left(\frac{\mathbf{x}}{k_2} - \frac{\theta\varepsilon\mathbf{y}}{2k_2}\right) d\mathbf{y},$$

which satisfies the exact equation

$$\mathbf{p}\cdot\nabla W - F = \frac{i}{2\varepsilon\theta}(\nu_1 - \nu_2^*)W + \frac{1}{\sqrt{\varepsilon}}\mathcal{L}W, \tag{68}$$

where the operator $\mathcal{L}$ is defined as

$$\begin{aligned}\mathcal{L}W(\mathbf{x},\mathbf{p}) = &\frac{i}{2\theta}\int \hat{V}_1(d\mathbf{q})e^{i\frac{\mathbf{q}\cdot\mathbf{x}}{\varepsilon k_1}} W\left(\mathbf{x},\mathbf{p} - \frac{\theta\mathbf{q}}{2k_1}\right)\\ &-\frac{i}{2\theta}\int \hat{V}_2^*(d\mathbf{q})e^{-i\frac{\mathbf{q}\cdot\mathbf{x}}{\varepsilon k_2}} W\left(\mathbf{x},\mathbf{p} - \frac{\theta\mathbf{q}}{2k_2}\right)\end{aligned}$$

and the function

$$\begin{aligned}F = &-\frac{i}{2(2\pi)^3}\int e^{-i\mathbf{p}\cdot\mathbf{y}} f_1\left(\frac{\mathbf{x}}{k_1} + \frac{\mathbf{y}}{2k_1}\right) U_2^*\left(\frac{\mathbf{x}}{k_2} - \frac{\mathbf{y}}{2k_2}\right) d\mathbf{y}\\ &+\frac{i}{2(2\pi)^3}\int e^{-i\mathbf{p}\cdot\mathbf{y}} U_1\left(\frac{\mathbf{x}}{k_1} + \frac{\mathbf{y}}{2k_1}\right) f_2^*\left(\frac{\mathbf{x}}{k_2} - \frac{\mathbf{y}}{2k_2}\right) d\mathbf{y}\end{aligned} \tag{69}$$

depends linearly on U_1 and U_2.

To capture the cross-frequency correlation in the radiative transfer regime we also need to restrict the frequency difference range

$$\lim_{\varepsilon\to 0} k_1 = \lim_{\varepsilon\to 0} k_2 = k, \quad \frac{k_2 - k_1}{\varepsilon\theta k} = \beta, \tag{70}$$

where $k, \beta > 0$ are independent of ε and θ. Assuming the differentiability of the mean refractive index's dependence on the wavenumber we write

$$\frac{\nu_2^* - \nu_1}{2\varepsilon\theta} = \nu', \tag{71}$$

where ν' is independent of ε, θ.

Using the multi-scale expansion we derive the two-frequency radiative transfer equation for the averaged Wigner distribution $\bar{W}$

$$\mathbf{p} \cdot \nabla_{\mathbf{x}} \bar{W} + i\nu' \bar{W} - \mathbb{E}F \qquad (72)$$
$$= \frac{\pi k^3}{\theta^4} \int d\mathbf{q} \Phi\left(\frac{k}{\theta}(\mathbf{p} - \mathbf{q})\right) \delta(|\mathbf{p}|^2 - |\mathbf{q}|^2) \left[e^{i\mathbf{x} \cdot (\mathbf{p}-\mathbf{q})\beta} \bar{W}(\mathbf{x}, \mathbf{q}) - \bar{W}(\mathbf{x}, \mathbf{p})\right].$$

The δ-function in the scattering kernel is due to elastic scattering which preserve the wavenumber. When $\beta = 0$ (then $\nu_1 = \nu_2$ and $i\nu' \sim$ the imaginary part of ν), eq. (72) reduce to the standard form of radiative transfer equation for the phase space energy density [7, 45]. For $\beta > 0$, the wave featue is retained in (72). When $\beta \to \infty$, the first term in the bracket on the right-hand side of (72) drops out, due to rapid phase fluctuation, so the random scattering effect is pure damping:

$$\mathbf{p} \cdot \nabla_{\mathbf{x}} \bar{W} + i\nu' \bar{W} - \mathbb{E}F = -\frac{\pi k^3}{\theta^4} \int d\mathbf{q} \Phi\left(\frac{k}{\theta}(\mathbf{p} - \mathbf{q})\right) \delta(|\mathbf{p}|^2 - |\mathbf{q}|^2) \bar{W}(\mathbf{x}, \mathbf{p}).$$

5.2.1   Geometrical radiative transfer

Let us consider the further limit $\theta \ll 1$ when the wavelength is much shorter than the correlation length of the medium fluctuation. To this end, the following form is more convenient to work with

$$\mathbf{p} \cdot \nabla_{\mathbf{x}} \bar{W} + i\nu' \bar{W} - \mathbb{E}F \qquad (73)$$
$$= \frac{\pi k}{2\theta^2} \int d\mathbf{q} \Phi(\mathbf{q}) \delta\left(\mathbf{q} \cdot \left(\mathbf{p} - \frac{\theta \mathbf{q}}{2k}\right)\right) \left[e^{i\mathbf{x} \cdot \mathbf{q}\beta\theta/k} \bar{W}\left(\mathbf{x}, \mathbf{p} - \frac{\theta \mathbf{q}}{k}\right) - \bar{W}(\mathbf{x}, \mathbf{p})\right],$$

which is obtained from eq. (72) after a change of variables. We expand the right-hand side of (73) in θ and pass to the limit $\theta \to 0$ to obtain

$$\mathbf{p} \cdot \nabla_{\mathbf{x}} \bar{W} + i\nu' \bar{W} - \mathbb{E}F = \frac{1}{4k} (\nabla_{\mathbf{p}} - i\beta\mathbf{x}) \cdot \mathbf{D} \cdot (\nabla_{\mathbf{p}} - i\beta\mathbf{x}) \bar{W} \quad (74)$$

with the (momentum) diffusion coefficient

$$\mathbf{D}(\mathbf{p}) = \pi \int \Phi(\mathbf{q}) \delta(\mathbf{p} \cdot \mathbf{q}) \mathbf{q} \otimes \mathbf{q} d\mathbf{q}. \qquad (75)$$

The symmetry $\Phi(\mathbf{p}) = \Phi(-\mathbf{p})$ plays an explicit role here in rendering the right-hand side of eq.(73) a second-order operator in the limit $\theta \to 0$. Equation (74) can be rigorously derived from geometrical optics by a probabilistic method [20].

5.2.2   Spatial (frequency) spread and coherence bandwidth

Through dimensional analysis, eq. (74) yields qualitative information about important physical parameters of the stochastic medium. To show this, let us assume for simplicity the isotropy of the medium, i.e. $\Phi(\mathbf{p}) = \Phi(|\mathbf{p}|)$, so that $\mathbf{D} = C|\mathbf{p}|^{-1}\Pi(\mathbf{p})$ where

$$C = \frac{\pi}{3}\int \delta\left(\frac{\mathbf{p}}{|\mathbf{p}|}\cdot\frac{\mathbf{q}}{|\mathbf{q}|}\right)\Phi(|\mathbf{q}|)|\mathbf{q}|d\mathbf{q} \tag{76}$$

is a constant and $\Pi(\mathbf{p})$ the orthogonal projection onto the plane perpendicular to $\mathbf{p}$. In view of (74) C (and $\mathbf{D}$) has the dimension of inverse length while the variables $\mathbf{x}$ and $\mathbf{p}$ are dimensionless.

Now consider the following change of variables

$$\mathbf{x} = \sigma_x k\tilde{\mathbf{x}}, \quad \mathbf{p} = \sigma_p\tilde{\mathbf{p}}/k, \quad \beta = \beta_c\tilde{\beta}, \tag{77}$$

where σ_x and σ_p are respectively the spreads in position and spatial frequency, and β_c is the coherence bandwidth. Let us substitute (77) into eq. (74) and aim for the standard form

$$\tilde{\mathbf{p}}\cdot\nabla_{\tilde{\mathbf{x}}}\bar{W} + i\nu'\bar{W} - \langle F\rangle = \left(\nabla_{\tilde{\mathbf{p}}} - i\tilde{\beta}\tilde{\mathbf{x}}\right)\cdot|\tilde{\mathbf{p}}|^{-1}\Pi(\tilde{\mathbf{p}})\left(\nabla_{\tilde{\mathbf{p}}} - i\tilde{\beta}\tilde{\mathbf{x}}\right)\bar{W}. \tag{78}$$

The 1-st term on the left side yields the first duality relation

$$\sigma_x/\sigma_p \sim 1/k^2. \tag{79}$$

The balance of terms in each pair of parentheses yields the second duality relation

$$\sigma_x\sigma_p \sim \frac{1}{\beta_c}, \tag{80}$$

whose left-hand side is the *space-spread-bandwidth product*. Finally the removal of the constant C determines

$$\sigma_p \sim k^{2/3}C^{1/3} \tag{81}$$

from which σ_x and β_c can be determined by using (79) and (80):

$$\sigma_x \sim k^{-4/3}C^{1/3}, \quad \beta_c \sim k^{2/3}C^{-2/3}.$$

We do not know if, as it stands, eq. (78) is analytically solvable but we can solve analytically for its boundary layer behavior.

5.2.3   Small-scale asymptotics

Consider the propagation distance less than the transport mean-free-path. The corresponding two-frequency Wigner distribution would be highly concentrated at the longitudinal momentum, say, $p = 1$. Hence we can assume that the projection $\Pi(\mathbf{p})$ in (78) is effectively just the projection onto the transverse plane coordinated by $\mathbf{x}_\perp$ and approximate eq. (74) by

$$\left[\partial_z + \mathbf{p}_\perp \cdot \nabla_{\mathbf{x}_\perp}\right]\bar{W} + i\nu'\bar{W} - \mathbb{E}F = \frac{C_\perp}{4k|p|}\left(\nabla_{\mathbf{p}_\perp} - i\beta\mathbf{x}_\perp\right)^2\bar{W}, \quad (82)$$

where the constant $C_\perp$ is the paraxial approximation of (75) for $|p| = 1$:

$$C_\perp = \frac{\pi}{2}\int \Phi(0, \mathbf{q}_\perp)|\mathbf{q}_\perp|^2 d\mathbf{q}_\perp.$$

Note that the longitudinal (momentum) diffusion vanishes and that the longitudinal momentum p plays the role of a parameter in eq.(82) which then can be solved in the direction of increasing z as an evolution equation with initial data given at a fixed z.

Let σ_* be the spatial spread in the transverse coordinates $\mathbf{x}_\perp$, ℓ_c the coherence length in the transverse dimensions and β_c the coherence bandwidth. Let L be the scale of the boundary layer. We then seek the following change of variables

$$\tilde{\mathbf{x}}_\perp = \frac{\mathbf{x}_\perp}{\sigma_* k}, \quad \tilde{\mathbf{p}}_\perp = \mathbf{p}_\perp k\ell_c, \quad \tilde{z} = \frac{z}{Lk}, \quad \tilde{\beta} = \frac{\beta}{\beta_c} \quad (83)$$

to remove all the physical parameters from (82) and to aim for the form

$$\partial_{\tilde{z}}\bar{W} + \tilde{\mathbf{p}}_\perp \cdot \nabla_{\tilde{\mathbf{x}}_\perp}\bar{W} + Lki\nu'\bar{W} - Lk\mathbb{E}F = \left(\nabla_{\tilde{\mathbf{p}}_\perp} - i\tilde{\beta}\tilde{\mathbf{x}}_\perp\right)^2\bar{W}. \quad (84)$$

The same reasoning as above now leads to

$$\ell_c\sigma_* \sim L/k, \quad \sigma_*/\ell_c \sim 1/\beta_c, \quad \ell_c \sim k^{-1}L^{-1/2}C_\perp^{-1/2}$$

and hence

$$\sigma_* \sim L^{3/2}C_\perp^{1/2}, \quad \beta_c \sim k^{-1}C_\perp^{-1}L^{-2}.$$

The layer thickness L may be determined by $\ell_c \sim 1$, i.e. $L \sim k^{-2}C_\perp^{-1}$.

After the inverse Fourier transform eq. (84) becomes

$$\partial_{\tilde{z}}\Gamma - i\nabla_{\tilde{\mathbf{y}}_\perp} \cdot \nabla_{\tilde{\mathbf{x}}_\perp}\Gamma + Lki\nu'\Gamma - Lk\mathbb{E}F = -\left|\tilde{\mathbf{y}}_\perp + \tilde{\beta}\tilde{\mathbf{x}}_\perp\right|^2\Gamma, \quad (85)$$

which is the governing equation for the two-frequency mutual coherence in the normalized variables. With data given on $\tilde{z} = 0$ and vanishing far-field boundary condition in the transverse directions, Eq. (85) can be solved analytically and its Green function is analogous to (65):

$$\frac{e^{-iLk\nu'}(i4\tilde{\beta})^{1/2}}{(2\pi)^2\tilde{z}\sinh\left[(i4\tilde{\beta})^{1/2}\tilde{z}\right]}\exp\left[\frac{1}{i4\tilde{\beta}\tilde{z}}\left|\tilde{\mathbf{y}}_\perp - \tilde{\beta}\tilde{\mathbf{x}}_\perp - \mathbf{y}'_\perp + \tilde{\beta}\mathbf{x}'_\perp\right|^2\right] \quad (86)$$

$$\times \exp\left[-\frac{\coth\left[(i4\tilde{\beta})^{1/2}\tilde{z}\right]}{(i4\tilde{\beta})^{1/2}}\left|\tilde{\mathbf{y}}_\perp + \tilde{\beta}\tilde{\mathbf{x}}_\perp - \frac{\mathbf{y}'_\perp + \tilde{\beta}\mathbf{x}'_\perp}{\cosh\left[(i4\tilde{\beta})^{1/2}\tilde{z}\right]}\right|^2\right]$$

$$\times \exp\left[-\frac{\tanh\left[(i4\tilde{\beta})^{1/2}\tilde{z}\right]}{(i4\tilde{\beta})^{1/2}}\left|\mathbf{y}'_\perp + \tilde{\beta}\mathbf{x}'_\perp\right|^2\right].$$

6   Application: time reversal

Time reversal is the process of recording the signal from a remote source, time-reversing and back-propagating it to retrofocus around the source (Figure 15). Time reversal of acoustic waves has been demonstrated to hold exciting technological potentials in subwavelength focusing, dispersion compensation, communications, imaging, remote-sensing and target detection in unknown environments (see [29], [30], [41] and references therein). The same should hold for the electromagnetic waves as well. Time reversal of electromagnetic waves is closely related to optical phase conjugation [34].

6.1   Spherical wave

In the simplest version of time reversal, a compactly supported, monochromatic source f emits a wave field u which is then recorded at the boundary ∂D enclosing the support of f. u satisfies the inhomogeneous Helmholtz equation

$$\Delta u + k^2 u = f.$$

Suppose both u and $\partial u/\partial n$ are recorded at ∂D, phase-conjugated and back-propagated into the domain D by using, respectively, $\partial G_0/\partial n$ and G_0. For a monochromatic wave, phase conjugation is equivalent to time reversal. We obtain as a result

$$v(\mathbf{r}) \equiv \int_{\partial D} u^*(\mathbf{r}')\frac{\partial G_0(\mathbf{r} - \mathbf{r}')}{\partial n}d\sigma(\mathbf{r}') - \int_{\partial D}\frac{\partial u^*(\mathbf{r}')}{\partial n}G_0(\mathbf{r} - \mathbf{r}')d\sigma(\mathbf{r}'). \quad (87)$$

Clearly, $v(\mathbf{r})$ is a solution of the homogeneous Helmholtz equation.

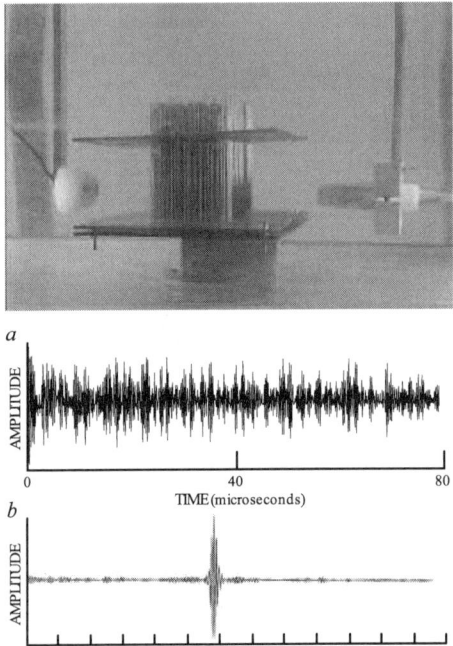

Figure 15   Acoustic chaotic pinball occurs when an underwater ultrasonic pulse emitted by the transducer (at left in photograph) ricochets among 2,000 randomly placed steel rods before reaching the 96-element time-reversing mirror at right. Each element of the array receives a chaotic-seeming sound signal (a portion of one is shown in the middle plot) lasting much longer than the original one-microsecond pulse. When the mirror plays back the chaotic signals, reversed and in synchrony, they ricochet back through the maze of rods and combine to re-create a well-defined pulse, shown in the bottom plot, at the transducer (adapted from [28]).

By the second Green identity

$$
\int_{\partial D} u^*(\mathbf{r}') \frac{\partial G_0(\mathbf{r}-\mathbf{r}')}{\partial n} d\sigma(\mathbf{r}') - \int_{\partial D} \frac{\partial u^*(\mathbf{r}')}{\partial n} G_0(\mathbf{r}-\mathbf{r}') d\sigma(\mathbf{r}')
$$
$$
= \int_D u^*(\mathbf{r}')(\Delta + k^2)G(\mathbf{r}-\mathbf{r}')d\mathbf{r}' - \int_D (\Delta + k^2)u^*(\mathbf{r}')G_0(\mathbf{r}-\mathbf{r}')d\mathbf{r}', \quad (88)
$$

we have

$$
v(\mathbf{r}) = u^*(\mathbf{r}) - \int_D f^*(\mathbf{r}')G_0(\mathbf{r}-\mathbf{r}')d\mathbf{r}'.
$$

Therefore

$$
\begin{aligned}
v(\mathbf{r}) &= \int_D G_0^*(\mathbf{r} - \mathbf{r}')f(\mathbf{r}')d\mathbf{r}' - \int_D G_0(\mathbf{r} - \mathbf{r}')f^*(\mathbf{r}')d\mathbf{r}' \\
&= \int_D \left[G_0^*(\mathbf{r} - \mathbf{r}') - G_0(\mathbf{r} - \mathbf{r}') \right] f^*(\mathbf{r}')d\mathbf{r}' \\
&= \frac{1}{2\pi i} \int_D \frac{\sin\left(k|\mathbf{r} - \mathbf{r}'|\right)}{|\mathbf{r} - \mathbf{r}'|} f^*(\mathbf{r}')d\mathbf{r}'.
\end{aligned}
\tag{89}
$$

In particular, for a point source $f(\mathbf{r}) = \delta(\mathbf{r} - \mathbf{r}_0)$, we have

$$
v(\mathbf{r}) = \frac{1}{2\pi i} \frac{\sin\left(k|\mathbf{r} - \mathbf{r}_0|\right)}{|\mathbf{r} - \mathbf{r}_0|}.
$$

In this case, the time reversal resolution is proportional to λ. Note that (89) is independent of the domain D as long as it contains the support of f.

6.2   Paraxial wave

Next, we consider time reversal in a random medium. For simplicity we use the Markovian model.

Here, a source $\Psi_0(\mathbf{x})$ located at $z = L$ emits a signal with the carrier wavenumber k toward the time reversal mirror (TRM) of aperture A located at $z = 0$ through a turbulent medium. The transmitted field is captured and time reversed at the TRM and then sent back toward the source point through the same turbulent medium, [25], [26].

The time-reversed, back-propagated wave field at $z = L$ can be expressed as

$$
\begin{aligned}
\Psi_{\mathrm{tr}}(\mathbf{x}) &= \int G(L, \mathbf{x}, \mathbf{x}_m)G^*(L, \mathbf{x}_s, \mathbf{x}_m)\Psi_0^*(\mathbf{x}_s)\mathbb{I}_A(\mathbf{x}_m)d\mathbf{x}_m d\mathbf{x}_s \\
&= \int e^{i\mathbf{p}\cdot(\mathbf{x}-\mathbf{x}_s)/\gamma}W\left(L, \frac{\mathbf{x} + \mathbf{x}_s}{2}, \mathbf{p}\right)\Psi_0^*(\mathbf{x}_s)d\mathbf{p}d\mathbf{x}_s,
\end{aligned}
\tag{90}
$$

where $\mathbb{I}_A$ is the indicator function of the TRM, G the propagator of the Schrödinger equation and W the mixed-state Wigner distribution function

$$
W(z, \mathbf{x}, \mathbf{p})
$$
$$
= \int W(z, \mathbf{x}, \mathbf{p}; \mathbf{x}_m)\mathbb{I}_A(\mathbf{x}_m)d\mathbf{x}_m
$$
$$
W(z, \mathbf{x}, \mathbf{p}; \mathbf{x}_m)
$$
$$
= \frac{1}{(2\pi)^2} \int e^{-i\mathbf{p}\cdot\mathbf{y}}G(z, \mathbf{x} + \gamma\mathbf{y}/2, \mathbf{x}_m)G^*(z, \mathbf{x} - \gamma\mathbf{y}/2, \mathbf{x}_m)d\mathbf{y}
$$

which is the convex combination of the pure-state Wigner distributions $W(\cdot; \mathbf{x}_m)$. Here we have used the fact that time reversing of the signal is equivalent to the phase conjugating of its spatial component.

Let us consider a point source located at $(z, 0)$ by substituting the Dirac-delta function $\delta(\mathbf{x})$ for Ψ_0 in (90) and calculate $\mathbb{E}\Psi_{\mathrm{tr}}$ with the Green function (45). We then obtain the point-spread function for the time reversed, refocused wave field written as $\mathcal{P}_{tr}(\mathbf{x}) = \mathcal{P}_0(\mathbf{x})T_{tr}(\mathbf{x})$ with

$$\mathcal{P}_0(\mathbf{x}) \equiv \left(\frac{1}{z\gamma}\right)^2 \exp\left[i\frac{|\mathbf{x}|^2}{2\gamma z}\right]\hat{\mathbb{I}}_A\left(\frac{\mathbf{x}}{\gamma z}\right)$$

$$T_{tr}(\mathbf{x}) \equiv \exp\left[-\frac{z}{\gamma^2}\int_0^1 D_*(-s\mathbf{x})ds\right]. \tag{91}$$

In the absence of random inhomogeneity the function T_{tr} is unity and the resolution scale ρ_0 is determined solely by $\mathcal{P}_0$:

$$\rho_0 \sim 2\pi\frac{\gamma z}{A}. \tag{92}$$

In view of definition of γ this is evidently the classical Rayleigh resolution formula. Note that we have used the dimensionless variables.

6.3 Anomalous focal spot

We shall see here that a turbulent medium such as the turbulent atmosphere can significantly reduce the focal spot size below the Rayleigh limit.

To this end we assume the inertial range asymptotic:

$$D_*(r) \approx C_*^2 r^{2H_*}, \quad \ell_0 \ll r \ll L_0 = 1, \tag{93}$$

where the effective Hölder exponent H_* is given by

$$H_* = \begin{cases} H + 1/2, & \text{for } H \in (0, 1/2), \\ 1, & \text{for } H \in (1/2, 1) \end{cases} \tag{94}$$

and the structure parameter C_* is proportional to $\sigma_H^{1/2}$. Outside of the inertial range we have instead $D_*(r) \sim r^2, r \ll \ell_0$ and $D_*(r) \to D_*(\infty)$ for $r \to \infty$ where $D_*(\infty) > 0$ is a finite constant. As in (35) we have chosen the correlation length L_0 as the reference length L_x.

First we consider the situation where there may be an inertial range behavior. This requires from (91) that

$$\gamma^{-2}D_*(\infty) \gg 1. \tag{95}$$

In the presence of random inhomogeneities the retrofocal spot size is determined by $\mathcal{P}_0$ or T_{tr} depending on which has a smaller support. For the power-law spectrum we have the inertial range asymptotic

$$T_{\mathrm{tr}}(\mathbf{x}) \sim \exp\left[-C_*^2 \gamma^{-2} z |\mathbf{x}|^{2H_*}(4H_* + 2)^{-1}\right] \tag{96}$$

for $\ell_0 \ll |\mathbf{x}| \ll 1$. We define the turbulence-induced time-reversal resolution as

$$\rho_{\mathrm{tr}} = \sqrt{\int |\mathbf{x}|^2 T_{\mathrm{tr}}^2(\mathbf{x}) d\mathbf{x} / \int T_{\mathrm{tr}}^2(\mathbf{x}) d\mathbf{x}}, \tag{97}$$

which by (96) has the inertial range asymptotic

$$\rho_{\mathrm{tr}} \sim \left(\frac{\gamma \lambda}{C_* \sqrt{z}}\right)^{1/H_*}, \quad \ell_0 \ll \rho_{\mathrm{tr}} \ll 1. \tag{98}$$

The nonlinear law (98) is valid only down to the inner scale ℓ_0 below which the linear law prevails $\rho_{\mathrm{tr}} \sim \gamma \lambda z^{-1/2}$.

We see that under (95) ρ_{tr} is independent of the aperture, has a superlinear dependence on the wavelength in the inertial range and the resolution is further enhanced as the distance z and random inhomogeneities (C_*) increase. This effect can be explained by the notion of turbulence-induced aperture which enlarges as z and C_* increase as the TRM is now able to capture signals initially propagating in the more oblique directions.

To recover the linear law previously reported in [4], let us consider the situation where $\rho_{\mathrm{tr}} = O(\gamma)$ and take the limit of vanishing Fresnel number $\gamma \to 0$ in eq. (46) by setting $\mathbf{x} = \gamma \mathbf{y}$. Then we have

$$\lim_{\gamma \to 0} \gamma^{-2} D_*(\gamma \mathbf{y}) = D_0 |\mathbf{y}|^2, \quad D_0 = \frac{1}{2} \int \Phi(0, \mathbf{q}) |\mathbf{q}|^2 d\mathbf{q}.$$

The resulting mean retrofocused field $\mathbb{E} \Psi_{\mathrm{tr}}(\gamma \mathbf{y})$ is Gaussian in the offset variable $\mathbf{y}$ and the refocal spot size on the original scale is given by

$$\rho_{\mathrm{tr}} \sim \gamma \lambda (D_0 z)^{-1/2}. \tag{99}$$

Hence the linear law prevails in the sub-inertial range.

6.4   Duality and turbulence-induced aperture

Intuitively speaking, the turbulence-induced aperture referred to in the previous section is closely related to how a wave is spread in the course of propagation through the turbulent medium. A quantitative estimation can be given by analyzing the spread of wave energy.

To this end let us calculate the mean energy density with the Gaussian initial wave amplitude

$$\Psi(0, \mathbf{x}) = \exp\left[-|\mathbf{x}|^2/(2\alpha^2)\right]. \tag{100}$$

We obtain

$$\mathbb{E}|\Psi(z, \mathbf{x})|^2 = \left(\frac{\alpha}{2\sqrt{\pi}}\right)^d \int \exp\left[-|\mathbf{q}|^2[\alpha^2/4 + \gamma^2 z^2/(4\alpha^2)]\right]$$
$$\times \exp\left[i\mathbf{q}\cdot\mathbf{x}\right]\exp\left[-\frac{z}{\gamma^2}\int_0^1 D_*(\mathbf{q}s\gamma)ds\right]d\mathbf{q}.$$

The reason we do not consider the point source right away is that for a point source $\mathbb{E}|\Psi|^2 \sim$ const. so to see the effect of the random diffraction we need to consider an extensive source.

From the above the turbulence-induced spread can be identified as convolution with the kernel which is the inverse Fourier transform $\mathcal{F}^{-1}T$ of the transfer function

$$T(\mathbf{q}) = \exp\left[-\frac{z}{\gamma^2}\int_0^1 D_*(\mathbf{q}s\gamma)ds\right].$$

In view of (91), we obtain that

$$\mathcal{F}^{-1}T(\mathbf{x}) = \frac{1}{\gamma^2 z^2}\mathcal{F}^{-1}T_{\mathrm{tr}}(\frac{\mathbf{x}}{\gamma z}). \tag{101}$$

In this case it is reasonable to define the turbulence-induced forward spread σ_* as

$$\sigma_* = \sqrt{\int |\mathbf{x}|^2 \left|\mathcal{F}^{-1}T\right|^2(\mathbf{x})d\mathbf{x}\bigg/\int \left|\mathcal{F}^{-1}T\right|^2(\mathbf{x})d\mathbf{x}},$$

which, in view of (97) and (101), then satisfies the uncertainty inequality (see also [14])

$$\sigma_*\rho_{\mathrm{tr}} \geqslant \gamma z. \tag{102}$$

The equality holds when T_{tr} is Gaussian, i.e. when $H^* = 1$ or in the subinertial range. This strongly suggests the definition of the turbulence-induced aperture as $A_* = 2\pi\gamma z/\rho_{\mathrm{tr}}$ in complete analogy to (92). And we have the inequality

$$A_* \leqslant 2\pi\sigma_*,$$

where equality holds true for a Gaussian wave structure function.

6.5   Coherence length

Another physical variable that is naturally dual to the wave spread is the coherence length. The physical intuition is that the larger the spread the smaller the coherence length.

In the Markovian model with the Gaussian data (100) the coherence length has the following expression:

$$\mathbb{E}\Psi(z, \mathbf{x} + \mathbf{y}/2)\Psi(z, \mathbf{x} - \mathbf{y}/2) \tag{103}$$

$$= \left(\frac{\alpha}{\sqrt{2\pi}}\right)^2 \int \exp\left[-|\mathbf{q}|^2\alpha^2/4\right] \exp\left[-\frac{|\mathbf{y} - \gamma z \mathbf{q}|^2}{4\alpha^2}\right]$$

$$\times \exp\left[i\mathbf{q} \cdot \mathbf{x}\right] \exp\left[-\frac{1}{\gamma^2}\int_0^z D_*(-\mathbf{y} + \gamma \mathbf{q}s)ds\right] d\mathbf{q}.$$

In the point-source limit $\alpha \to 0$, we have

$$\mathbb{E}\Psi(z, \mathbf{x} + \mathbf{y}/2)\Psi(z, \mathbf{x} - \mathbf{y}/2) \tag{104}$$

$$\approx \left(\frac{\sqrt{2}\alpha^2}{\gamma z}\right)^2 \exp\left[i\frac{1}{\gamma z}\mathbf{y} \cdot \mathbf{x}\right] \exp\left[-\frac{z}{\gamma^2}\int_0^1 D_*(-\mathbf{y}s)ds\right].$$

In view of (104) let us define the turbulence-induced coherence length δ_* as

$$\delta_* = \sqrt{\int |\mathbf{y}|^2 T_2^2(\mathbf{y})d\mathbf{y} / \int T_2^2(\mathbf{y})d\mathbf{y}}, \ \ T_2(\mathbf{y}) = \exp\left[-\frac{z}{\gamma^2}\int_0^1 D_*(-\mathbf{y}s)ds\right].$$

Since $T_2 = T_{\mathrm{tr}}$, δ_* is equal to the turbulence-induced time-reversal resolution ρ_{tr} and is related to the wave spread as

$$\sigma_*\delta_* \geqslant \gamma z,$$

where the equality holds for a Gaussian wave structure function. Because of the identity of δ_* and ρ_{tr} the time reversal refocal spot size can be used to estimate the coherence length of the wave field which is more difficult to measure directly.

6.6   Broadband time reversal communications

Now we would like to discuss time reversal with broadband signals as a means of communication in random media (Figure 16). We consider the multiple-input-multiple-output (MIMO) broadcast channel described in [19], [18]. We assume that the random medium is described by the Markovian model.

Let the M receivers located at $(L, \mathbf{r}_j), j = 1, \cdots, M$ first send a pilot signal $\int e^{i\frac{kt}{\gamma}}g(k)dk\delta(\mathbf{r}_j - \mathbf{a}_i)$ to the N-element TRA located at $(0, \mathbf{a}_i), i =$

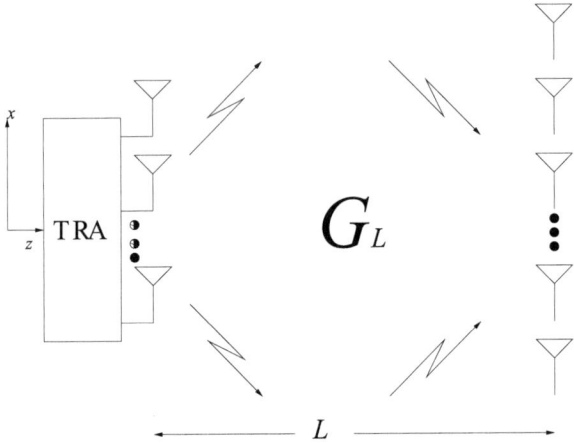

Figure 16   MIMO Broadcast Channel.

$1, \cdots, N$ which then use the time-reversed version of the received signals $\int e^{i\frac{kt}{\gamma}} g(k) G_L(\mathbf{r}_j, \mathbf{a}_i; k) dk$ to modulate streams of symbols and send them back to the receivers. Here G_L is the Green function of

$$i\frac{\partial \Psi_z}{\partial z} + \frac{\gamma}{2k}\Delta_\mathbf{x}\Psi_z + \frac{k}{\gamma}\chi_z \circ \Psi_z = 0, \quad \mathbf{x} \in \mathbb{R}^d, \qquad (105)$$

and $g^2(k)$ is the power density at k. For simplicity we take $g^2(k) = \exp\left(-\frac{|k-1|^2}{2B^2\gamma^2}\right)$. As shown in [4], [10], when the TRA has an infinite time-window, the signal arriving at the receiver plane with delay $L+t$ is given by

$$S(\mathbf{r}, t) = \sum_{l=1}^{T}\sum_{i=1}^{N}\sum_{j=1}^{M} m_j(\tau_l) \int e^{-i\frac{k}{\gamma}(t-\tau_l)} g(k)$$
$$\times G_L(\mathbf{r}, \mathbf{a}_i; k) G_L^*(\mathbf{r}_j, \mathbf{a}_i; k) dk, \qquad (106)$$

where $m_j(\tau_l), l = 1, \cdots, T \leqslant \infty$ are a stream of T symbols intended for the j-th receiver transmitted at times $\tau_1 < \tau_2 < \cdots < \tau_T$. We assume for simplicity that $|m_j(\tau_l)| = 1, \forall j, l$.

Consider the mean $E(\mathbf{r}, t) = \gamma^{-d} \int f((\mathbf{x} - \mathbf{r})/\ell_c)\mathbb{E}S(\mathbf{x}, t)d\mathbf{x}$ and the variance

$$V(\mathbf{r}, t) = \gamma^{-2d}\mathbb{E}\Big[\int f((\mathbf{x} - \mathbf{r})/\ell_c)S(\mathbf{x}, t)d\mathbf{x}\Big]^2 - E^2(\mathbf{r}, t),$$

where the coupling with the test function f can be viewed as the averaging induced by measurement. We have made the test function f act on the scale of the coherence length ℓ_c, the smallest spatial scale of interest (the speckle size) in the present context. Different choices of scale would not affect the conclusion of our analysis.

The primary object of our analysis is

$$\rho = \frac{E^2(\mathbf{r}_j, \tau_l)}{V(\mathbf{r}, t)}, \quad j = 1, \cdots, M, \; l = 1, \cdots, T, \tag{107}$$

which is the signal-to-interference ratio (SIR) if $\mathbf{r} = \mathbf{r}_j, t = \tau_l$ and the signal-to-sidelobe ratio (SSR) if $|\mathbf{r} - \mathbf{r}_j| \gg \ell_c, \forall j$ (spatial sidelobes) or $|t - \tau_l| \gg B^{-1}, \forall l$ (temporal sidelobes) (as $V(\mathbf{r}, \tau) \approx E^2(\mathbf{r}, \tau)$ as we will see below). We shall refer to it as the signal-to-interference-or-sidelobe ratio (SISR). In the special case of $\mathbf{r} = \mathbf{r}_j$ and $|t - \tau_l| \gg B^{-1}, \forall l$, ρ^{-1} is a measure of intersymbol interference. To show stability and resolution, we shall find the precise conditions under which $\rho \to \infty$ and $\mathbb{E}S(\mathbf{r}, t)$ is asymptotically $\sum_{l=1}^{T} \sum_{j=1}^{M} m_j(\tau_l) S_{jl}(\mathbf{r}, t)$ where

$$S_{jl}(\mathbf{r}, t) \approx \sum_{i=1}^{N} \int e^{-i\frac{k(t-\tau_l)}{\gamma}} g(k) \mathbb{E}\left[G_L(\mathbf{r}, \mathbf{a}_i; k) G_L^*(\mathbf{r}_j, \mathbf{a}_i; k) \right] dk \tag{108}$$

is a sum of δ-like functions around $\mathbf{r}_j$ and $\tau_l = 0, \forall l$. In other words, we employ the TRA as a multiplexer to transmit the M scrambled data-streams to the receivers and we would like to turn the medium into a demultiplexer by employing the broadband time reversal technique.

Provided that the antenna spacing is greater than the coherence length and the frequency separation is greater than the coherence bandwidth, the sufficient condition for achieving the desired goal (i.e. stability and refocusing) is the multiplexing condition

$$NB \gg MC, \tag{109}$$

where C is the number of symbols per unit time in the datum streams intended for each receiver [19], [18]. The proof makes a nontrivial use of the asymptotic solution (65).

In terms of resolution, the best experimental result in this direction so far has been achieved by [42].

7   Application: imaging in random media

7.1   Imaging of phase objects

Consider the electron transmission microscope whose image formation is based on the interaction of electrons with the object. Two kinds of scattering are involved: elastic and inelastic scattering. The former involves no transfer of energy and give rise to high-resolution information. The latter involves transfer of energy and produces low resolution information. The electron microscopy mainly uses the former for imaging.

In most applications, the elastic scattering interaction can be described as phase shift as in (21). For weak phase object, the Born approximation $\Psi = \Psi_0(1 + i\Phi)$ is valid [32]. The Fraunhofer diffraction can be used since the observation is always made in the far distance from the object and close to the optical axis. In that approximation the wave function in the back focal plane of the objective lens is – in the absence of aberration – the Fourier transform of u. However, the lens aberrations and the defocusing have the effect of shifting the phase of the scattered wave by an amount expressed by $2\pi\chi(\mathbf{k})$ where χ is called the wave aberration function. In a polar coordinate system $(|\mathbf{k}|, \phi)$ we have [32]

$$\chi(k, \phi) = -\frac{1}{2}\lambda\left[\Delta z + \frac{z_0}{2}\sin(2\phi)\right]|\mathbf{k}|^2 + \frac{1}{4}\lambda^3 C_s|\mathbf{k}|^4,$$

where λ is the electron wavelength; Δz the defocus of the objective lens; z_a the focal difference due to axial astigmatism; C_s the third-order spherical aberration constant.

An ideal lens will transform an incoming plane wave into a spherical wave front converging into a single point on the back focal plane. Lens aberrations have the effect of deforming the spherical wave front. In particular, the spherical aberration term C_s acts in a way that the outer zones of the wave front are curved more than the inner zones, leading to a decreased focal length in the outer zones.

The above discussion leads to the wave function

$$\Psi_b(\mathbf{k}) = \mathfrak{F}[\Psi](\mathbf{k})e^{i2\pi\chi(\mathbf{k})}$$

in the back focal plane of the objective lens. Next, the wave function in the image plane is obtained from the wave in the back focal plane, after modification by the aperture function $A(\mathbf{k})$, through an inverse Fourier transform

$$\Psi_i(\mathbf{y}) = \mathfrak{F}^{-1}[\mathfrak{F}[\Psi](\mathbf{k})A(\mathbf{k})e^{i2\pi\chi(\mathbf{k})}],$$

where $A(\mathbf{k})$ can be taken as the indicator function of the aperture:

$$A(\mathbf{k}) = \begin{cases} 1, & \text{for } |\mathbf{k}| \leqslant \theta_1/\lambda, \\ 0, & \text{else}, \end{cases}$$

where θ_1 is the angle corresponding to the radius of the objective aperture. Finally, the observed intensity in the image plane is

$$I(\mathbf{y}) = |\Psi_i|^2(\mathbf{y}).$$

If we apply the Born approximation, assume Φ is real-valued and subtract the constant background to consider only the contrast of the image intensity, then we obtain a linear relationship between $O(\mathbf{k}) = \mathfrak{F}[\Phi](\mathbf{k})$ and the Fourier transform of the image contrast $\mathfrak{F}[I](\mathbf{k})$:

$$\mathfrak{F}[I](\mathbf{k}) = O(\mathbf{k})A(\mathbf{k})\sin\left(2\pi\chi(\mathbf{k})\right)$$

or equivalently

$$I(\mathbf{y}) = \int \Phi(\mathbf{y}')h(\mathbf{y} - \mathbf{y}')d\mathbf{y}',$$

where $h(\mathbf{y}) = \mathfrak{F}[A(\mathbf{k})\sin\left(2\pi\chi\right)]$ is called the point-spread function of the imaging system. The function $\sin 2\pi\chi$ is known as the phase contrast transfer function and the function $A(\mathbf{k})\sin\left(2\pi\chi\right)$ is the optical transfer function.

Optical imaging systems are often built out of lens, pinholes and mirrors. For optical waves many objects can be treated as phase objects such as thin sheets or organic specimens, air flows, vortices and shock waves, strains in transparent materials, density changes in heating. The basic configuration of, for example, a microscope has 2 lenses of $4f$ geometry which is equivalent to iterated (windowed) Fourier transforms and produces an inverted image. Then the imaging quality of the system is determined by the point-spread function. The intensity of the image is the convolution of the object intensity and the point-spread function.

7.2   Long-exposure imaging

The refractive index fluctuation in the turbulent atmosphere restricts the angular resolution of large, ground-based telescopes to the seeing limit of 0.5 arcsec. On the other hand, the theoretical resolution of a 5m telescope is about 0.02 arcsec at wavelength 0.5μm. This is more than 20 times of reduction in resolving power.

The seeing quality can be rated by Pickering's scale which ranges from P-1(worst) to P-10 (best), Figure 17. The Pickering scale is based on what a highly magnified star looks like when carefully focused, in a small telescope. A star at high magnification, under perfect seeing (P-10) looks like a bull's eye. A small central disk surrounded by one or more concentric rings. At P-1, it is just an amorphous blob. The central disk is known as the Airy disk and it's size in inversely proportional to the size of the telescope objective.

Consider an object such as a faraway star or galaxy. The wave field incident on the top of the atmosphere is Ψ_0. For simplicity we consider spatially incoherent object, i.e. $\langle\Psi(\mathbf{x}_1)\Psi^*(\mathbf{x}_2)\rangle = |\Psi(\mathbf{x}_1)|^2\delta(\mathbf{x}_1 - \mathbf{x}_2)$, where $\langle\cdot\rangle$ denotes the averaging with respect to the random phase of the object and is independent of the averaging with respect to the medium

PICKERING'S SCALE

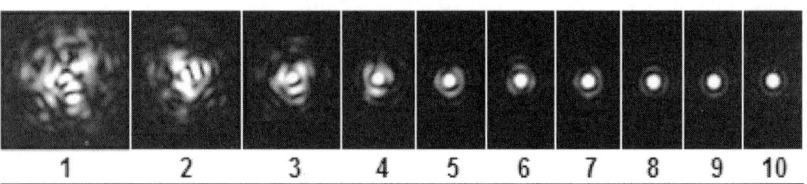

| 1 | 2 | 3 | 4 | 5 | 6 | 7 | 8 | 9 | 10 |

Figure 17   Pickering's scale of rating atmospheric turbulence: the photographs show the image (intensity distribution) of a star under various atmospheric conditions.

ensemble. In the case of a single star the wave field is nearly a plane wave. Let $G_L(\mathbf{x}, \bar{\mathbf{x}})$ be the Green function for the turbulent medium of thickness L. The wave field impinging on the lens is $\int G_L(\mathbf{x}, \bar{\mathbf{x}})\Psi_0(\bar{\mathbf{x}})d\bar{\mathbf{x}}$. The lens introduces a phase factor of the form $e^{-\frac{i}{2\gamma f}|\mathbf{x}|^2}$ and on the focal plane the wave field is given by

$$\Psi(\mathbf{x}) = e^{-\frac{i}{2\gamma f}|\mathbf{x}|^2} \int \Psi_0(\bar{\mathbf{x}})G_L(\mathbf{x}', \bar{\mathbf{x}})\mathbb{I}_A(\mathbf{x}')e^{\frac{i}{f\gamma}\mathbf{x}\cdot\mathbf{x}'}d\mathbf{x}'d\bar{\mathbf{x}}, \quad (110)$$

where $\mathbb{I}_A(\mathbf{x}')$ is the indicator function of the lens. The observed intensity is then

$$I(\mathbf{x}) =$$
$$\int |\Psi_0(\bar{\mathbf{x}})|^2 G_L(\mathbf{x}_1', \bar{\mathbf{x}})G_L^*(\mathbf{x}_2', \bar{\mathbf{x}})\mathbb{I}_A(\mathbf{x}_1')\mathbb{I}_A(\mathbf{x}_2')e^{\frac{i}{f\gamma}\mathbf{x}\cdot(\mathbf{x}_1'-\mathbf{x}_2')}d\mathbf{x}_1'd\mathbf{x}_2'd\bar{\mathbf{x}}. (111)$$

Without the complete knowledge of G_L it is difficult to solve the basic imaging equation (111) and recover the impinging wave field Ψ_0 from the observed intensity I.

Now let us consider the long-exposure imaging equation for a spatially incoherent object. Assuming the ergodicity of the turbulent medium, after sufficiently long exposure, the intensity in the focal plane is the statistical average:

$$\mathbb{E}I(\mathbf{x}) =$$
$$\int |\Psi_0(\bar{\mathbf{x}})|^2 e^{\frac{i}{f\gamma}\mathbf{x}\cdot(\mathbf{x}_1'-\mathbf{x}_2')}\Gamma_2(L, \mathbf{x}_1'-\mathbf{x}_2'; 0)\mathbb{I}_A(\mathbf{x}_1')\mathbb{I}_A(\mathbf{x}_2')d\mathbf{x}_1'd\mathbf{x}_1'd\bar{\mathbf{x}}. \quad (112)$$

Let

$$I_0(\mathbf{x}) = \frac{1}{L^2} \int e^{\frac{i}{2L\gamma}(\mathbf{x}_1-\mathbf{x}_2)\cdot(\mathbf{x}_1+\mathbf{x}_2-2\bar{\mathbf{x}})}$$
$$\times |\Psi_0(\bar{\mathbf{x}})|^2 e^{\frac{i}{f\gamma}\mathbf{x}\cdot(\mathbf{x}_1'-\mathbf{x}_2')}\mathbb{I}_A(\mathbf{x}_1')\mathbb{I}_A(\mathbf{x}_2')d\mathbf{x}_1'd\bar{\mathbf{x}}$$

be the intensity in the absence of the turbulent medium. Then in view of (47) we can write eq. (112) in the form

$$\mathbb{E}I(\mathbf{x}) = \int S(\mathbf{x} - \mathbf{x}')I_0(\mathbf{x}')d\mathbf{x}',$$

where

$$S(\mathbf{x}) = \int e^{-\frac{L}{\gamma^2}\int_0^1 D_*(s(\mathbf{x}'_2 - \mathbf{x}'_1))ds} e^{\frac{i}{f\gamma}\mathbf{x}\cdot(\mathbf{x}'_1 - \mathbf{x}'_2)}d\mathbf{x}'_1 d\mathbf{x}'_2$$

represents the turbulence-induced pattern of a point source. For the purpose of imaging, the entire propagation modeling is to supply this function S.

More realistically the inversion problem should be posed with inclusion of noise:

$$\mathbb{E}I(\mathbf{x}) = \int S(\mathbf{x} - \mathbf{x}')I_0(\mathbf{x}')d\mathbf{x}' + N(\mathbf{x})$$

or in the Fourier domain

$$\mathbb{E}\hat{I}(\mathbf{k}) = \hat{S}(\mathbf{k})\hat{I}_0(\mathbf{k}) + \hat{N}(\mathbf{k}).$$

If we write the solution of the inversion problem as

$$\tilde{I}_0(\mathbf{k}) = \hat{T}(\mathbf{k})\mathbb{E}I(\mathbf{k}),$$

then T can be determined from minimizing the mean-squared error

$$E = \int \mathbb{E}|\hat{I}_0(\mathbf{k}) - \tilde{I}_0(\mathbf{k})|^2 d\mathbf{k},$$

where $\langle \cdot \rangle$ stands for the average with respect to noise. The minimizer is called the Wiener fileter and is given by

$$\hat{T}(\mathbf{k}) = \frac{\hat{S}^*(\mathbf{k})}{|\hat{S}|^2(\mathbf{k}) + \text{SSR}^{-1}},$$

where SSR stands for the signal-to-noise ratio

$$\text{SSR} = \frac{|\mathbb{E}\hat{I}(\mathbf{k})|^2}{\mathbb{E}|\hat{N}(\mathbf{k})|^2}.$$

In the limit of vanishing noise, the Wiener filter reduces to the inverse filter while in the large noise limit it reduces to the matched filter. Other solutions to the inverse problem in the presence of noise can be obtained by the maximum likelihood method which seeks to maximize the likelihood function $\mathbb{P}(\mathbb{E}I|\hat{I}_0)$ under the assumption of independent Poisson or Gaussian noise [1].

7.3 Short-exposure imaging

The limitations on long-exposure imaging with a thin lens may be quite severe. In the case of atmospheric imaging, this is mainly due to the phase distortion which causes imperfect focus. A natural approach for circumventing the problem is to use an imaging method which is insensitive to phase distortion. This is the essence of interferometric imaging techniques which in their simplest forms, produce images of the object autocorrelation function rather than the object itself. One technique is called the amplitude interferometry explainable in terms of the Michelson stellar interferometer, see Figure 18.

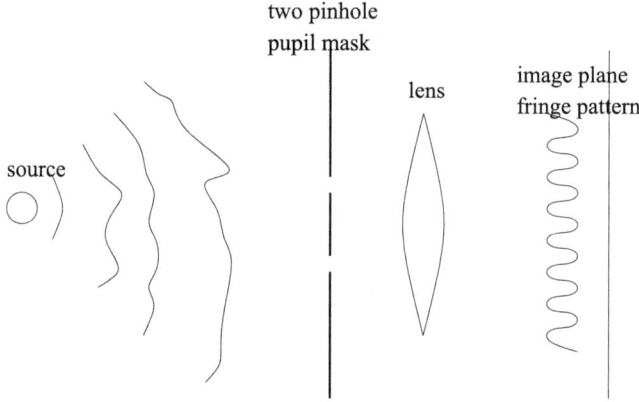

Figure 18   Michelson interferometry.

A related technique is called the speckle interferometry. The general procedure is to take a long series of short exposures and find the spatial power spectrum of the image [1]. The advantage of the speckle interferometry over the simple two-pinhole amplitude interferometry is that the former measures all angular frequencies simultaneously.

The assumptions are that the *phase* of the light wave from the source obeys the Gaussian statistics and the object is small enough to be contained in an atmospheric coherence area. The latter ensures the isoplanicity condition to be satisfied and implies the following simpler form than (111)

$$I(\mathbf{x}) = \int S_s(\mathbf{x} - \mathbf{x}')I_0(\mathbf{x}')d\mathbf{x}', \tag{113}$$

where S_s changes with each realization of turbulent media. After Fourier transform, we have

$$\hat{I}(\mathbf{k}) = \hat{S}_s(\mathbf{k})\hat{I}_0(\mathbf{k}).$$

Squaring modulus and averaging over a large number of short exposures we then have

$$\mathbb{E}|\hat{I}(\mathbf{k})|^2 = \mathbb{E}|\hat{S}_s(\mathbf{k})|^2|\hat{I}_0(\mathbf{k})|^2, \tag{114}$$

which can be solved by various inversion techniques once $\mathbb{E}|\hat{S}_s(\mathbf{k})|^2$ is known. In view of the discussion on the general linear inversion problem, we see that the isoplanicity assumption can be relaxed. To recover the phase of $\hat{I}_0$ we need to employ a phase retrieval technique, see [9] and [54].

The success of the method depends on the relative insensitivity of $\mathbb{E}|\hat{S}_s(\mathbf{k})|^2$ to the random phase distortion, i.e. $\mathbb{E}|\hat{S}_s(\mathbf{k})|^2$ is significantly greater than $|\mathbb{E}\hat{S}_s(\mathbf{k})|^2$. When the phase of the light wave obeys the Gaussian statistics it can be shown that $\mathbb{E}|\hat{S}_s(\mathbf{k})|^2$ is proportional to that of the homogeneous case for large value of $\mathbf{k}$. As a consequence, the speckle interferometry can yield near-diffraction-limited resolution. The effect of the large intensity fluctuations on the resolution of speckle interferometry is less clear [1].

Speckle masking is a triple correlation imaging technique which retains the phase information [54]. The quantity considered is the bispectrum

$$\mathbb{E}\hat{I}^{(3)}(\mathbf{k}, \mathbf{k}') = \mathbb{E}\left[\hat{I}(\mathbf{k})\hat{I}(\mathbf{k}')\hat{I}(-\mathbf{k} - \mathbf{k}')\right].$$

It follows that

$$\mathbb{E}\hat{I}^{(3)}(\mathbf{k}, \mathbf{k}') = \mathbb{E}\hat{S}_s^{(3)}(\mathbf{k}, \mathbf{k}')\hat{I}_0^{(3)}(\mathbf{k}, \mathbf{k}'). \tag{115}$$

The function $\hat{I}_0^{(3)}(\mathbf{k}, \mathbf{k}')$ is the bispectrum of the object in the absence of the turbulent medium provided that we have infinite lens aperture. The function $\mathbb{E}\hat{S}_s^{(3)}(\mathbf{k}, \mathbf{k}')$ is known as the speckle masking transfer function and can be derived from the speckle interferograms of a point source or it can be calculated theoretically.

By (115) the phase $\phi_0^{(3)}$ of $\hat{I}_0^{(3)}$ can be determined from $\mathbb{E}\hat{I}^{(3)}(\mathbf{k}, \mathbf{k}')$ and $\mathbb{E}\hat{S}_s^{(3)}(\mathbf{k}, \mathbf{k}')$. Writing

$$\hat{I}_0(\mathbf{k}) = |\hat{I}_0(\mathbf{k})|e^{i\phi_0(\mathbf{k})},$$

we obtain

$$\phi_0^{(3)}(\mathbf{k}, \mathbf{k}') = \phi_0(\mathbf{k}) + \phi_0(\mathbf{k}') + \phi_0(-\mathbf{k} - \mathbf{k}') \tag{116}$$

from which a recursion relation relating ϕ_0 to $\phi^{(3)}$ can be developed. Equation (116) is called a closure phase relation. Since the speckle masking transfer function is greater than zero up to the diffraction cutoff frequency, the diffraction-limited resolution can be achieved by the speckle masking method [54].

7.4 Coherent imaging of multiple point targets in Rician media

A Rician (fading) medium is a random medium whose mean or coherent component is non-vanishing and whose fluctuations obey a Gaussian distribution. Such a model is widely used in wireless literature to describe certain wireless communication channels. The Rician factor K of a Rician medium is the ratio of signal power in coherent component over the fluctuating power. Typically a Rician channel arises when there is a line-of-sight between the antennas and the targets. On the other hand, for a richly scattering environment, the coherent component is so dim that $K \approx 0$ effectively. Such a medium is called Rayleigh fading. Clearly Rayleigh media pose a greater challenge to imaging obscured targets than Rician media. In this section, we discuss the theory and practice of imaging multiple point targets in a Rician medium. For more details, the reader is referred to [27] and the references therein. We discuss briefly the case with Rayleigh media in the next section.

There are two main ingredients in this theory: the first is time reversal and averaging with the mean (coherent) Green function at various frequencies; the second is the method of differential fields. We consider two kinds of arrays: the passive array when the targets are point sources and the active array when the targets are scattering objects. The differential field method is used only in the active case.

7.4.1 Differential scattered field in clutter

Consider the reduced wave equation with a randomly heterogeneous background

$$\Delta u_0(\mathbf{x}) + k^2 \mu_0 \varepsilon_0(\mathbf{x}) u_0 = S, \tag{117}$$

where μ_0 is the magnetic permeability assumed to be unity, ε_0 the dielectric constant representing the random medium and S the source of illumination.

Suppose there is intrusion of a foreign object and as a result, the total dielectric constant $\varepsilon(\mathbf{x})$ is given by $\varepsilon_0(\mathbf{x}) + \tilde{\epsilon}(\mathbf{x})$ where $\tilde{\epsilon}(\mathbf{x})$ is a localized function representing the intrusion. Then with the same illumination the resulting electric field u satisfies

$$\Delta u(\mathbf{x}) + k^2 \big(\varepsilon_0(\mathbf{x}) + \tilde{\epsilon}(\mathbf{x})\big) u = S. \tag{118}$$

The differential (scattered) field, defined as $\tilde{u} = u - u_0$, then satisfies

$$\Delta \tilde{u}(\mathbf{x}) + k^2 \varepsilon_0(\mathbf{x}) \tilde{u} = -k^2 \tilde{\epsilon} u. \tag{119}$$

Let $\mathcal{H}_0$ and $\mathcal{H}$ be the transfer operators associated with eqs. (117) and (118) respectively. Namely, $u_0 = \mathcal{H}_0 S$ and $u = \mathcal{H}S$. By (119) we can write

$$\tilde{u} = -k^2 \mathcal{H}_0 \left[\tilde{\epsilon}\mathcal{H}S\right].$$

One can visualize the multiple scattering events by noting the following perturbation expansion

$$\mathcal{H} = \left(1 - k^2 \mathcal{H}_0 \tilde{\epsilon} + k^4 \left(\mathcal{H}_0 \tilde{\epsilon}\right)^2 - k^6 \left(\mathcal{H}_0 \tilde{\epsilon}\right)^3 + \cdots\right)\mathcal{H}_0 \qquad (120)$$

from which we obtain the expansion for the differential field

$$\tilde{u} = -k^2 \mathcal{H}_0 \tilde{\epsilon}\mathcal{H}_0 S + k^4 \mathcal{H}_0 \left(\tilde{\epsilon}\mathcal{H}_0 \tilde{\epsilon}\right)\mathcal{H}_0 S - k^6 \mathcal{H}_0 \left(\tilde{\epsilon}\mathcal{H}_0 \tilde{\epsilon}\mathcal{H}_0 \tilde{\epsilon}\right)\mathcal{H}_0 S + \cdots .(121)$$

The terms in parentheses of (121) correspond the multiple scattering events between the target and the clutter, Figure 19.

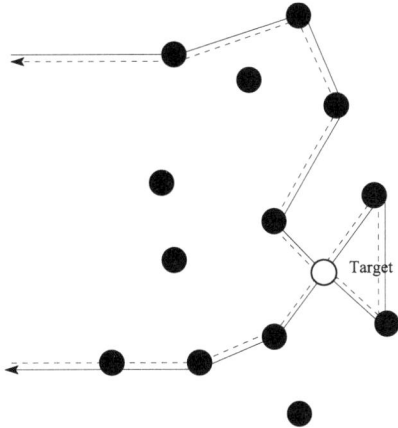

Figure 19   Double-passage interpretation of differential field.

For the simplicity we assume a weakly scattering target such that the multiple scattering between the target and the medium is negligible. This leads to the following simplification for the differential field

$$\tilde{u} = -k^2 \mathcal{H}_0 \tilde{\epsilon}\mathcal{H}_0 S. \qquad (122)$$

In this setting, the imaging problem is to determine $\tilde{\epsilon}$ from the information about $\tilde{u}$ without detailed information about $\mathcal{H}_0$.

7.4.2 Imaging functions

For a passive array, the signals are sampled by the antenna array and phase conjugated. The imaging method consists of back-propagating the resulting signals in computation domain by using an imaging filter $P(\omega)$ related to the mean Green function at frequency ω. Let $\tau_i(\omega)$ be the strength of i-th point sources $i = 1, \cdots, M$. The resulting imaging field is

$$u(\mathbf{x}) = \sum_{l=1}^{B} \sum_{j=1}^{N} \sum_{i=1}^{M} \tau_i(\omega_l) P(\mathbf{x}, \mathbf{y}_j; \omega_l) H_{ij}(\omega_l).$$

We write $\mathbf{P}(\omega) = [P_{ij}(\omega)]$ with $P_{ij}(\omega) = P(\mathbf{x}_i, \mathbf{y}_j; \omega)$.

In the case of an active array, we apply the method of the differential (response) field. In this approach, probing signals of various frequencies are first used to survey the random media in the absence of targets. Then in the presence of targets (with unknown locations) the same set of probing signals are used again to survey the media which is assumed to be fixed. The differences between these two responses is called the differential response which is then used to image the targets.

Let $\tau_i(\omega)$ be the scattering strength (reflectivity) of the i-th target, $i = 1, \cdots, M$ at frequency ω and let $\mathbf{H} = [H_{ij}]$ be the transfer matrix between the point targets and the antenna array. We shall assume weakly scattering targets so that the multiple scattering between the targets and the clutter is negligible and the only multiple scattering effect is in the clutter. In this approximation, the differential responses are given by $\sum_{i=1}^{M} \tau_i(\omega_l) H_{ij}(\omega_l) H_{in}(\omega_l), j = 1, \cdots, N, l = 1, \cdots, B$ where the index $n = 1, \cdots, N$ indicates the array elements emitting the probing signals. The imaging field in this case is given by

$$u(\mathbf{x}) = \sum_{l=1}^{B} \sum_{i=1}^{M} \sum_{j,n=1}^{N} \tau_i(\omega_l) P(\mathbf{x}, \mathbf{y}_j; \omega_l) H_{ij}(\omega_l) H_{in}(\omega_l) P(\mathbf{y}_n, \mathbf{x}; \omega_l)$$

or more generally

$$u(\mathbf{x}) = \sum_{l=1}^{B} f(\omega_l) \sum_{i=1}^{M} \sum_{j,n=1}^{N} \tau_i(\omega_l) P(\mathbf{x}, \mathbf{y}_j; \omega_l) H_{ij}(\omega_l) H_{in}(\omega_l) P(\mathbf{y}_n, \mathbf{x}; \omega_l)$$

with a weight function f of the frequency.

One of the central questions with imaging of cluttered targets is the statistical stability. Namely, how to construct an imaging functional that is independent of a certain class of unknown, random media? For any imaging function u a useful metric of stability is the signal-to-interference ratio (SIR) of the imaging functional at the location $\mathbf{x}$ given by

$$\Re(\mathbf{x}) \equiv \frac{|\mathbb{E}u(\mathbf{x})|^2}{\mathbb{E}(|u|^2(\mathbf{x})) - |\mathbb{E}(u(\mathbf{x}))|^2}.$$

A statistically stable imaging function corresponds to $\mathfrak{R} \gg 1$ whenever $\mathbb{E}u(\mathbf{x}) \neq 0$ and, in particular, in the neighborhood of every target point. Under such condition, the peaks of the amplitude $|u(\mathbf{x})|$ correspond to the point targets.

Our aim is to achieve stable imaging with as little information on the full Green function as possible. A natural choice for P is the phase factor of the mean Green function of the clutter. Our main assumptions are that the antenna elements and the point targets are sufficiently separated from one another, and that the multiple frequencies used are also sufficiently separated (see [27]). Under these conditions, we show that a sufficient condition for imaging stability is $KBN \gg M$ where B is the number of frequencies, N the number of antenna elements and M the number of point targets.

7.4.3   Numerical simulation with a Rician medium

Consider a discrete medium consisting of many randomly distributed point scatterers. Multiple scattering of waves in such a medium can be conveniently simulated by using the Foldy-Lax formulation of the Lippmann-Schwinger equation [27].

In the simulations, either 1000 or 3000 point scatterers are uniformly randomly distributed in the domain $[2000, 4000] \times [0, 5000]$, while the whole computation domain is $[-5000, 5000] \times [0, 5000]$. The transverse and longitudinal profiles of the Green function are shown in Figure 20. With 1000 particles, the K-factor in the clutter is on the order of unity.

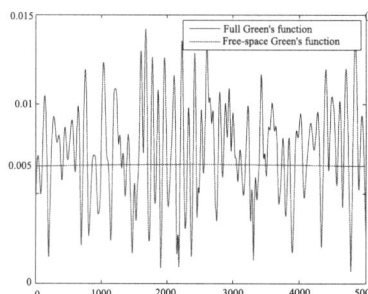 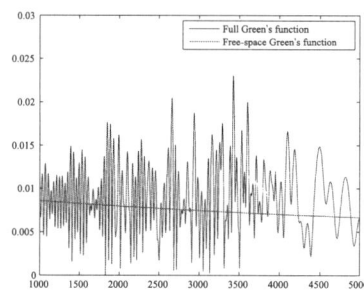

Figure 20   Typical transverse (left) and longitudinal (right) profiles of the intensity of the Green function with 1000 randomly distributed particles corresponding to $K \sim 1$.

For uniformly distributed scatterers, with a constant number density ρ, the mean Green function $\bar{H}$ in the high-frequency, forward scattering approximation satisfies the effective equation

$$\left(\nabla^2 + K_{\text{eff}}^2\right)\bar{G} = 0$$

with the effective wavenumber

$$K_{\text{eff}} = k + 2\pi f(\omega)\rho/k, \tag{123}$$

where $f(\omega)$ is the forward scattering amplitude at frequency ω. By the forward scattering theorem [37], the total extinction cross section σ_t is given by

$$\sigma_t = \begin{cases} \frac{4\pi}{k}\Im[f(\omega)], & d = 3, \\ 4\Im[f(\omega)], & d = 2. \end{cases} \tag{124}$$

Therefore the mean Green function in three dimensions has the form

$$\bar{G}(\mathbf{x}, \mathbf{y}; k) = -e^{-\rho\sigma_t r/2}\frac{e^{i\Re[K_{\text{eff}}]r}}{4\pi r}, \quad r = |\mathbf{x} - \mathbf{y}|. \tag{125}$$

From Figure 21 we see that $\Re[K_{\text{eff}}]$ is linearly proportional to k with a constant only slightly larger than one.

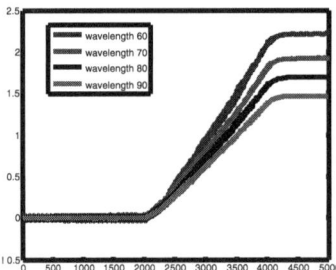 

Figure 21   The difference between the phase of the *mean* Green function, calculated with 3000 particles, and that of the free-space Green function along the longitudinal at various frequencies. The figure on the left shows linear growth of the phase difference with the propagation distance in the clutter for longer wavelengths while the figure on the right displays large fluctuations of the phase difference for wavelength 50.

To demonstrate the robustness of our approach, we will just use the phase factor e^{ikr} of the free-space Green function as the back-propagator P for imaging. Figures 22 and 23 show the results with 7 obscured point targets and $f(\omega) = 1$ and $f(\omega) = \omega^{-1}$, respectively.

This simple imaging method can be easily extended to the case of extended targets. Figure 24 shows the result with 5 line segments in both the passive and active array cases.

7.5   Coherent imaging in a Rayleigh medium

A Rayleigh medium is characterized by a zero-mean (i.e. zero K-factor) Gaussian distributed Green function G. Physically speaking, this assumption can be expressed by $g \gg 1$, where the dimensionless conductance g is the ensemble average of the transmittance [2], [47].

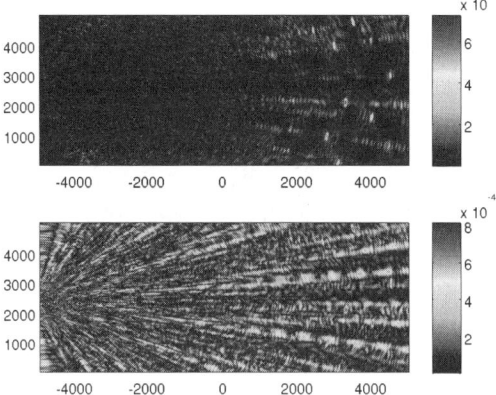

Figure 22   Active array imaging $|u(\mathbf{x})|$ of 7 point targets located at $[3100, 100]$, $[2800, 1000]$, $[4000, 1600]$, $[3300, 2100]$, $[4500, 3000]$, $[3000, 4000]$, and $[3500, 4800]$: the top and bottom plots are simulated with 1000 and 3000 point scatterers, respectively, randomly distributed in $x \in [2000, 4000]$. The scattering strengths of an individual scatterer to a target are 70 : 1. The 6 equally spaced antenna elements are on $y \in [1500, 3500]$, $x = -5000$. We use 20 equally spaced wavelengths from 52 to 90 and the weight function $f(\omega) = 1$.

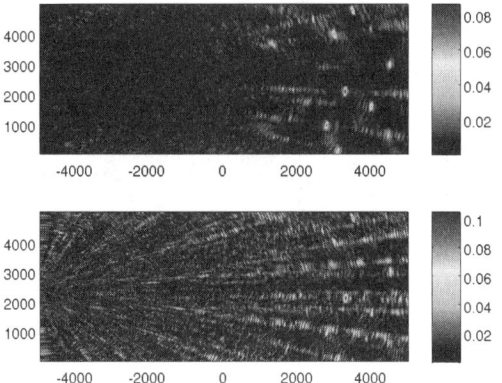

Figure 23   The same setting as in Figure 22 except with $f(\omega) = |\omega|^{-1}$. Compared to Figure 22, the resolution worsens but the stability is improved.

When $g \gg 1$, the transfer function G possesses a dominant short-range correlation on the scale of the transport mean-free-path. If the field point separations are larger than ℓ_t we can make the approximation

$$\mathbb{E}\left[G(\mathbf{x}, \mathbf{y})G^*(\mathbf{x}', \mathbf{y}')\right] \approx \mathbb{E}\left|G(\mathbf{x}, \mathbf{y})\right|^2 \delta(\mathbf{x} - \mathbf{x}')\delta(\mathbf{y} - \mathbf{y}'), \qquad (126)$$

where $\mathbf{x}, \mathbf{x}'$ are points on the front surface X of the wall while $\mathbf{y}, \mathbf{y}'$ are points on the back surface Y.

By (126) and the rule of computing Gaussian moments we obtain the

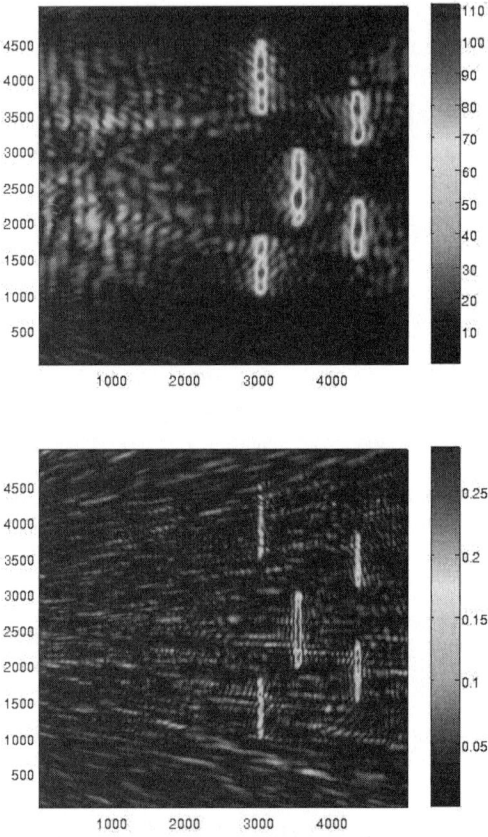

Figure 24   Passive (left) and active (right) array imaging with 51 antennas of 5 line segments cluttered in 3000 particles. The 51 equally spaced antennas are placed on the boundary $x = -5000, y \in [0, 5000]$ (not shown). The other imaging parameters are the same as in Figure 22.

fourth-order coherence function

$$\mathbb{E}\left[G(\mathbf{x}_1, \mathbf{y}_1)G^*(\mathbf{x}_1', \mathbf{y}_1')G(\mathbf{x}_2, \mathbf{y}_2)G^*(\mathbf{x}_2', \mathbf{y}_2')\right] \qquad (127)$$

$$\approx \mathbb{E}\left[\left|G(\mathbf{x}_1, \mathbf{y}_1)\right|^2\right] \mathbb{E}\left[\left|G(\mathbf{x}_2, \mathbf{y}_2)\right|^2\right]$$

$$\times \Big[\delta(\mathbf{x}_1 - \mathbf{x}_1')\delta(\mathbf{y}_1 - \mathbf{y}_1')\delta(\mathbf{x}_2 - \mathbf{x}_2')\delta(\mathbf{y}_2 - \mathbf{y}_2')$$

$$+ \delta(\mathbf{x}_1 - \mathbf{x}_2')\delta(\mathbf{y}_1 - \mathbf{y}_2')\delta(\mathbf{x}_2 - \mathbf{x}_1')\delta(\mathbf{y}_2 - \mathbf{y}_1')\Big].$$

For a statistically homogeneous medium the mean angular transmission coefficient $\mathbb{E}\left|G(\mathbf{x}, \mathbf{y})\right|^2$ is a function of $|\mathbf{x} - \mathbf{y}|$ and may be expressed as

$$\mathbb{E}\left|G(\mathbf{x}, \mathbf{y})\right|^2 = \hat{T} f_T(|\mathbf{x} - \mathbf{y}|),$$

where the mean (total) transmission coefficient $\hat{T}$ is proportional to ℓ_t/L_c and the angular transmission density function f_G is nonnegative and normalized

$$\int f_T(r)dr = \frac{1}{2\pi}.$$

Now let us describe the imaging geometry, Figure 25. The array elements are assumed to have the capability of making coherent measurement as well as transmitting coherent signals. Furthermore, the array elements are assumed to be a point transmitter/receiver and separated by more than one coherence length ℓ_c of the channel, which is typically comparable to the wavelength λ, so as to form a non-redundant aperture.

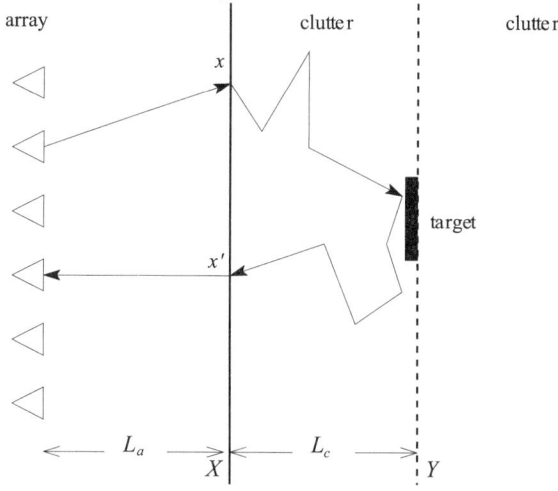

Figure 25   Geometry for imaging the front of the cluttered target.

The differential scattered field from the front of the target is given by

$$u(\mathbf{a};\mathbf{s}) = \int d\mathbf{x}'dr d\mathbf{x} G_0(\mathbf{s},\mathbf{x})G(\mathbf{x},\mathbf{r})\tau(\mathbf{r})G(\mathbf{x}',\mathbf{r})G_0(\mathbf{a},\mathbf{x}').$$

With this we can compute the mutual coherence function of the differential scattered field

$$\mathbb{E}\left[u(\mathbf{a}_1;\mathbf{s}_1)u^*(\mathbf{a}_2;\mathbf{s}_2)\right]$$
$$= \hat{T}^2 \int d\mathbf{r} |\tau(\mathbf{r})|^2 \int d\mathbf{x} f_T(|\mathbf{x}-\mathbf{r}|)G_0(\mathbf{s}_1,\mathbf{x})G_0^*(\mathbf{s}_2,\mathbf{x})$$
$$\times \int d\mathbf{x}' f_T(|\mathbf{x}'-\mathbf{r}|)G_0(\mathbf{a}_1,\mathbf{x}')G_0^*(\mathbf{a}_2,\mathbf{x}')$$
$$+ \hat{T}^2 \int d\mathbf{y} |\tau(\mathbf{r})|^2 \int d\mathbf{x} G_0^*(\mathbf{a}_2,\mathbf{x})f_T(|\mathbf{x}-\mathbf{r}|)G_0(\mathbf{s}_1,\mathbf{x})$$
$$\times \int d\mathbf{x}' G_0^*(\mathbf{s}_2,\mathbf{x}')f_T(|\mathbf{x}'-\mathbf{r}|)G_0(\mathbf{a}_1,\mathbf{x}'). \tag{128}$$

The second term in (128) represents the interference effect missing in radiative transfer theory.

We then have from (128) that

$$[u(\mathbf{a}_1;\mathbf{s}_1)u^*(\mathbf{a}_2;\mathbf{s}_2)]$$
$$= \int d\mathbf{r} |\tau(\mathbf{r})|^2 \Big[F_T(\mathbf{r};\mathbf{s}_1,\mathbf{s}_2)F_T(\mathbf{r};\mathbf{a}_1,\mathbf{a}_2)$$
$$+ F_T(\mathbf{r};\mathbf{s}_1,\mathbf{a}_2)F_T(\mathbf{r};\mathbf{a}_1,\mathbf{s}_2) \Big], \tag{129}$$

where

$$F_T(\mathbf{y};\mathbf{a},\mathbf{a}') = \int d\mathbf{x} f_T(|\mathbf{x}-\mathbf{y}|)G_0(\mathbf{a},\mathbf{x})G_0^*(\mathbf{a}',\mathbf{x}).$$

To bring (129) into more familiar form, let us assume that the clutter is in the far field such that the incident field is nearly a plane wave. Under such condition the Fraunhofer diffraction is valid:

$$G_0(\mathbf{a},\mathbf{x}) \sim e^{-ik\mathbf{a}\cdot\mathbf{x}/L_a},$$

where L_a is the distance between the array and the wall. Unimportant constants have been ignored. Then

$$F_T(\mathbf{y};\mathbf{a},\mathbf{a}') \sim \mathfrak{F}[f_T]\Big(\frac{\mathbf{a}-\mathbf{a}'}{\lambda L_a}\Big)e^{-ik(\mathbf{a}-\mathbf{a}')\cdot\mathbf{y}/L_a},$$

where $\mathfrak{F}$ stands for the Fourier transform. As a consequence we obtain

$$\mathbb{E}\left[u(\mathbf{a}_1;\mathbf{s}_1)u^*(\mathbf{a}_2;\mathbf{s}_2)\right] \sim$$
$$\hat{T}^2 \mathfrak{F}[|\tau|^2]\Big(\frac{\mathbf{s}_1-\mathbf{s}_2+\mathbf{a}_1-\mathbf{a}_2}{\lambda L_a}\Big)\Big[\mathfrak{F}[f_T]\Big(\frac{\mathbf{s}_1-\mathbf{s}_2}{\lambda L_a}\Big)\mathfrak{F}[f_T]\Big(\frac{\mathbf{a}_1-\mathbf{a}_2}{\lambda L_a}\Big)$$
$$+\mathfrak{F}[f_T]\Big(\frac{\mathbf{s}_1-\mathbf{a}_2}{\lambda L_a}\Big)\mathfrak{F}[f_T]\Big(\frac{\mathbf{a}_1-\mathbf{s}_2}{\lambda L_a}\Big)\Big]. \tag{130}$$

The function $\mathfrak{F}[f_T]$ is typically exponentially decaying

$$\mathfrak{F}[f_T](\mathbf{p}) \sim \frac{L_c|\mathbf{p}|}{\sinh\left(L_c|\mathbf{p}|\right)}$$

and limits the number of accessible modes of $\mathfrak{F}[|\tau|^2]$ to up to $1/L_c$ [47]. In other words, the resolution of $|\tau|^2$ is $\sim L_c$.

Acknowledgement

I am grateful to Professors Thomas Hou, Ta-Tsien Li and Chun Liu for inviting me to participate *The Second Shanghai Summer School in Applied Mathematics, 2006* and their hospitality while I was visiting Fudan University. I thank my PhD student, Mike Yan, for performing the numerical calculations shown in Figures 20–24.

References

[1] M.J. Beran and J. Oz-Vogt, *Imaging through turbulence in the atmosphere, Prog. Opt.* 33 (1994), 319-388.

[2] R. Berkovits and S. Feng, *Correlations in coherent multiple scattering. Phys. Rep.* 238 (1994), 135-172.

[3] M. Born and E. Wolf, *Principles of Optics: Electromagnetic Theory of Propagation, Interference and Diffraction of Light* (7th Edition), Cambridge University Press, 1999.

[4] P. Blomgren, G. Papanicolaou and H. Zhao, *Super-resolution in time-reversal acoustics. J. Acoust. Soc. Am.* 111(2002), 230.

[5] Z. Bouchal, J. Wagner and M. Chlup, *Self-reconstruction of a distorted nondiffracting beam, Opt. Comm.* 151 (1998) 207-211.

[6] Bruesselbach, D.C. Jones and D.A. Rockwell and R.C. Lind, *J. Opt. Soc. Am.* B 12(1995), 1434-1447.

[7] S. Chandrasekhar, *Radiative Transfer* (Dover Publications, New York, 1960).

[8] Courant and D. Hilbert, *Methods for Mathematical Physics, Vol. II.*

[9] J.C. Dainty and J.R. Fienup, *Phase retrieval and image reconstruction for astronomy,* in *Image Recovery,* ed. H. Stark, Academic Press, New York, pp, 231-275, 1987.

[10] A. Derode, E. Larose, M. Tanter, J. de Rosny, A. Tourin, M. Campillo and M. Fink, *J. Acoust. Soc. Am.*113 (2003), 2973.

[11] A. Derode, A. Tourin, J. de Rosny, M. Tanter, S. Yon, and M. Fink, *Phys. Rev. Lett.*90(2003), 014301.

[12] D. Dragoman, *The Wigner distribution function in optics and optoelectronics*, *Prog. Opt.*37(1997), 1-56.

[13] J. Durnin, *Exact solutions for nondiffracting beams. I. The scalar theory*,*J. Opt. Soc. Am. A* 4, 651-654 (1987).

[14] A. C. Fannjiang, *Parabolic approximation, radiative transfer and time reversal* , *J. Phys. A. Conf. Ser.* 7 (2005), 121-137.

[15] A. C. Fannjiang, *Self-averaging scaling limits for random parabolic waves. Arch. Rat. Mech. Anal.* 175:3 (2005), 343 - 387

[16] A. C. Fannjiang, *White-Noise and Geometrical Optics Limits of Wigner-Moyal Equation for Wave Beams in Turbulent Media*, *Comm. Math. Phys.* 254:2 (2005), pp. 289-322.

[17] A.C. Fannjiang, *White-Noise and Geometrical Optics Limits of Wigner-Moyal Equation for Wave Beams in Turbulent Media II. Two-Frequency Wigner Distribution Formulation*, *J. Stat. Phys.* 120 (2005), 543-586.

[18] A. Fannjiang, *Time Reversal Communication in Rayleigh Fading Broadcast Channels with Pinholes*, *Phys. Lett. A* 353/5 (2006), pp 389-397

[19] A. Fannjiang, *Information Transfer in Disordered Media by Broadband Time-Reversal: Stability, Resolution and Capacity*, *Nonlinearity* 19 (2006) 2425-2439.

[20] A.C. Fannjiang, *Space-frequency correlation of classical waves in disordered media: high-frequency and small-scale asymptotics*, *Europhys. Lett.* 80 (2007) 14005.

[21] A.C. Fannjiang, *Two-frequency radiative transfer and asymptotic solution*, *J. Opt. Soc. Am. A* 24 (2007), 2248-2256.

[22] A.C. Fannjiang, *Self-Averaging Scaling Limits of Two-Frequency Wigner Distribution for Random Paraxial Waves*, *J. Phys. A: Math. Theor.* 40 (2007) 5025-5044.

[23] A.C. Fannjiang, *Mutual Coherence of Polarized Light in Disordered Media: Two-Frequency Method Extended*, *J. Phy. A: Math. Theor.* 40 (2007) 13667-13683.

[24] A.C. Fannjiang, *Two-Frequency Radiative Transfer. II: Maxwell Equations in Random Dielectrics*, *J. Opt. Soc. Am. A* 24 (2007) 3680-3690.

[25] A.C. Fannjiang and K. Solna *Superresolution and Duality for Time-Reversal of Waves in Random Media*, *Phys. Lett. A*352:1-2 (2005), pp. 22-29.

[26] A. C. Fannjiang and K. Solna, *Propagation and Time-reversal of Wave Beams in Atmospheric Turbulence*, *SIAM Multiscale Mod. Sim.* 3:3 (2005), pp. 522-558.

[27] A.C. Fannjiang and M. Yan, *Multi-frequency imaging of multiple targets in Rician fading channels: stability and resolution*, *Inverse Problems* 23 (2007) 1801-1819.

[28] M. Fink, *Time-reversed acoustics*, *Sci. Amer.* November 1999, 91-97.

[29] M. Fink, D. Cassereau, A. Derode, C. Prada, P. Roux, M. Tanter, J.L. Thomas and F. Wu, *Rep. Progr. Phys.* 63(2000), 1933-1995.

[30] M. Fink and C. Prada, *Inverse Problems* 17 (2001), R1-R38.

[31] G.W. Forbes, V.I. Man'ko, H.M. Ozaktas, R. Simon, K.B. Wolf eds., *Wigner Distributions and Phase Space in Optics*, 2000 (feature issue, *J. Opt. Soc. Am. A* 17 No. 12).

[32] J. Frank, *Three-Dimensional Electron Microscopy of Macromolecular Assemblies*, Oxford University Press, New York, New York, 2006.

[33] A. Ghatak and K. Thyagarajan, *Graded index optical waveguide: a review*. *Prog. Opt.*18 (1980), 3-126.

[34] M. Gower and D. Proch, *Optical Phase Conjugation*. Springer-Verlag, Berlin, Heidelberg, 1994.

[35] F. Hlawatsch, *Time-Frequency Analysis and Synthesis of Linear Signal Spaces*, Kluwer Academic Publishers, Boston, 1998.

[36] M. Hegner, *Optics: The light fantastic*, *Nature* 419 (2002), 125-127.

[37] A. Ishimaru, *Wave Propagation and Scattering in Random Media. Vol. 1*, Academic Press, New York, 1978.

[38] A. Ishimaru, *Wave Propagation and Scattering in Random Media. Vol. 2*, Academic Press, New York, 1978.

[39] S. Jaffard, Y. Meyer and R.D. Ryan, *Wavelets – Tools for Science & Technology*, SIAM, Philadelphia, 2001.

[40] G. S. Kino, *Acoustic Waves – Devices, Imaging & Analog Signal Processing*, Prentice-Hall, Englewood Cliffs, New Jersey, 1987.

[41] W.A. Kuperman, W. Hodgkiss, H.C. Song, T. Akal, C. Ferla and D.R. Jackson, *Phase conjugation in the ocean: experimental demonstration of an acoustic time-reversal mirror*. *J. Acoust. Soc. Amer.* 102 (1997), 1-16.

[42] G. Lerosey, J. de Rosny, A. Tourin, M. Fink, *Focusing Beyond the Diffraction Limit with Far-Field Time Reversal*, *Science* 315 (2007), 1120.

[43] L. Mandel and E. Wolf, *Optical Coherence and Quantum Optics* (Cambridge University Press, 1995).

[44] C.A. McQueen, J. Arit and K. Dholakia, *An experiment to study a "nondiffracting" light beam*, Am. J. Phys. **67** (1999) 912-915.

[45] M. Mishchenko, L. Travis, A. Lacis, *Multiple Scattering of Light by Particles: Radiative Transfer and Coherent Backscattering* (Cambridge University Press, Cambridge, 2006).

[46] F. Natterer, *The Mathematics of Computerized Tomography*, Teubner, Stuttgart, 1986.

[47] M.C.W. van Rossum and Th. M. Nieuwenhuizen, *Multiple scattering of classical waves: microscopy, mesoscopy and diffusion.* Rev. Mod. Phys. **71** (1999), 313-371.

[48] Spohn H., *Derivation of the transport equation for electrons moving through random impurities*, J. Stat. Phys. **17**(1977), 385-412.

[49] J.W. Strohbehn, *Laser Beam Propagation in the Atmosphere*, Springer-Verlag, Berlin, 1978.

[50] F. Tappert: *The Parabolic approximation method*, in *Wave Propagation and Underwater Acoustics*, J.B. Keller &J.S. Papadakis, eds., Lect. Notes in Phys. Vol. 70, Springer-Verlag, New York, 1977.

[51] V.I. Tatarskii, A. Ishimaru and V.U. Zavorotny, *Wave Propagation in Random Media*, SPIE, Bellingham, Washington, 1993.

[52] V.I. Tatarskii and V.U. Zavorotny, *Strong fluctuations in light propagation in a randomly inhomogeneous medium*, Prog. Opt. **18** (1980), 205-254.

[53] A.D. Wheelon, *Electromagnetic Scintillation. Vol. II*, Cambridge University Press, Cambridge, UK, 2001.

[54] G. Weigelt, *Triple-correlation imaging in optical astronomy*, Prog. Opt. **29** (1991), 293-319.

Multiscale Computations for Flow and Transport in Porous Media

Thomas Y. Hou

Applied Mathematics, 217-50, Caltech
Pasadena, CA 91125, USA
E-mail: hou@acm.caltech.edu

Abstract

Many problems of fundamental and practical importance have multiple scale solutions. The direct numerical solution of multiple scale problems is difficult to obtain even with modern supercomputers. The major difficulty of direct solutions is the scale of computation. The ratio between the largest scale and the smallest scale could be as large as 10^5 in each space dimension. From an engineering perspective, it is often sufficient to predict the macroscopic properties of the multiple-scale systems, such as the effective conductivity, elastic moduli, permeability, and eddy diffusivity. Therefore, it is desirable to develop a method that captures the small scale effect on the large scales, but does not require resolving all the small scale features. This paper reviews some of the recent advances in developing systematic multiscale methods with particular emphasis on multiscale finite element methods with applications to flow and transport in heterogeneous porous media. This manuscript is not intended to be a general survey paper on this topic. The discussion is limited by the scope of the lectures and expertise of the author.

1 Introduction

Many problems of fundamental and practical importance have multiple scale solutions. Composite materials, porous media, and turbulent transport in high Reynolds number flows are examples of this type. A complete analysis of these problems is extremely difficult. For example, the difficulty in analyzing groundwater transport is mainly caused by the heterogeneity of subsurface formations spanning over many scales. This heterogeneity is often represented by the multiscale fluctuations in the permeability of media. For composite materials, the dispersed phases (particles or fibers), which may be randomly distributed in the matrix, give rise to fluctuations in the thermal or electrical conductivity; moreover, the conductivity is usually discontinuous across the phase boundaries. In turbulent transport problems, the convective velocity field fluctuates randomly and contains many scales depending on the Reynolds number of the flow.

The direct numerical solution of multiple scale problems is difficult even with the advent of supercomputers. The major difficulty of direct solutions is the scale of computation. For groundwater simulations, it is common that millions of grid blocks are involved, with each block having a dimension of tens of meters, whereas the permeability measured from cores is at a scale of several centimeters. This gives more than 10^5 degrees of freedom per spatial dimension in the computation. Therefore, a tremendous amount of computer memory and CPU time are required, and this can easily exceed the limit of today's computing resources. The situation can be relieved to some degree by parallel computing; however, the size of the discrete problem is *not* reduced. The load is merely shared by more processors with more memory. Whenever one can afford to resolve all the small scale features of a physical problem, direct solutions provide quantitative information of the physical processes at all scales. On the other hand, from an engineering perspective, it is often sufficient to predict the macroscopic properties of the multiscale systems, such as the effective conductivity, elastic moduli, permeability, and eddy diffusivity. Therefore, it is desirable to develop a method that captures the small scale effect on the large scales, but does not require resolving all the small scale features. Upscaling procedures have been commonly applied for this purpose and are effective in many cases. More recently, a number of multiscale techniques have been developed and successfully applied to various areas, e.g., porous media flows. The main idea of upscaling techniques is to form coarse-scale equations with a prescribed analytical form that may differ from the underlying fine-scale equations. In multiscale methods, the fine-scale information is carried throughout the simulation and the coarse-scale equations are generally not expressed analytically, but rather formed and solved numerically.

The purpose of this lecture note is to review some recent advances in developing multiscale finite element (volume) methods for flow and transport in strongly heterogeneous porous media. Extra effort is made in developing a multiscale computational method that can be potentially used for practical multiscale for problems with a large range of nonseparable scales. Substantial progress has been made in recent years by combining modern mathematical techniques such as multiscale analysis, random sampling, and adaptivity and multiresolution. The lectures can be roughly divided into six parts. In Section 2, we will review some homogenization theory for elliptic and hyperbolic equations as well as for incompressible flows. This homogenization theory provides the critical guideline for designing effective multiscale methods. In Section 3, we discuss numerical homogenization based on sampling techniques. In Section 4, we discuss some recent developments of multiscale finite element methods. We also discuss the issue of upscaling one-phase, two-phase flows through heterogeneous porous media and the use of limited global information in multiscale finite element methods. In Section 5, we discuss the generalization of the multiscale finite element methods to nonlinear partial differential equations. In Section 6, we will consider multiscale simulations of two-phase flow immiscible flows using a flow-based adaptive coordinate system. There are many other multiscale methods which we will not cover due to the limited scope of these lectures. The above methods are chosen because they are similar philosophically and the materials complement each other very well. This paper is not intended to be a detailed survey of all available multiscale methods. The discussion is limited by scope of the lectures and expertise of the author.

2 Review of homogenization theory

In this section, we will review some classical homogenization theory for elliptic and hyperbolic PDEs. This homogenization theory will play an essential role in designing effective multiscale numerical methods for partial differential equations with multiscale solutions.

2.1 Homogenization theory for elliptic problems

Consider the second order elliptic equation

$$\mathcal{L}(u_\varepsilon) \equiv -\frac{\partial}{\partial x_i}\left(a_{ij}\left(x/\varepsilon\right)\frac{\partial}{\partial x_j}\right)u_\varepsilon + a_0(x/\varepsilon)u_\varepsilon = f, \ u_\varepsilon|_{\partial\Omega} = 0, \quad (2.1)$$

where $a_{ij}(y)$ and $a_0(y)$ are 1-periodic in both variables of y, and satisfy $a_{ij}(y)\xi_i\xi_j \geqslant \alpha\xi_i\xi_i$, with $\alpha > 0$, and $a_0 > \alpha_0 > 0$. Here we have used the Einstein summation notation, i.e., repeated index means summation with respect to that index.

This model equation represents a common difficulty shared by several physical problems. For porous media, it is the pressure equation through Darcy's law, the coefficient a_ε representing the permeability tensor. For composite materials, it is the steady heat conduction equation and the coefficient a_ε represents the thermal conductivity. For steady transport problems, it is a symmetrized form of the governing equation. In this case, the coefficient a_ε is a combination of transport velocity and viscosity tensor.

Homogenization theory is to study the limiting behavior $u_\varepsilon \to u$ as $\varepsilon \to 0$. The main task is to find the homogenized coefficients, a_{ij}^* and a_0^*, and the *homogenized equation* for the limiting solution u,

$$-\frac{\partial}{\partial x_i}\left(a_{ij}^* \frac{\partial}{\partial x_j}\right)u + a_0^* u = f, \quad u|_{\partial\Omega} = 0. \tag{2.2}$$

Define the L^2 and H^1 norms over Ω as follows:

$$\|v\|_0^2 = \int_\Omega |v|^2 \, dx, \quad \|v\|_1^2 = \|v\|_0^2 + \|\nabla v\|_0^2. \tag{2.3}$$

Further, we define the bilinear form

$$a^\varepsilon(u, v) = \int_\Omega a_{i,j}^\varepsilon(x)\frac{\partial u}{\partial x_j}\frac{\partial v}{\partial x_i}\, dx + \int_\Omega a_0^\varepsilon uv \, dx. \tag{2.4}$$

It is easy to show that

$$c_1\|u\|_1^2 \leqslant a^\varepsilon(u, u) \leqslant c_2\|u\|_1^2, \tag{2.5}$$

with $c_1 = \min(\alpha, \alpha_0)$, $c_2 = \max(\|a_{ij}\|_\infty, \|a_0\|_\infty)$.

The elliptic problem can also be formulated as a variational problem: find $u_\varepsilon \in H_0^1$,

$$a^\varepsilon(u_\varepsilon, v) = (f, v) \quad \text{for all } v \in H_0^1(\Omega), \tag{2.6}$$

where (f, v) is the usual L^2 inner product, $\int_\Omega fv \, dx$.

Special case: one-dimensional problem. Let $\Omega = (x_0, x_1)$ and take $a_0 = 0$. We have

$$-\frac{d}{dx}\left(a(x/\varepsilon)\frac{du_\varepsilon}{dx}\right) = f \quad \text{in } \Omega, \tag{2.7}$$

where $u_\varepsilon(x_0) = u_\varepsilon(x_1) = 0$, and $a(y) > \alpha_0 > 0$ is y-periodic with period y_0.

By taking $v = u_\varepsilon$ in the bilinear form, we have

$$\|u_\varepsilon\|_1 \leqslant c.$$

Therefore, one can extract a subsequence, still denoted by u_ε, such that

$$u_\varepsilon \rightharpoonup u \quad \text{in } H_0^1(\Omega) \text{ weakly}. \tag{2.8}$$

On the other hand, we notice that

$$a^\varepsilon \rightharpoonup m(a) = \frac{1}{y_0}\int_0^{y_0} a(y) \, dy \quad \text{in } L^\infty(\Omega) \text{ weak star}. \tag{2.9}$$

It is tempting to conclude that u satisfies

$$-\frac{d}{dx}\left(m(a)\frac{du}{dx}\right) = f,$$

where $m(a) = \frac{1}{y_0}\int_0^{y_0} a(y)\,dy$ is the arithmetic mean of a. However, this is *not* true. To derive the correct answer, we introduce

$$\xi^\varepsilon = a^\varepsilon\frac{du^\varepsilon}{dx}.$$

Since a^ε is bounded, and u_x^ε is bounded in $L^2(\Omega)$, so ξ^ε is bounded in $L^2(\Omega)$. Moreover, since $-\frac{d\xi^\varepsilon}{dx} = f$, we have $\xi^\varepsilon \in H^1(\Omega)$. Thus we get

$$\xi^\varepsilon \to \xi \quad \text{in } L^2(\Omega) \text{ strongly,}$$

so that

$$\frac{1}{a^\varepsilon}\xi^\varepsilon \to m(1/a)\xi \quad \text{in } L^2(\Omega) \text{ weakly.}$$

Further, we note that $\frac{1}{a^\varepsilon}\xi^\varepsilon = \frac{du^\varepsilon}{dx}$. Therefore, we arrive at

$$\frac{du}{dx} = m(1/a)\xi.$$

On the other hand, $-\frac{d\xi^\varepsilon}{dx} = f$ implies $-\frac{d\xi}{dx} = f$. This gives

$$-\frac{d}{dx}\left(\frac{1}{m(1/a)}\frac{du}{dx}\right) = f. \tag{2.10}$$

This is the correct homogenized equation for u. Note that $a^* = \frac{1}{m(1/a)}$ is the harmonic average of a^ε. It is in general not equal to the arithmetic average $\overline{a^\varepsilon} = m(a)$.

Multiscale asymptotic expansions. The above analysis does not generalize to multi-dimensions. In this subsection, we introduce the multiscale expansion technique in deriving homogenized equations. This technique is very effective and can be used in a number of applications.

We shall look for $u_\varepsilon(x)$ in the form of asymptotic expansion

$$u_\varepsilon(x) = u_0(x, x/\varepsilon) + \varepsilon u_1(x, x/\varepsilon) + \varepsilon^2 u_2(x, x/\varepsilon) + \cdots, \tag{2.11}$$

where the functions $u_j(x, y)$ are double periodic in y with period 1.

Denote by A^ε the second order elliptic operator:

$$A^\varepsilon = -\frac{\partial}{\partial x_i}\left(a_{ij}(x/\varepsilon)\frac{\partial}{\partial x_j}\right). \tag{2.12}$$

When differentiating a function $\phi(x, x/\varepsilon)$ with respect to x, we have

$$\frac{\partial}{\partial x_j} = \frac{\partial}{\partial x_j} + \frac{1}{\varepsilon}\frac{\partial}{\partial y_j},$$

where y is evaluated at $y = x/\varepsilon$. With this notation, we can expand A^ε as follows:

$$A^\varepsilon = \varepsilon^{-2} A_1 + \varepsilon^{-1} A_2 + \varepsilon^0 A_3, \tag{2.13}$$

where

$$A_1 = -\frac{\partial}{\partial y_i}\left(a_{ij}(y)\frac{\partial}{\partial y_j}\right), \tag{2.14}$$

$$A_2 = -\frac{\partial}{\partial y_i}\left(a_{ij}(y)\frac{\partial}{\partial x_j}\right) - \frac{\partial}{\partial x_i}\left(a_{ij}(y)\frac{\partial}{\partial y_j}\right), \tag{2.15}$$

$$A_3 = -\frac{\partial}{\partial x_i}\left(a_{ij}(y)\frac{\partial}{\partial x_j}\right) + a_0. \tag{2.16}$$

Substituting the expansions for u_ε and A^ε into $A^\varepsilon u_\varepsilon = f$, and equating the terms of the same power, we get

$$A_1 u_0 = 0, \tag{2.17}$$
$$A_1 u_1 + A_2 u_0 = 0, \tag{2.18}$$
$$A_1 u_2 + A_2 u_1 + A_3 u_0 = f. \tag{2.19}$$

Equation (2.17) can be written as

$$-\frac{\partial}{\partial y_i}\left(a_{ij}(y)\frac{\partial}{\partial y_j}\right) u_0(x, y) = 0, \tag{2.20}$$

where u_0 is periodic in y. The theory of second order elliptic PDEs [57] implies that $u_0(x, y)$ is independent of y, i.e., $u_0(x, y) = u_0(x)$. This simplifies equation (2.18) for u_1,

$$-\frac{\partial}{\partial y_i}\left(a_{ij}(y)\frac{\partial}{\partial y_j}\right) u_1 = \left(\frac{\partial}{\partial y_i}a_{ij}(y)\right)\frac{\partial u}{\partial x_j}(x).$$

Define $\chi^j = \chi^j(y)$ as the solution to the following *cell problem*:

$$\frac{\partial}{\partial y_i}\left(a_{ij}(y)\frac{\partial}{\partial y_j}\right)\chi^j = \frac{\partial}{\partial y_i}a_{ij}(y), \tag{2.21}$$

where χ^j is double periodic in y. The general solution of equation (2.18) for u_1 is then given by

$$u_1(x, y) = -\chi^j(y)\frac{\partial u}{\partial x_j}(x) + \tilde{u}_1(x). \tag{2.22}$$

Finally, we note that the equation for u_2 is given by

$$\frac{\partial}{\partial y_i}\left(a_{ij}(y)\frac{\partial}{\partial y_j}\right)u_2 = A_2 u_1 + A_3 u_0 - f. \tag{2.23}$$

The solvability condition implies that the right-hand side of (2.23) must have mean zero in y over one periodic cell $Y = [0,1] \times [0,1]$, i.e.,

$$\int_Y (A_2 u_1 + A_3 u_0 - f)\, dy = 0.$$

This solvability condition for second order elliptic PDEs with periodic boundary condition [57] requires that the right-hand side of equation (2.23) have mean zero with respect to the fast variable y. This solvability condition gives rise to the homogenized equation for u:

$$-\frac{\partial}{\partial x_i}\left(a_{ij}^*\frac{\partial}{\partial x_j}\right)u + m(a_0)u = f, \tag{2.24}$$

where $m(a_0) = \frac{1}{|Y|}\int_Y a_0(y)\, dy$ and

$$a_{ij}^* = \frac{1}{|Y|}\int_Y \left(a_{ij} - a_{ik}\frac{\partial \chi^j}{\partial y_k}\right) dy. \tag{2.25}$$

Justification of formal expansions. The above multiscale expansion is based on a formal asymptotic analysis. However, we can justify its convergence rigorously.

Let $z_\varepsilon = u_\varepsilon - (u + \varepsilon u_1 + \varepsilon^2 u_2)$. Applying A^ε to z_ε, we get

$$A^\varepsilon z_\varepsilon = -\varepsilon r_\varepsilon,$$

where $r_\varepsilon = A_2 u_2 + A_3 u_1 + \varepsilon A_3 u_2$. If f is smooth enough, so is u_2. Thus we have $\|r_\varepsilon\|_\infty \leqslant c$.

On the other hand, we have

$$z_\varepsilon|_{\partial\Omega} = -(\varepsilon u_1 + \varepsilon^2 u_2)|_{\partial\Omega}.$$

Thus, we obtain

$$\|z_\varepsilon\|_{L^\infty(\partial\Omega)} \leqslant c\varepsilon.$$

It follows from the maximum principle [57] that

$$\|z_\varepsilon\|_{L^\infty(\Omega)} \leqslant c\varepsilon$$

and therefore we conclude that

$$\|u_\varepsilon - u\|_{L^\infty(\Omega)} \leqslant c\varepsilon.$$

Boundary corrections. The above asymptotic expansion does not take into account the boundary condition of the original elliptic PDEs. If we add a boundary correction, we can obtain higher order approximations.

Let $\theta_\varepsilon \in H^1(\Omega)$ denote the solution to

$$\nabla_x \cdot a^\varepsilon \nabla_x \theta_\varepsilon = 0 \text{ in } \Omega, \quad \theta_\varepsilon = u_1(x, x/\varepsilon) \text{ on } \partial\Omega.$$

Then we have

$$(u_\varepsilon - (u + \varepsilon u_1(x, x/\varepsilon) - \varepsilon \theta_\varepsilon)) |_{\partial\Omega} = 0.$$

Moskow and Vogelius [84] have shown that

$$
\begin{aligned}
\|u_\varepsilon - u - \varepsilon u_1(x, x/\varepsilon) + \varepsilon \theta_\varepsilon\|_0 &\leqslant C_\omega \varepsilon^{1+\omega} \|u\|_{2+\omega}, \\
\|u_\varepsilon - u - \varepsilon u_1(x, x/\varepsilon) + \varepsilon \theta_\varepsilon\|_1 &\leqslant C\varepsilon \|u\|_2,
\end{aligned}
\tag{2.26}
$$

where we assume $u \in H^{2+\omega}(\Omega)$ with $0 \leqslant \omega \leqslant 1$, and Ω is assumed to be a bounded, convex curvilinear polygon of class C^∞. This improved estimate will be used in the convergence analysis of the multiscale finite element method to be presented in Section 4.

2.2   Homogenization for hyperbolic problems

In this subsection, we will review some homogenization theory for semilinear hyperbolic systems. As we will see below, homogenization for hyperbolic problems is very different from that for elliptic problems. The phenomena are also very rich.

Consider the semilinear Carleman equations [23]:

$$
\begin{aligned}
\frac{\partial u_\varepsilon}{\partial t} + \frac{\partial u_\varepsilon}{\partial x} &= v_\varepsilon^2 - u_\varepsilon^2, \\
\frac{\partial v_\varepsilon}{\partial t} - \frac{\partial v_\varepsilon}{\partial x} &= u_\varepsilon^2 - v_\varepsilon^2,
\end{aligned}
$$

with oscillatory initial data, $u_\varepsilon(x, 0) = u_0^\varepsilon(x)$, $v_\varepsilon(x, 0) = v_0^\varepsilon(x)$.

Assume that the initial conditions are positive and bounded. Then it can be shown that there exists a unique *bounded* solution for all times. Thus we can extract a subsequence of u_ε and v_ε such that $u_\varepsilon \rightharpoonup u$ and $v_\varepsilon \rightharpoonup v$ as $\varepsilon \to 0$.

Denote u_m as the weak limit of u_ε^m, and v_m as the weak limit of v_ε^m. By taking the weak limit of both sides of the equations, we get

$$
\begin{aligned}
\frac{\partial u_1}{\partial t} + \frac{\partial u_1}{\partial x} &= v_2 - u_2, \\
\frac{\partial v_1}{\partial t} - \frac{\partial v_1}{\partial x} &= u_2 - v_2.
\end{aligned}
$$

By multiplying the Carleman equations by u_ε and v_ε respectively, we get

$$\frac{\partial u_\varepsilon^2}{\partial t} + \frac{\partial u_\varepsilon^2}{\partial x} = 2u_\varepsilon v_\varepsilon^2 - 2u_\varepsilon^3,$$

$$\frac{\partial v_\varepsilon^2}{\partial t} + \frac{\partial v_\varepsilon^2}{\partial x} = 2v_\varepsilon u_\varepsilon^2 - 2v_\varepsilon^3.$$

Thus the weak limit of u_ε^2 depends on the weak limit of u_ε^3 and the weak limit of $u_\varepsilon v_\varepsilon^2$.

Denote by $\overline{w_\varepsilon}$ as the weak limit of w_ε. To obtain a closure, we would like to express $\overline{u_\varepsilon v_\varepsilon^2}$ in terms of the product $\overline{u_\varepsilon}$ and $\overline{v_\varepsilon^2}$. This is *not* possible in general. In this particular case, we can use the Div-Curl Lemma [85, 86, 98] to obtain a closure.

The Div-Curl Lemma. *Let Ω be an open set of $\mathbb{R}^N$ and u_ε and v_ε be two sequences such that*

$$u_\varepsilon \rightharpoonup u \quad in \ \left(L^2(\Omega)\right)^N \ weakly,$$

$$v_\varepsilon \rightharpoonup v \quad in \ \left(L^2(\Omega)\right)^N \ weakly.$$

Further, we assume that

$$\mathrm{div}\, u_\varepsilon \ is \ bounded \ in \ L^2(\Omega) \ (or compact in H^{-1}(\Omega)),$$

$$\mathbf{curl}\, v_\varepsilon \ is \ bounded \ in \ \left(L^2(\Omega)\right)^{N^2} \ (or \ compact \ in \ \left(H^{-1}(\Omega)\right)^{N^2}).$$

Let $\langle \cdot, \cdot \rangle$ denote the inner product in $\mathbb{R}^N$, i.e.,

$$\langle \mathbf{u}, \mathbf{v} \rangle = \sum_{i=1}^{N} u_i v_i.$$

Then we have

$$\langle \mathbf{u}_\varepsilon \cdot \mathbf{v}_\varepsilon \rangle \rightharpoonup \langle \mathbf{u} \cdot \mathbf{v} \rangle \quad weakly. \tag{2.27}$$

Remark 2.1. We remark that the Div-Curl Lemma is the simplest form of the more general compensated compactness theory developed by Tartar [98] and Murat [85, 86].

Applying the Div-Curl Lemma to $(u_\varepsilon, u_\varepsilon)$ and $(v_\varepsilon^2, v_\varepsilon^2)$ in the space-time domain, one can show that $\overline{u_\varepsilon v_\varepsilon^2} = \overline{u_\varepsilon}\; \overline{v_\varepsilon^2}$. Similarly, one can show that $\overline{u_\varepsilon^2 v_\varepsilon} = \overline{u_\varepsilon^2}\; \overline{v_\varepsilon}$. Using this fact, Tartar [99] obtained the following infinite hyperbolic system for u_m and v_m [99]:

$$\frac{\partial u_m}{\partial t} + \frac{\partial u_m}{\partial x} = m u_{m-1} v_2 - m u_{m+1},$$

$$\frac{\partial v_m}{\partial t} - \frac{\partial v_m}{\partial x} = m v_{m-1} u_2 - m v_{m+1}.$$

Note that the weak limit of u_ε^m, u_m, depends on the weak limit of u_ε^{m+1}, u_{m+1}. Similarly, v_m depends on v_{m+1}. Thus one cannot obtain a closed system for the weak limits u_ε and v_ε by a finite system. This is a generic phenomenon for nonlinear partial differential equations with microstructure. It is often referred to as the closure problem. On the other hand, for the Carleman equations, Tartar showed that the infinite system is hyperbolic and the system is well-posed.

The situation is very different for a 3×3 system of Broadwell type [22]:

$$\frac{\partial u_\varepsilon}{\partial t} + \frac{\partial u_\varepsilon}{\partial x} = w_\varepsilon^2 - u_\varepsilon v_\varepsilon, \qquad (2.28)$$

$$\frac{\partial v_\varepsilon}{\partial t} - \frac{\partial v_\varepsilon}{\partial x} = w_\varepsilon^2 - u_\varepsilon v_\varepsilon, \qquad (2.29)$$

$$\frac{\partial w_\varepsilon}{\partial t} + \alpha\frac{\partial w_\varepsilon}{\partial x} = u_\varepsilon v_\varepsilon - w_\varepsilon^2, \qquad (2.30)$$

with oscillatory initial data, $u_\varepsilon(x,0) = u_0^\varepsilon(x)$, $v_\varepsilon(x,0) = v_0^\varepsilon(x)$ and $w_\varepsilon(x,0) = w_0^\varepsilon(x)$. When $\alpha = 0$, the above system reduces to the original Broadwell model. We will refer to the above system as the generalized Broadwell model.

Note that in the generalized Broadwell model, the right-hand side of the w-equation depends on the product of uv. If we try to obtain an evolution equation for w_ε^2, it will depend on the triple product $u_\varepsilon v_\varepsilon w_\varepsilon$. The Div-Curl Lemma cannot be used here to characterize the weak limit of this triple product in terms of the weak limits of u_ε, v_ε and w_ε.

Assume that the initial oscillations are periodic, i.e.,

$$u_0^\varepsilon = u_0(x, x/\varepsilon), \ v_0^\varepsilon = v_0(x, x/\varepsilon), \ w_0^\varepsilon = w_0(x, x/\varepsilon),$$

where $u_0(x,y)$, $v_0(x,y)$, $w_0(x,y)$ are 1-periodic in y.

There are two cases to consider.

Case 1. $\alpha = m/n$ is a rational number. Let $\{U(x,y,t), V(x,y,t), W(x, y,t)\}$ be the homogenized solution which satisfies

$$U_t + U_x = \int_0^1 W^2\, dy - U\int_0^1 V\, dy,$$

$$V_t - V_x = \int_0^1 W^2\, dy - U\int_0^1 V\, dy,$$

$$W_t + \alpha W_x = -W^2 + \frac{1}{n}\int_0^n U(x, y + (\alpha-1)z, t)V(x, y + (\alpha+1)z, t)\, dz,$$

where $U|_{t=0} = u_0(x,y)$, $V|_{t=0} = v_0(x,y)$ and $W|_{t=0} = w_0(x,y)$. Then we have

$$\|u_\varepsilon(x,t) - U(x, \tfrac{x-t}{\varepsilon}, t)\|_{L^\infty} \leqslant C\varepsilon,$$

$$\|v_\varepsilon(x,t) - V(x, \tfrac{x+t}{\varepsilon}, t)\|_{L^\infty} \leqslant C\varepsilon,$$

$$\|w_\varepsilon(x,t) - W(x, \tfrac{x-\alpha t}{\varepsilon}, t)\|_{L^\infty} \leqslant C\varepsilon.$$

Case 2. α is an irrational number. Let $\{U(x,y,t), V(x,y,t), W(x,y,t)\}$ be the homogenized solution which satisfies

$$U_t + U_x = \int_0^1 W^2 \, dy - U \int_0^1 V \, dy,$$

$$V_t - V_x = \int_0^1 W^2 \, dy - U \int_0^1 V \, dy,$$

$$W_t + \alpha W_x = -W^2 + \left(\int_0^1 U \, dy \right) \left(\int_0^1 V \, dy \right),$$

where $U|_{t=0} = u_0(x,y)$, $V|_{t=0} = v_0(x,y)$ and $W|_{t=0} = w_0(x,y)$. Then we have

$$\|u_\varepsilon(x,t) - U(x, \tfrac{x-t}{\varepsilon}, t)\|_{L^\infty} \leqslant C\varepsilon,$$

$$\|v_\varepsilon(x,t) - V(x, \tfrac{x+t}{\varepsilon}, t)\|_{L^\infty} \leqslant C\varepsilon,$$

$$\|w_\varepsilon(x,t) - W(x, \tfrac{x-\alpha t}{\varepsilon}, t)\|_{L^\infty} \leqslant C\varepsilon.$$

We refer the reader to [59] for the proof of the above results.

Note that when α is a rational number, the interaction of u_ε and v_ε can generate a high frequency contribution to w_ε. This is *not* the case when α is an irrational number. The rational α case corresponds to a resonance interaction.

The derivation and analysis of the above results rely on the following two lemmas:

Lemma 2.1. *Let* $f(x), g(x,y) \in C^1$. *Assume that* $g(x,y)$ *is n-periodic in* y. *Then we have*

$$\int_a^b f(x)g(x,x/\varepsilon) \, dx = \int_a^b f(x) \left(\frac{1}{n} \int_0^n g(x,y) \, dy \right) dx + O(\varepsilon).$$

Lemma 2.2. *Let* $f(x,y,z) \in C^1$. *Assume that* $f(x,y,z)$ *is 1-periodic in* y *and* z. *If* γ_2/γ_1 *is an irrational number, then we have*

$$\int_a^b f\left(x, \tfrac{x_1 + \gamma_1 x}{\varepsilon}, \tfrac{x_2 + \gamma_1 x}{\varepsilon} \right) dx = \int_a^b \left(\int_0^1 \int_0^1 f(x,y,z) \, dy \, dz \right) dx + O(\varepsilon).$$

The proof uses some basic ergodic theory. It can be seen easily by expanding in Fourier series in the periodic variables [59]. For the sake of completeness, we present a simple proof of the above homogenization result for the case of $\alpha = 0$ in the next subsection.

Homogenization of the Broadwell model. In this subsection, we give a simple proof of the homogenization result in the special case of $\alpha = 0$. The homogenized equations can be derived by multiscale asymptotic expansions [82].

Consider the Broadwell model

$$\partial_t u + \partial_x u = w^2 - uv \quad \text{in } \mathbb{R} \times (0, T), \tag{2.31}$$

$$\partial_t v - \partial_x v = w^2 - uv \quad \text{in } \mathbb{R} \times (0, T), \tag{2.32}$$

$$\partial_t w = uv - w^2 \quad \text{in } \mathbb{R} \times (0, T), \tag{2.33}$$

with oscillatory initial values

$$u(x,0) = u_0(x, \tfrac{x}{\varepsilon}), \quad v(x,0) = v_0(x, \tfrac{x}{\varepsilon}), \quad w(x,0) = w_0(x, \tfrac{x}{\varepsilon}), \tag{2.34}$$

where $u_0(x,y), v_0(x,y), w_0(x,y)$ are 1-periodic in y. We introduce an extra variable, y, to describe the fast variable, x/ε. Let the solution of the homogenized equation be $\{U(x,y,t), V(x,y,t), W(x,y,t)\}$ which satisfies

$$\partial_t U + \partial_x U + U \int_0^1 V \, dy - \int_0^1 W^2 \, dy = 0 \text{ in } \mathbb{R} \times (0, T), \tag{2.35}$$

$$\partial_t V - \partial_x V + V \int_0^1 U \, dy - \int_0^1 W^2 \, dy = 0 \text{ in } \mathbb{R} \times (0, T), \tag{2.36}$$

$$\partial_t W + W^2 - \int_0^1 U(x, y-z, t) V(x, y+z, t) \, dz = 0 \text{ in } \mathbb{R} \times (0, T), \tag{2.37}$$

with initial values given by

$$U(x,y,0) = u_0(x,y), \quad V(x,y,0) = v_0(x,y), \quad W(x,y,0) = w_0(x,y). \tag{2.38}$$

Note that $U(x,y,t), V(x,y,t), W(x,y,t)$ are 1-periodic in y and system (2.35)–(2.18) is a set of partial differential equations in (x,t) with $y \in [0,1]$ as a parameter. The global existence of systems (2.31)–(2.34) and (2.35)–(2.38) has been established, see the references cited in [49].

Theorem 2.1. *Let (u, v, w) and (U, V, W) be the solutions of systems (2.31)–(2.34) and (2.35)–(2.38), respectively. Then we have the following error estimate:*

$$\max_{0 \leqslant t \leqslant T} E(t) \leqslant \left[5(M(T)^2 + 2TK(T)M(T)) \exp(6M(T)T) \right] \varepsilon := C_1(T)\varepsilon, \tag{2.39}$$

where the error function $E(t)$ is given by

$$E(t) = \max_{x \in \mathbb{R}} \left\{ \left| u(x,t) - U(x, \tfrac{x-t}{\varepsilon}, t) \right| + \left| v(x,t) - V(x, \tfrac{x+t}{\varepsilon}, t) \right| \right.$$

$$\left. + \left| w(x,t) - W(x, \tfrac{x}{\varepsilon}, t) \right| \right\}$$

and the constants $M(T)$ and $K(T)$ are given by

$$M(T) = \max_{(x,y,t) \in \mathbb{R} \times [0,1] \times [0,T]} \left(|u|, |v|, |w|, |U|, |V|, |W| \right), \tag{2.40}$$

$$N(T) = \max_{(x,y,t) \in \mathbb{R} \times [0,1] \times [0,T]}$$

$$\left(|\partial_x U|, |\partial_t U|, |\partial_x V|, |\partial_t V|, |\partial_x W|, |\partial_t W| \right). \tag{2.41}$$

This homogenization result was first obtained by McLaughlin, Papanicolaou and Tartar using an L^p norm estimate ($0 < p < \infty$) [82]. Since we need an L^∞ norm estimate in the convergence analysis of our particle method, we give another proof of this result in L^∞ norm. As a first step, we prove the following lemma.

Lemma 2.3. *Let $g(x,y) \in C^1(\mathbb{R} \times [0,1])$ be 1-periodic in y and satisfy the relation $\int_0^1 g(x,y)\,dy = 0$. Then for any $\varepsilon > 0$ and for any constants a and b, the following estimate holds:*

$$\left| \int_a^b g(x, \tfrac{x}{\varepsilon})\,dx \right| \leqslant B(g)\varepsilon + |b - a| B(\partial_x g)\varepsilon, \tag{2.42}$$

where $B(\zeta) = \max_{(x,y) \in \mathbb{R} \times [0,1]} |\zeta(x,y)|$ for any function ζ defined on $\mathbb{R} \times [0,1]$.

Proof. Estimate (2.42) is a direct consequence of the identity

$$g(x, \tfrac{x}{\varepsilon}) = \frac{d}{dx} \int_a^x g(x, \tfrac{s}{\varepsilon})\,ds - \int_a^x \frac{\partial g}{\partial x}(x, \tfrac{s}{\varepsilon})\,ds$$

and the estimates

$$\left| \int_a^b g(x, \tfrac{s}{\varepsilon})\,ds \right| \leqslant B(g)\varepsilon, \quad \left| \int_a^x \frac{\partial g}{\partial x}(x, \tfrac{s}{\varepsilon})\,ds \right| \leqslant B(\partial_x g)\varepsilon,$$

which follow from the 1-periodicity of $g(x,y)$ in y and that $\int_0^1 g(x,y)\,dy = 0$. This completes the proof. $\qquad\square$

Proof of Theorem 2.1. Subtracting (2.35) from (2.31) and integrating the resulting equation along the characteristics from 0 to t, we get

$$u(x,t) - U(x, \tfrac{x-t}{\varepsilon}, t)$$

$$= \int_0^t \left[w(x-t+s, s)^2 - W(x-t+s, \tfrac{x-t+s}{\varepsilon}, s)^2 \right] ds$$

$$+ \int_0^t \left[W(x-t+s, \tfrac{x-t+s}{\varepsilon}, s)^2 - \int_0^1 W(x-t+s, y, s)^2 \, dy \right] ds$$

$$- \int_0^t \left[u(x-t+s, s)v(x-t+s, s) \right.$$

$$\left. - U(x-t+s, \tfrac{x-t}{\varepsilon}, s)V(x-t+s, \tfrac{x-t+2s}{\varepsilon}, s) \right] ds$$

$$- \int_0^t U(x-t+s, \tfrac{x-t}{\varepsilon}, s) \left[V(x-t+s, \tfrac{x-t+2s}{\varepsilon}, s) \right.$$

$$\left. - \int_0^1 V(x-t+s, y, s) \, dy \right] ds$$

$$:= (\mathrm{I})_1 + \cdots + (\mathrm{I})_4. \tag{2.43}$$

It is clear from the definition of $E(t)$ and $M(T)$ that

$$|(\mathrm{I})_1 + (\mathrm{I})_3| \leqslant 2M(T) \int_0^t E(s) \, ds.$$

To estimate $(\mathrm{I})_2$, we define for fixed $(x,t) \in \mathbb{R} \times [0,T]$,

$$g_{(x,t)}(s,y) = W(x-t+s, \tfrac{x-t}{\varepsilon} + y, s)^2.$$

Since the 1-periodicity of $W(x,y,t)$ in y implies

$$\int_0^1 W(x-t+s, y, s)^2 \, dy = \int_0^1 W(x-t+s, \tfrac{x-t}{\varepsilon} + y, s)^2 \, dy,$$

we obtain by applying Lemma 2.1 that

$$|(\mathrm{I})_2| = \left| \int_0^t \left[g_{(x,t)}(s, \tfrac{s}{\varepsilon}) - \int_0^1 g_{(x,t)}(s,y) \, dy \right] ds \right|$$

$$\leqslant M(T)^2 \varepsilon + 2M(T)K(T)T\varepsilon.$$

Similarly, we have

$$(\mathrm{I})_4 \leqslant M(T)^2 \varepsilon + 2M(T)K(T)T\varepsilon.$$

Substituting these estimates into (2.43) we get

$$\left| u(x,t) - U(x, \tfrac{x-t}{\varepsilon}, t) \right|$$

$$\leqslant 2M(T) \int_0^t E(s) \, ds + 2M(T)^2 \varepsilon + 4M(T)K(T)T\varepsilon. \tag{2.44}$$

Similarly, we conclude from (2.36)–(2.37) and (2.32)–(2.33) that

$$\left| v(x,t) - V(x, \tfrac{x+t}{\varepsilon}, t) \right|$$
$$\leqslant 2M(T) \int_0^t E(s)\, ds + 2M(T)^2 \varepsilon + 4M(T)K(T)T\varepsilon,$$

(2.45)

$$\left| w(x,t) - W(x, \tfrac{x}{\varepsilon}, t) \right|$$
$$\leqslant 2M(T) \int_0^t E(s)\, ds + M(T)^2 \varepsilon + 2M(T)K(T)T\varepsilon. \quad (2.46)$$

Now the desired estimate (2.39) follows from summing (2.44)–(2.46) and using the Gronwall inequality. ☐

Remark 2.2. The homogenization theory tells us that the initial oscillatory solutions propagate along their characteristics. The nonlinear interaction can generate only low frequency contributions to the u and v components. On the other hand, the nonlinear interaction of u, v on w can generate both low and high frequency contribution to w. That is, even if w has no oscillatory component initially, the dynamical interaction of u, v and w can generate a high frequency contribution to w at later time. This is not the case for the u and v components. Due to this resonant interaction of u, v and w, the weak limit of $u_\varepsilon v_\varepsilon w_\varepsilon$ is not equal to the product of the weak limits of u_ε, v_ε, w_ε. This explains why the compensated compactness result does not apply to this 3×3 system [99].

Although it is difficult to characterize the weak limit of the triple product, $u_\varepsilon v_\varepsilon w_\varepsilon$ for arbitrary oscillatory initial data, it is possible to say something about the weak limit of the triple product for oscillatory initial data that have periodic structure, such as the ones studies here. Depending on α being rational or irrational, the limiting behavior is very different. In fact, one can show that $\overline{u_\varepsilon v_\varepsilon w_\varepsilon} = \overline{u_\varepsilon}\,\overline{v_\varepsilon}\,\overline{w_\varepsilon}$ when α is equal to an irrational number. This is not true in general when α is a rational number.

2.3 Convection of microstructure

It is most interesting to see if one can apply homogenization technique to obtain an averaged equation for the large scale quantity for incompressible Euler or Navier-Stokes equations. In 1985, McLaughlin, Papanicolaou and Pironneau [83] attempted to obtain a homogenized equation for the 3-D incompressible Euler equations with highly oscillatory velocity field. More specifically, they considered the following initial value problem:

$$u_t + (u \cdot \nabla)u = -\nabla p,$$

with $\nabla \cdot u = 0$ and highly oscillatory initial data

$$u(x,0) = U(x) + W(x, x/\varepsilon).$$

They then constructed multiscale expansions for both the velocity field and the pressure. In doing so, they made an important assumption that the microstructure is convected by the mean flow. Under this assumption, they constructed a multiscale expansion for the velocity field as follows:

$$u^\varepsilon(x,t) = u(x,t) + w(\tfrac{\theta(x,t)}{\varepsilon}, \tfrac{t}{\varepsilon}, x, t) + \varepsilon u_1(\tfrac{\theta(x,t)}{\varepsilon}, \tfrac{t}{\varepsilon}, x, t) + O(\varepsilon^2).$$

The pressure field p^ε is expanded similarly. From this ansatz, one can show that θ is convected by the mean velocity:

$$\theta_t + u \cdot \nabla \theta = 0, \qquad \theta(x,0) = x.$$

It is a very challenging problems to develop a systematic approach to study the large scale solution in three dimensional Euler and Navier-Stokes equations. The work of McLaughlin, Papanicolaou and Pironneau provided some insightful understanding into how small scales interact with large scale and how to deal with the closure problem. However, the problem is still not completely resolved since the cell problem obtained this way does not have a unique solution. Additional constraints need to be enforced in order to derive a large scale averaged equation. With additional assumptions, they managed to derive a variant of the $k - \varepsilon$ model in turbulence modeling.

Remark 2.3. One possible way to improve the work of [83] is take into account the oscillation in the Lagrangian characteristics, θ_ε. The oscillatory part of θ_ε in general could have order one contribution to the mean velocity of the incompressible Euler equation. In [65–67], Hou and Yang and co-workers have studied convection of microstructure of the 2-D and 3-D incompressible Euler equations using a new approach. They do not assume that the oscillation is propagated by the mean flow. In fact, they found that it is crucial to include the effect of oscillations in the characteristics on the mean flow. Using this new approach, they can derive a well-posed cell problem which can be used to obtain an effective large scale average equation.

More can be said for a passive scalar convection equation

$$v_t + \frac{1}{\varepsilon} \nabla \cdot (u(x/\varepsilon)v) = \alpha \Delta v,$$

with $v(x,0) = v_0(x)$. Here $u(y)$ is a known incompressible periodic (or stationary random) velocity field with zero mean. Assume that the initial condition is smooth.

Expand the solution v^ε in powers of ε

$$v^\varepsilon = v(t, x) + \varepsilon v_1(t, x, x/\varepsilon) + \varepsilon^2 v_2(t, x, x/\varepsilon) + \cdots .$$

The coefficients of ε^{-1} lead to

$$\alpha \Delta_y v_1 - u \cdot \nabla_y v_1 - u \cdot \nabla_x v = 0.$$

Let e_k, $k = 1, 2, 3$, be the unit vectors in the coordinate directions and let $\chi^k(y)$ satisfy the cell problem:

$$\alpha \Delta_y \chi^k - u \cdot \nabla_y \chi^k - u \cdot e_k = 0.$$

Then we have

$$v_1(t, x, y) = \sum_{k=1}^{3} \chi^k(y) \frac{v(t, x)}{\partial x_k}.$$

The coefficients of ε^0 give

$$\alpha \Delta_y v_2 - u \cdot \nabla_y v_2 = u \cdot \nabla_x v_1 - 2\alpha \nabla_x \cdot \nabla_y v_1 - \alpha \Delta_x v + v_t.$$

The solvability condition for v_2 requires that the right-hand side has zero mean with respect to y. This gives rise to the equation for homogenized solution v,

$$v_t = \alpha \Delta_x v - \overline{u \cdot \nabla_x v_1}.$$

Using the cell problem, McLaughlin, Papanicolaou, and Pironneau obtained [83]

$$v_t = \sum_{i,j=1}^{3} (\alpha \delta_{ij} + \alpha_{T_{ij}}) \frac{\partial^2 v}{\partial x_i \partial x_j},$$

where $\alpha_{T_{ij}} = -\overline{u_i \chi^j}$.

Nonlocal memory effect of homogenization. It is interesting to note that for certain degenerate problem, the homogenized equation may have a nonlocal memory effect.

Consider the simple 2-D linear convection equation:

$$\frac{\partial u_\varepsilon(x, y, t)}{\partial t} + a_\varepsilon(y) \frac{\partial u_\varepsilon(x, y, t)}{\partial x} = 0,$$

with initial condition $u_\varepsilon(x, y, 0) = u_0(x, y)$. Note that $y = x_2$ is not a fast variable here.

We assume that a_ε is bounded and u_0 has compact support. While it is easy to write down the solution explicitly,

$$u_\varepsilon(x, y, t) = u_0(x - a_\varepsilon(y)t, y),$$

it is not an easy task to derive the homogenized equation for the weak limit of u_ε.

Using Laplace transform and measure theory, Luc Tartar 100 showed that the weak limit u of u_ε satisfies

$$\frac{\partial}{\partial t}u(x,y,t) + A_1(y)\frac{\partial}{\partial x}u(x,y,t)$$
$$= \int_0^t \int \frac{\partial^2}{\partial x^2}u(x - \lambda(t-s),y,s)d\mu_y(\lambda)\,ds,$$

with $u(x,y,0) = u_0(x,y)$, where $A_1(y)$ is the weak limit of $a_\varepsilon(y)$, and μ_y is a probability measure of y and has support in $[\min(a_\varepsilon), \max(a_\varepsilon)]$.

As we can see, the degenerate convection induces a nonlocal history dependent diffusion term in the propagating direction (x). The homogenized equation is *not* amenable to computation since the measure μ_y cannot be expressed explicitly in terms of a_ε.

3   Numerical homogenization based on sampling techniques

Homogenization theory provides a critical guideline for us to design effective numerical methods to compute multiscale problems. Whenever homogenized equations are applicable they are very useful for computational purposes. There are, however, many situations for which we do not have well-posed effective equations or for which the solution contains different frequencies such that effective equations are not practical. In these cases we would like to approximate the original equations directly. In this part of my lectures, we will investigate the possibility of approximating multiscale problems using particle methods together with sampling technique. The classes of equations we consider here include semilinear hyperbolic systems and the incompressible Euler equation with oscillatory solutions.

When we talk about convergence of an approximation to an oscillatory solution, we need to introduce a new definition. The traditional convergence concept is too weak in practice and does not discriminate between solutions which are highly oscillatory and those which are smooth. We need the error to be small essentially independent of the wavelength in the oscillation when the computational grid size is small. On the other hand we cannot expect the approximation to be well behaved pointwise. It is enough if the continuous solution and its discrete approximation have similar local or moving averages.

Definition 3.1 (Engquist [48]). Let v^n be the numerical approximation to u at time t_n $(t_n = n\Delta t)$, ε represent the wave length of oscillation in the solution. The approximation v^n converges to u as $\Delta t \to 0$, essentially independent of ε, if for any $\delta > 0$ and $T > 0$, there exists a set $s(\varepsilon, \Delta t_0) \in (0, \Delta t_0)$ with measure $(s(\varepsilon, \Delta t_0)) \geqslant (1 - \delta)\Delta t_0$ such that

$$||u(\cdot, t_n) - v^n|| \leqslant \delta, \qquad 0 \leqslant t_n \leqslant T$$

is valid for all $\Delta t \in s(\varepsilon, \Delta t_0)$ and where Δt_0 is independent of ε.

The convergence concept of "essentially independent of ε" is strong enough to mimic the practical case where the high frequency oscillations are not well resolved on the grid. A small set of values of Δt has to be removed in order to avoid resonance between Δt and ε. Compare the almost always convergence for the Monte Carlo methods [87].

It is natural to compare our problem with the numerical approximation of discontinuous solutions of nonlinear conservation laws. Shock capturing methods do not produce the correct shock profiles but the overall solution may still be good. For this the scheme must satisfy certain conditions such as conservation form. We are here interested in analogous conditions on algorithms for oscillatory solutions. These conditions should ideally guarantee that the numerical approximation in some sense is close to the solution of the corresponding effective equation when the wave length of the oscillation tends to zero.

There are three central sources of problems for discrete approximations of highly oscillatory solutions.

(i) The first one is the sampling of the computational mesh points $(x_j = j\Delta x, j = 0, 1, ...)$. There is the risk of resonance between the mesh points and the oscillation. For example, if Δx equals the wave length of the periodic oscillation, the discrete initial data may only get values from the peaks of a curve like the upper envelope of the oscillatory solution. We can never expect convergence in that case. Thus Δx cannot be completely independent of the wave length.

(ii) Another problem comes from the approximation of advection. The group velocity for the differential equation and the corresponding discretization are often very different [51]. This means that an oscillatory pulse which is not well resolved is not transported correctly even in average by the approximation. Furthermore, dissipative schemes do not advect oscillations correctly. The oscillations are damped out very fast in time.

(iii) Finally, the nonlinear interaction of different high frequency components in a solution must be modeled correctly. High frequency interactions may produce lower frequencies that influence the averaged solution. We can show that this nonlinear interaction is well approximated by certain particle methods applied to a class of semilinear differential equations. The problem is open for the approximation of more general nonlinear equations.

In [49, 50], we studied a particle method approximation to the nonlinear discrete Boltzmann equations in kinetic theory of discrete velocity with multiscale initial data. In such equations, high frequency components can be transformed into lower frequencies through nonlinear interactions, thus affecting the average of solutions. We assume that the initial data are of the form $a(x, x/\varepsilon)$ with $a(x, y)$ 1-periodic in each component of y. As we see from the homogenization theory in the previous section, the behavior of oscillatory solutions for the generalized Broadwell model is very sensitive to the velocity coefficients. It depends on whether a certain ratio among the velocity components is a rational number or an irrational number.

It is interesting to note that the structure of oscillatory solutions for the generalized Broadwell model is quite stable when we perturb the velocity coefficient α around irrational numbers. In this case, the resonance effect of u and v on w vanishes in the limit of $\varepsilon \to 0$. However, the behavior of oscillatory solutions for the generalized Broadwell model becomes singular when perturbing around integer velocity coefficients. There is a strong interaction between the high frequency components of u and v, and the interaction in the uv term would create an oscillation of order $O(1)$ on the w component. In [99], Tartar showed that for the Carleman model the weak limit of all powers of the initial data will uniquely determine the weak limit of the oscillatory solutions at later times, using the Compensated Compactness Theorem. We found that this is no longer true for the generalized Broadwell model with integer-values velocity coefficients [59].

In [49, 50], we showed that this subtle behavior for the generalized Broadwell model with oscillatory initial data can be captured correctly by a particle method even on a coarse grid. The particle method converges to the effective solution essentially independent of ε. For the Broadwell model, the hyperbolic part is solved exactly by the particle method. No averaging is therefore needed in the convergence result. We also analyze a numerical approximation of the Carleman equations with variable coefficients. The scheme is designed such that particle interaction can be accounted for without introducing interpolation. There are errors in the particle method approximation of the linear part of the system. As a result, the convergence can only be proved for moving averages. The convergence proofs for the Carleman and the Broadwell equations have one feature in common. The local truncation errors in both cases are of order $O(\Delta t)$. In order to show convergence, we need to take into account cancellation of the local errors at different time levels. This is very different from the conventional convergence analysis for finite difference methods. This is also the place where numerical sampling becomes crucial in order to obtain error cancellation at different time levels.

In the next two subsections, we present a careful study of the Broadwell model with highly oscillatory initial data in order to demonstrate the basic idea of the numerical homogenization based on sampling techniques.

3.1 Convergence of the particle method

Now we consider how to capture this oscillatory solution on a coarse grid using a particle method. Since the discrete velocity coefficients are integers for the Broadwell model, we can express a particle method in the form of a special finite difference method by choosing $\Delta x = \Delta t$. Denote by u_i^n, v_i^n, w_i^n the approximations of $u(x_i, t^n), v(x_i, t^n)$ and $w(x_i, t^n)$ respectively with $x_i = i\Delta x$ and $t^n = n\Delta t$. Our particle scheme is given by

$$u_i^n = u_{i-1}^{n-1} + \Delta t(w^2 - uv)_{i-1}^{n-1}, \tag{3.1}$$

$$v_i^n = v_{i+1}^{n-1} + \Delta t(w^2 - uv)_{i+1}^{n-1}, \tag{3.2}$$

$$w_i^n = w_i^{n-1} - \Delta t(w^2 - uv)_i^{n-1}, \tag{3.3}$$

with the initial conditions given by

$$u_i^0 = u(x_i, 0), \quad v_i^0 = v(x_i, 0), \quad w_i^0 = w(x_i, 0). \tag{3.4}$$

To study the convergence of the particle scheme (3.1)–(3.4) we need the following lemma, which is a discrete analogue of Lemma 2.1.

Lemma 3.1. *Let* $g(x, y) \in C^3([0, T] \times [0, 1])$ *be 1-periodic in* y *and satisfy the relation* $\int_0^1 g(x, y) \, dy = 0$. *Let* $x_k = kh$ *and* $r = h/\varepsilon$. *If* $h \in S(\varepsilon, h_0)$, *where*

$$S(\varepsilon, h_0) = \left\{ 0 < h \leqslant h_0 : \frac{kh}{\varepsilon} \notin \left(i - \frac{\tau}{|k|^{3/2}}, i + \frac{\tau}{|k|^{3/2}} \right), \right.$$

$$\left. \text{for } i = 1, 2, \cdots, \left[\frac{kh_0}{\varepsilon} \right] + 1, 0 \neq k \in Z, 0 < \varepsilon \leqslant 1 \right\},$$

then we have

$$\left| \sum_{k=0}^{n-1} g\left(x_k, \frac{x_k}{\varepsilon} \right) h \right| \leqslant \frac{C_0(1 + T)L(g)h}{\tau}, \quad \forall n = 1, 2, \cdots, \left[\frac{T}{h} \right],$$

where C_0 *is a constant independent of* h, ε, T, τ *and* g, *and*

$$L(g) = \max_{(x,y) \in [0,T] \times [0,1]} \left(|\partial_y^3 g(x, y)|, |\partial_x \partial_y^3 g(x, y)| \right).$$

Moreover, it is obvious that

$$|S(\varepsilon, h_0)| \geqslant h_0 \left(1 - \tau \sum_{k=1}^{\infty} k^{-3/2}\right) \geqslant h(1 - 3\tau).$$

Proof. Since g is 1-periodic in y with mean zero, it can be expanded in a Fourier series

$$g(x, y) = \sum_{m \neq 0} a_m(x) e^{2\pi i m y}, \text{ where } a_m(x) = \int_0^1 g(x, y) e^{-2\pi i m y} \, dy.$$

Simple integration by parts yields that

$$|a_m(x)| \leqslant \frac{1}{(2\pi|m|)^3} L(g), \quad |a'_m(x)| \leqslant \frac{1}{(2\pi|m|)^3} L(g).$$

Thus we have

$$\left| \sum_{k=0}^{n-1} g(x_k, \frac{x_k}{\varepsilon}) h \right| = \left| \sum_{k=0}^{n-1} \sum_{m \neq 0} a_m(x_k) e^{2\pi i m x_k/\varepsilon} h \right|$$

$$= \left| \sum_{m \neq 0} \sum_{k=0}^{n-1} a_m(x_k) e^{2\pi i m k h/\varepsilon} h \right|.$$

Summation by parts yields

$$\left| \sum_{k=0}^{n-1} a_m(x_k) e^{2\pi i k h/\varepsilon} \right| \leqslant \left| a_m(x_{n-1}) \sum_{k=0}^{n-1} e^{2\pi i k h/\varepsilon} \right|$$

$$+ \left| \sum_{k=0}^{n-1} \left(\sum_{j=1}^{k} e^{2\pi i m j h/\varepsilon} \right) (a_m(x_k) - a_m(x_{k+1})) \right|$$

$$\leqslant \frac{2(1+T)L(g)}{(2\pi|m|)^3 |1 - e^{2\pi i m h/\varepsilon}|}.$$

But for $h \in S(\varepsilon, h_0)$, we have

$$|1 - e^{2\pi i m h/\varepsilon}| = 2|\sin(\pi m h/\varepsilon)| \geqslant \frac{2\pi\tau}{|m|^{3/2}}.$$

Hence, for $h \in S(\varepsilon, h_0)$,

$$\sum_{k=0}^{n-1} \left| g(x_k, \frac{x_k}{\varepsilon}) h \right| \leqslant \frac{2(1+T)L(g)h}{(2\pi)^4 \tau} \sum_{m \neq 0} \frac{1}{|m|^{3/2}} =: \frac{C_0 h(1+T)L(g)}{\tau}.$$

This completes the proof.   □

Now we are ready to study the approximation property of the particle scheme (3.1)–(3.4). First denote by

$$E^n = \max_i \left(|u(x_i, t^n) - u_i^n|, \; |v(x_i, t^n) - v_i^n|, \; |w(x_i, t^n) - w_i^n| \right). \quad (3.5)$$

Integrating (2.28) from 0 to t^n along its characteristics, we get

$$u(x_i, t^n) = u(x_i - t^n, 0) + \int_0^{t^n} (w^2 - uv)(x_i - t^n + s, s) \, ds. \quad (3.6)$$

From (3.1) we know that

$$u_i^n = u_{i-n}^0 + \sum_{k=0}^{n-1} (w^2 - uv)_{i-k}^k \Delta t. \quad (3.7)$$

Subtracting (3.7) from (3.6) we obtain that

$$
\begin{aligned}
u(x_i, t^n) &- u_i^n \\
&= \int_0^{t^n} (w^2 - uv)(x_i - t^n + s, s) \, ds - \sum_{k=0}^{n-1} (w^2 - uv)(x_i - t^k, t^k) \Delta t \\
&\quad + \sum_{k=0}^{n-1} \Delta t \left[(w^2 - uv)(x_i - t^k, t^k) - (w^2 - uv)_{i-k}^k \right] \\
&:= (\mathrm{II}) + (\mathrm{III}). \quad (3.8)
\end{aligned}
$$

Let $M(T)$ be defined as in (2.40) and $N(T)$ be given by

$$N(T) = \max \left\{ |u_i^k|, |v_i^k|, |w_i^k| : i \in Z, 0 \leqslant k \leqslant [T/\Delta t] \right\}. \quad (3.9)$$

It can be shown that $N(T)$ is bounded for finite time independent of ε, see [49]. Then it is clear that

$$(\mathrm{III}) \leqslant (M(T) + N(T)) \sum_{k=0}^{n-1} \Delta t E^k. \quad (3.10)$$

It remains to estimate (II). For convenience, let $\theta = w^2 - uv$ and

$$\Theta(x, t) = W\left(x, \tfrac{x}{\varepsilon}, t\right)^2 - U\left(x, \tfrac{x-t}{\varepsilon}, t\right) V\left(x, \tfrac{x+t}{\varepsilon}, t\right).$$

Then we have

$$
\begin{aligned}
\text{(II)} = &\int_0^{t^n} \left[\theta(x_i - t^n + s, s)\, ds - \Theta(x_i - t^n + s, s) \right] ds \\
&+ \left[\int_0^{t^n} \Theta(x_i - t^n + s, s)\, ds - \sum_{k=0}^{n-1} \Delta t \Theta(x_i - t^k, t^k) \right] \\
&+ \sum_{k=0}^{n-1} \left[\Theta(x_i - t^k, t^k) - \theta(x_i - t^k, t^k) \right] \Delta t \\
:= &\ \text{(II)}_1 + \cdots + \text{(II)}_3.
\end{aligned} \tag{3.11}
$$

By Theorem 2.1 we get

$$
|\text{(II)}_1 + \text{(II)}_3| \leqslant 2TM(T)C_1(T)\varepsilon. \tag{3.12}
$$

To proceed further, let, for fixed (x_i, t^n),

$$
g_i^n(s, y) = W(x_i - t^n + s, \tfrac{x_i - t^n}{\varepsilon} + y, s)^2.
$$

It is clear that g_i^n is 1-periodic in y. Now by Lemmas 2.1 and 2.2, we have

$$
\begin{aligned}
&\int_0^{t^n} W(x_i - t^n + s, s)^2\, ds - \sum_{k=0}^{n-1} W(x_i - t^k, t^k)^2 \Delta t \\
&= \int_0^{t^n} \left[g_i^n(s, \tfrac{s}{\varepsilon}) - \int_0^1 g_i^n(s, y)\, dy \right] ds \\
&\quad + \int_0^{t^n} \int_0^1 g_i^n(s, y)\, dy\, ds - \sum_{k=0}^{n-1} \Delta t \int_0^1 g_i^n(t^{n-k}, y)\, dy \\
&\quad - \sum_{k=0}^{n-1} \Delta t \left[g(t^{n-k}, \tfrac{t^{n-k}}{\varepsilon}) - \int_0^1 g_i^n(t^{n-k}, y)\, dy \right] \leqslant C(T)(\varepsilon + \Delta t), \tag{3.13}
\end{aligned}
$$

where we have used standard methods to estimate the second term, since the derivative of g_i^n with respect to s is independent of ε. Here and in the remainder of this section, we will always denote by $C(T)$ the various constants which are independent of ε and Δt. Now similar to the reasoning leading to (3.13) we can obtain

$$
|\text{(II)}_3| \leqslant C(T)(\varepsilon + \Delta t). \tag{3.14}
$$

From (3.8)–(3.12) and (3.14) we finally get

$$
|u(x_i, t^n) - u_i^n| \leqslant C(T)(\varepsilon + \Delta t) + (M(T) + N(T)) \sum_{k=0}^{n-1} \Delta t E^k. \tag{3.15}
$$

Similarly, we have

$$|v(x_i, t^n) - v_i^n| \leqslant C(T)(\varepsilon + \Delta t) + (M(T) + N(T)) \sum_{k=0}^{n-1} \Delta t E^k, \quad (3.16)$$

$$|w(x_i, t^n) - w_i^n| \leqslant C(T)(\varepsilon + \Delta t) + (M(T) + N(T)) \sum_{k=0}^{n-1} \Delta t E^k. \quad (3.17)$$

To summarize, we have the following theorem by summing (3.15)–(3.17) and applying the Gronwall inequality.

Theorem 3.1. *Let (u, v, w) be the solution of (2.31)–(2.34) and (u_i^n, v_i^n, w_i^n) be the solution of the particle scheme (3.1)–(3.4). Assume that $\Delta t \in S(\varepsilon, \Delta t_0)$, where $S(\varepsilon, \Delta t_0)$ is defined in Lemma 3.1. Then the following estimate holds:*

$$\max_{1 \leqslant n \leqslant [T/\Delta t]} E^n \leqslant C(T)(\varepsilon + \Delta t),$$

where $C(T)$ is independent of ε and Δt, and E^n is defined as in (3.5).

Remark 3.1. It is important that we perform the error analysis globally in time in order to account for cancellation of local truncation errors at different time steps. As we can see from the analysis, the local truncation error is of order Δt in one time step. If we do not take into account the error cancellation in time, we would obtain an error bound of order $O(1)$ which is an over-estimate. The error cancellation is closely related to the sampling we choose. This is the place where we can see the difference between a good sampling and a resonant sampling.

Remark 3.2. As we can see from the error analysis, error cancellation along Lagrangian characteristics is essential in obtaining convergence independent of the oscillation. This idea can be generalized to hyperbolic systems with variable coefficient velocity fields. In the special case of the Carleman model with variable coefficients, we have analyzed the convergence of a particle method in [50]. However, the particle method analyzed in [50] does not generalize to multi-dimensions or 3×3 systems. Together Razvan Fetecau [54], we have designed a modified Lagrangian particle method. In this method, each component of the solution is updated along its own characteristic. So there is no fixed grid. When we update one component of the solution, say u, we need values of the other components (say v and w) along the u characteristic. We obtain these values by using some high order interpolation scheme (such as cubic spline). In the case of the Carleman model with variable coefficient and the non-resonant Broadwell model (α being irrational), we can prove rigorously the convergence of the modified particle method essentially independent of the small scale. Our numerical experiments show that the modified particle method works even for the original Broadwell model, which is surprising. This modified Lagrangian particle method in principle works for any number of families of characteristics and for multi-dimensions.

Below we describe briefly the results we obtain for the variable coefficient Carleman equations

$$u_t + a(x,t)u_x = v^2 - u^2, \tag{3.18}$$

$$v_t - b(x,t)v_x = u^2 - v^2, \tag{3.19}$$

with initial data $u(x,0) = u_0(x, x/\varepsilon)$, $v(x,0) = v_0(x, x/\varepsilon)$. In Figure 3.1, we illustrate the particle trajectories for the u and v components.

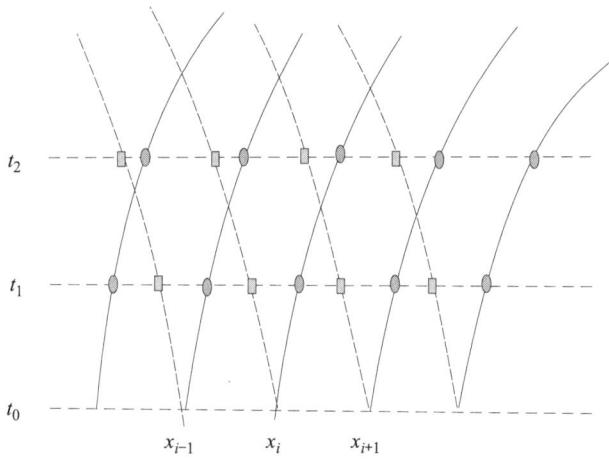

Figure 3.1. Schematic particle trajectories for different components.

We choose the oscillatory coefficients as follows:

$$a(x,t) = 1 + 0.5 \sin\left(\tfrac{xt}{\varepsilon}\right) \text{ and } b(x,t) = 1 + 0.2 \cos\left(\tfrac{xt}{\varepsilon}\right).$$

The initial conditions for u and v are chosen as

$$u_0\left(x, \frac{x}{\varepsilon}\right) = \begin{cases} 0.5\sin^4(\pi(x-3)/2)(1+\sin(2\pi(x-3)/\varepsilon)), & |x-4| < 1, \\ 0, & |x-4| \geqslant 1, \end{cases}$$
$$\tag{3.20}$$

$$v_0\left(x, \frac{x}{\varepsilon}\right) = \begin{cases} 0.5\sin^4(\pi(x-4)/2)(1+\sin(2\pi(x-4)/\varepsilon)), & |x-5| < 1, \\ 0, & |x-5| \geqslant 1. \end{cases}$$
$$\tag{3.21}$$

In our calculations, we choose $\Delta x = 0.01$, $\Delta t = \frac{\Delta x}{\sqrt{5}}$, and $\varepsilon = \Delta x\sqrt{2} \approx 0.014$. We plot the u-characteristic in Figure 3.2. The coarse grid solution for the u-component is plotted in Figure 3.3(a). We can see that it

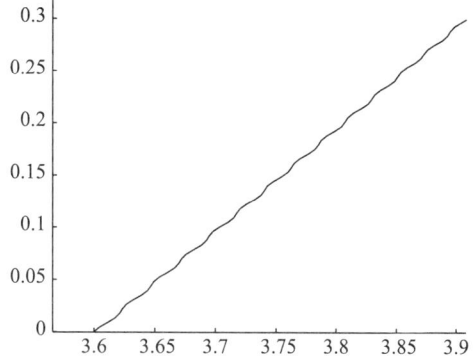

Figure 3.2. A typical u-characteristic trajectory.

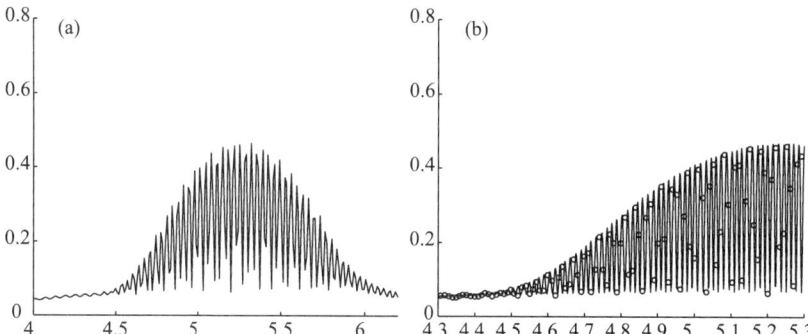

Figure 3.3. (a) Coarse grid solution u at time $t = 1.28$. (b) Putting the coarse grid solution u on top of a well-resolved computation (solid line).

captures very well the high frequency information. In Figure 3.3(b), we put the coarse grid solution on top of the corresponding well-resolved solution. The agreement is very good. We also check the accuracy of the moving average [50] of the solution and the average of its second order moments. The results are plotted in Figure 3.4. Again, we observe excellent agreement between the coarse grid calculations and the well-resolved calculations.

We have also performed the same calculations for the 3×3 Broadwell model with rational or irrational coefficient α. The subtle homogenization behavior is captured correctly for both rational α and irrational α. We do not present the results here.

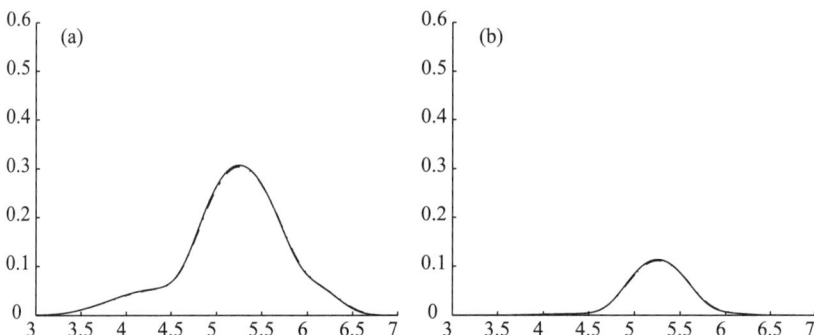

Figure 3.4. (a) The averaged solution $\bar{u}$ (dashdot line); the solid line represents a very well resolved computation. (b) The averaged second order moment $\overline{u^2}$ (dashdot line); the solid line represents a very well resolved computation.

3.2 Vortex methods for incompressible flows

The generalization of the particle method to the incompressible flows is the vortex method. In [38], we have analyzed the convergence of the vortex method for 2-D incompressible Euler equations with oscillatory vorticity field. Our analysis relies on the observation that there are tremendous cancellations among the local errors at different space locations in the velocity approximation. Thus the local errors do not add up to $O(1)$ as predicted by the classical error estimate in the case where the grid size is large compared to the oscillatory wavelength.

Consider the 2-D incompressible Euler equation in vorticity form:

$$\omega_t + (u \cdot \nabla)\omega = 0,$$

with oscillatory initial vorticity $\omega(x,0) = \omega_0(x, x/\varepsilon)$.

Define the particle trajectory, denoted as $X(t, \alpha)$,

$$\frac{dX(t, \alpha)}{dt} = u(X(t, \alpha), t), \quad X(0, \alpha) = \alpha.$$

Vorticity is conserved along characteristics:

$$\omega(X(t, \alpha), t) = \omega_0(\alpha).$$

On the other hand, velocity can be expressed in terms of vorticity by the Biot-Savart law:

$$u(X(t, \alpha), t) = \int K(X(t, \alpha) - X(t, \alpha'))\omega_0(\alpha')d\alpha',$$

with K given by $K(x) = (-x_2, x_1)/(2\pi|x|^2)$.

The Biot-Savart kernel K has a singularity at the origin. To regularize the kernel, Chorin introduced the vortex blob method (see, e.g. [26]), replacing K by $K_\delta = K * \zeta_\delta$,

$$\zeta_\delta = \frac{1}{\delta^2} \zeta \left(\frac{x}{\delta}\right), \quad \delta = h^\sigma, \text{ with } \sigma < 1,$$

ζ is typically chosen as a variant of Gaussian.

The vortex blob method is given by

$$\frac{dX_i^h(t)}{dt} = \sum_j K_\delta(X_i^h(t) - X_j^h(t))\omega_j h^2,$$

where $X_i^h(0) = \alpha_i$, and $w_j = w_0(\alpha_j, \alpha_j/\varepsilon)$.

Together with Weinan E, we have proved that the vortex method converges essentially independent of ε [38].

The case studied in [38] deals with bounded oscillatory vorticity. This assumption leads to strong convergence of the velocity field. It is more physical to consider homogenization for highly oscillatory velocity field. Would the vortex blob method still capture the correct large scale solution with a relatively coarse grid (or small number of particles)? This is still an open question.

4 Numerical upscaling based on multiscale finite element methods

It is natural to consider the possibility of generalizing the sampling technique to second order elliptic equations with highly oscillatory coefficients. In [8], we showed that finite difference approximations converge essentially independent of the small scale ε for one-dimensional elliptic problems. In several space dimensions we found that only in the case of rapidly oscillating periodic coefficients do the above results generalize, in a weaker form. In the case of almost periodic or random coefficients in several space dimensions we showed, both theoretically and with a simple counterexample, that numerical homogenization by sampling does not work efficiently. New ideas seem to be needed.

In order to overcome the difficulty we mentioned above for the sampling technique, we have introduced a multiscale finite element method (MsFEM) for solving partial differential equations with multiscale solutions, see [3, 25, 40, 42, 62–64, 101]. The central goal of this approach is to obtain the large scale solutions accurately and efficiently without resolving the small scale details. The main idea is to construct finite element base functions which capture the small scale information within each element. The small scale information is then brought to the large scales through the coupling of the global stiffness matrix. Thus, the effect of small scales on the large scales is correctly captured. In our method, the base functions are constructed from the leading order homogeneous elliptic equation in each element. As a consequence, the base functions are adapted to the local microstructure of the differential operator. In the case of two-scale periodic structures, we have proved that the multiscale method indeed converges to the correct solution independent of the small scale in the homogenization limit [64].

In practical computations, a large amount of overhead time comes from constructing the base functions. In general, these multiscale base functions are constructed numerically, except for certain special cases. Since the base functions are independent of each other, they can be constructed independently and can be done perfectly in parallel. This greatly reduces the overhead time in constructing these bases. In many applications, it is important to obtain a scale-up equation from the fine grid equation. For example, the high degree of variability and multiscale nature of formation properties in subsurface flows (such as permeability) pose significant challenges for subsurface flow modeling. Geological characterizations that capture these effects are typically developed at scales that are too fine for direct flow simulation, so techniques are required to enable the solution of flow problems in practice. Upscaling procedures have been commonly applied for this purpose and are effective in many cases (see e.g., [73] for reviews and discussion). Our multiscale finite element method can be used for a similar purpose and successfully applied for problems of this type.

As discussed in [73], upscaling methods and multiscale numerical techniques (as applied within the context of subsurface flow modeling) have many similarities and some important differences. Upscaling techniques provide coefficients, which are typically computed in a preprocessing step, for coarse scale equations of prescribed analytical forms. In multiscale methods, the coarse scale equations are formed numerically and fine scale information may be carried throughout the simulation and used at various stages. For example, in multiscale procedures for subsurface flow applications, different grids are often used for flow and transport computations. The advantage of deriving a scale-up equation or performing multiscale computations is that one can perform many useful tests on the coarse model with different boundary conditions or source terms. This would be very expensive if we have to perform all these tests on a fine grid. For time dependent problems, the coarse-scale equation also allows for larger time steps. This results in additional computational saving.

It should be mentioned that many numerical methods have been developed with goals similar to ours. These include generalized finite element methods [10, 11, 13], wavelet based numerical homogenization methods [18, 29, 31, 77], methods based on the homogenization theory (cf. [16, 28, 35, 52]), equation-free computations (e.g., [76]), variational multiscale methods [21, 71, 72], heterogeneous multiscale methods [37], matrix-dependent multigrid based homogenization [29, 77], generalized p-FEM in homogenization [79, 80], and some upscaling methods based on simple physical and/or mathematical motivations (cf. [33, 81]). The methods based on the homogenization theory have been successfully applied to determine the effective conductivity and permeability of certain composite materials and porous media. However, their range of applications is usually limited by restrictive assumptions on the media, such as scale separation and periodicity [15, 75]. They are also expensive to use for solving problems with many separate scales since the cost of computation grows exponentially with the number of scales. But for the multiscale method, the number of scales does not increase the overall computational cost exponentially. The upscaling methods are more general and have been applied to problems with random coefficients with partial success (cf. [33,81]). But the design principle is strongly motivated by the homogenization theory for periodic structures. Their application to nonperiodic structures is not always guaranteed to work.

Most multiscale methods presented to date have applied local calculations for the determination of basis functions. Though effective in many cases, global effects can be important for some problems. The importance of global information has been illustrated within the context of upscaling procedures as well as multiscale computations in recent investigations. These studies have shown that the use of limited global information in the calculation of the coarse-scale parameters (such as basis functions) can significantly improve the accuracy of the resulting coarse model. In this lecture note, we describe the use of limited global information in multiscale simulations.

We remark that the idea of using base functions governed by the differential equations has been applied to convection-diffusion equation with boundary layers (see, e.g., [14] and references therein). Babuska et al. applied a similar idea to 1-D problems [13] and to a special class of 2-D problems with the coefficient varying locally in one direction [11]. Most of these methods are based on the special property of one-dimensional properties of the coefficients. As indicated by our convergence analysis, there is a fundamental difference between one-dimensional problems and genuinely multi-dimensional problems. Special complications such as the resonance between the mesh scale and the physical scale never occur in the corresponding 1-D problems.

4.1 Multiscale finite element methods for elliptic PDEs

In this section we consider the multiscale finite element method applied to the following problem:

$$L_\varepsilon p := -\nabla \cdot (a(\tfrac{x}{\varepsilon})\nabla p) = f \ \text{ in } \Omega, \quad p = 0 \ \text{ on } \Gamma = \partial\Omega, \qquad (4.1)$$

where Ω is a convex polygon in $\mathbb{R}^2$. The unknown is changed to p, since it will be used later in porous media flow simulations, where the solution represents the pressure field. ε is assumed to be a small parameter, and $a(x) = (a_{ij}(x/\varepsilon))$ is symmetric and satisfies $\alpha|\xi|^2 \leqslant a_{ij}\xi_i\xi_j \leqslant \beta|\xi|^2$, for all $\xi \in \mathbb{R}^2$ and with $0 < \alpha < \beta$. Furthermore, $a_{ij}(y)$ are smooth periodic function in y in a unit cube Y. We will always assume that $f \in L^2(\Omega)$. In fact, the smoothness assumption on a_{ij} can be relaxed, which will be discussed later.

Let p_0 be the solution of the homogenized equation

$$L_0 p_0 := -\nabla \cdot (a^*\nabla p_0) = f \ \text{ in } \Omega, \quad p_0 = 0 \ \text{ on } \Gamma, \qquad (4.2)$$

where $\Gamma = \partial\Omega$ and

$$a_{ij}^* = \frac{1}{|Y|}\int_Y a_{ik}(y)\left(\delta_{kj} - \frac{\partial\chi^j}{\partial y_k}\right) dy,$$

and $\chi^j(y)$ is the periodic solution of the cell problem

$$\nabla_y \cdot (a(y)\nabla_y\chi^j) = \frac{\partial}{\partial y_i}a_{ij}(y) \ \text{ in } Y, \quad \int_Y \chi^j(y)\, dy = 0.$$

It is clear that $p_0 \in H^2(\Omega)$ since Ω is a convex polygon. Denote by $p_1(x, y) = -\chi^j(y)\frac{\partial p_0(x)}{\partial x_j}$ and let θ_ε be the solution of the problem

$$L_\varepsilon\theta_\varepsilon = 0 \ \text{ in } \Omega, \quad \theta_\varepsilon(x) = p_1(x, \tfrac{x}{\varepsilon}) \ \text{ on } \Gamma. \qquad (4.3)$$

Our analysis of the multiscale finite element method relies on the following homogenization result obtained by Moskow and Vogelius [84].

Lemma 4.1. *Let $p_0 \in H^2(\Omega)$ be the solution of (4.2), $\theta_\varepsilon \in H^1(\Omega)$ be the solution to (4.3) and $p_1(x) = -\chi^j(x/\varepsilon)\partial p_0(x)/\partial x_j$. Then there exists a constant C independent of u_0, ε and Ω such that*

$$\| p - p_0 - \varepsilon(u_1 - \theta_\varepsilon)\|_{1,\Omega} \leqslant C\varepsilon(|p_0|_{2,\Omega} + \|f\|_{0,\Omega}).$$

Now we are going to introduce the multiscale finite element methods. Let $\mathcal{T}_h$ be a regular partition of Ω into triangles. Let $\{x_j\}_{j=1}^J$ be the interior nodes of the mesh $\mathcal{T}_h$ and $\{\psi_j\}_{j=1}^J$ be the nodal basis of the standard linear finite element space $W_h \subset H_0^1(\Omega)$. Denote by $S_i = \operatorname{supp}(\psi_i)$ and define ϕ^i with support in S_i as follows:

$$L_\varepsilon \phi^i = 0 \text{ in } K, \quad \phi^i = \psi_i \text{ on } \partial K \; \forall \, K \in \mathcal{T}_h, K \subset S_i. \qquad (4.4)$$

It is obvious that $\phi^i \in H_0^1(S_i) \subset H_0^1(\Omega)$. Finally, let $V_h \subset H_0^1(\Omega)$ be the finite element space spanned by $\{\phi^i\}_{i=1}^J$.

With above notation we can introduce the following discrete problem: find $p_h \in V_h$ such that

$$(a(\tfrac{x}{\varepsilon})\nabla p_h, \nabla v_h) = (f, v_h) \; \forall \, v_h \in V_h, \qquad (4.5)$$

where and hereafter we denote by $(\cdot, \cdot)$ the L^2 inner product in $L^2(\Omega)$.

As we will see later, the choice of boundary conditions in defining the multiscale bases will play a crucial role in approximating the multiscale solution. Intuitively, the boundary condition for the multiscale base function should reflect the multiscale oscillation of the solution p across the boundary of the coarse grid element. By choosing a linear boundary condition for the base function, we will create a mismatch between the exact solution p and the finite element approximation across the element boundary. In the next section, we will discuss this issue further and introduce an over-sampling technique to alleviate this difficulty. The over-sampling technique plays an important role when we need to reconstruct the local fine grid velocity field from a coarse grid pressure computation for two-phase flows. This technique enables us to remove the artificial numerical boundary layer across the coarse grid boundary element.

We remark that the multiscale finite element method with linear boundary conditions for the multiscale base functions is similar in spirit to the residual-free bubbles finite element method [20] and the variational multiscale method [71, 21]. In a recent paper [94], G. Sangalli derives a multiscale method based on the residual-free bubbles formulation and compares it with the multiscale finite element method described here. There are many striking similarities between the two approaches.

To gain some insight into the multiscale finite element method, we next perform an error analysis for the multiscale finite element method in the simplest case, i.e., we use linear boundary conditions for the multiscale base functions.

4.2 Error estimates ($h < \varepsilon$)

The starting point is the well-known Cea's lemma.

Lemma 4.2. *Let p be the solution of* (4.1) *and p_h be the solution of* (4.5). *Then we have*

$$\| p - p_h \|_{1,\Omega} \leqslant C \inf_{v_h \in V_h} \| p - v_h \|_{1,\Omega}.$$

Let $\Pi_h : C(\bar{\Omega}) \to W_h \subset H_0^1(\Omega)$ be the usual Lagrange interpolation operator:

$$\Pi_h p(x) = \sum_{j=1}^{J} p(x_j)\psi_j(x) \quad \forall\, u \in C(\bar{\Omega})$$

and $I_h : C(\bar{\Omega}) \to V_h$ be the corresponding interpolation operator defined through the multiscale base function ϕ,

$$I_h p(x) = \sum_{j=1}^{J} p(x_j)\phi^j(x) \quad \forall\, u \in C(\bar{\Omega}).$$

From the definition of the basis function ϕ^i in (4.4) we have

$$L_\varepsilon(I_h p) = 0 \ \text{ in } K, \quad I_h p = \Pi_h p \ \text{ on } \partial K \qquad (4.6)$$

for any $K \in \mathcal{T}_h$.

Lemma 4.3. *Let $p \in H^2(\Omega)$ be the solution of (4.1). Then there exists a constant C independent of h, ε such that*

$$\| p - I_h p \|_{0,\Omega} + h \| p - I_h p \|_{1,\Omega} \leqslant Ch^2(| p |_{2,\Omega} + \| f \|_{0,\Omega}). \qquad (4.7)$$

Proof. At first it is known from the standard finite element interpolation theory that

$$\| p - \Pi_h p \|_{0,\Omega} + h \| p - \Pi_h p \|_{1,\Omega} \leqslant Ch^2(| p |_{2,\Omega} + \| f \|_{0,\Omega}). \qquad (4.8)$$

On the other hand, since $\Pi_h p - I_h p = 0$ on ∂K, the standard scaling argument yields

$$\| \Pi_h p - I_h p \|_{0,K} \leqslant Ch |\Pi_h p - I_h p|_{1,K} \quad \forall\, K \in \mathcal{T}_h. \qquad (4.9)$$

To estimate $|\Pi_h p - I_h p|_{1,K}$ we multiply the equation in (4.6) by $I_h p - \Pi_h p \in H_0^1(K)$ to get

$$(a(\tfrac{x}{\varepsilon})\nabla I_h p, \nabla(I_h p - \Pi_h p))_K = 0,$$

where $(\cdot, \cdot)_K$ denotes the L^2 inner product of $L^2(K)$. Thus, upon using the equation in (4.1), we get

$$(a(\tfrac{x}{\varepsilon})\nabla(I_h p - \Pi_h p), \nabla(I_h p - \Pi_h p))_K$$
$$= (a(\tfrac{x}{\varepsilon})\nabla(p - \Pi_h p), \nabla(I_h p - \Pi_h p))_K - (a(\tfrac{x}{\varepsilon})\nabla p, \nabla(I_h p - \Pi_h p))_K$$
$$= (a(\tfrac{x}{\varepsilon})\nabla(p - \Pi_h p), \nabla(I_h p - \Pi_h p))_K - (f, I_h p - \Pi_h p)_K.$$

This implies that

$$|I_h p - \Pi_h p|_{1,K} \leqslant Ch | p |_{2,K} + \| I_h p - \Pi_h p \|_{0,K} \| f \|_{0,K}.$$

Hence

$$|I_h p - \Pi_h p|_{1,K} \leqslant Ch(|p|_{2,K} + \|f\|_{0,K}), \tag{4.10}$$

where we have used (4.9). Now the lemma follows from (4.8)–(4.10). □

In conclusion, we have the following estimate by using Lemmas 4.2 and 4.3.

Theorem 4.1. *Let $p \in H^2(\Omega)$ be the solution of (4.1) and $p_h \in V_h$ be the solution of (4.5). Then we have*

$$\|p - p_h\|_{1,\Omega} \leqslant Ch(|p|_{2,\Omega} + \|f\|_{0,\Omega}). \tag{4.11}$$

Note that estimate (4.11) blows up like h/ε as $\varepsilon \to 0$ since $|p|_{2,\Omega} = O(1/\varepsilon)$. This is insufficient for practical applications. In next subsection we derive an error estimate which is uniform as $\varepsilon \to 0$.

4.3 Error estimates ($h > \varepsilon$)

In this section, we will show that the multiscale finite element method gives a convergence result uniform in ε as ε tends to zero. This is the main feature of this multiscale finite element method over the traditional finite element method. The main result in this section is the following theorem.

Theorem 4.2. *Let $p \in H^2(\Omega)$ be the solution of (4.1) and $p_h \in V_h$ be the solution of (4.5). Then we have*

$$\|p - p_h\|_{1,\Omega} \leqslant C(h + \varepsilon)\|f\|_{0,\Omega} + C\left(\frac{\varepsilon}{h}\right)^{1/2}\|p_0\|_{1,\infty,\Omega}, \tag{4.12}$$

where $p_0 \in H^2(\Omega) \cap W^{1,\infty}(\Omega)$ is the solution of the homogenized equation (4.2).

To prove the theorem, we first denote

$$p_I(x) = I_h p_0(x) = \sum_{j=1}^{J} p_0(x_j)\phi^j(x) \in V_h.$$

From (4.6) we know that $L_\varepsilon p_I = 0$ in K and $p_I = \Pi_h p_0$ on ∂K for any $K \in \mathcal{T}_h$. The homogenization theory (see (2.26)) implies that

$$\|p_I - p_{I0} - \varepsilon(p_{I1} - \theta_{I\varepsilon})\|_{1,K} \leqslant C\varepsilon(\|f\|_{0,K} + |p_{I0}|_{2,K}), \tag{4.13}$$

where p_{I0} is the solution of the homogenized equation on K:

$$L_0 p_{I0} = 0 \text{ in } K, \quad p_{I0} = \Pi_h p_0 \text{ on } \partial K, \tag{4.14}$$

$p_{\text{I}1}$ is given by the relation

$$p_{\text{I}1}(x, y) = -\chi^j(y)\frac{\partial p_{\text{I}0}}{\partial x_j} \quad \text{in } K, \tag{4.15}$$

and $\theta_{\text{I}\varepsilon} \in H^1(K)$ is the solution of the problem:

$$L_\varepsilon \theta_{\text{I}\varepsilon} = 0 \quad \text{in } K, \quad \theta_{\text{I}\varepsilon}(x) = p_{\text{I}1}(x, \tfrac{x}{\varepsilon}) \quad \text{on } \partial K. \tag{4.16}$$

It is obvious from (4.14) that

$$p_{\text{I}0} = \Pi_h p_0 \quad \text{in } K, \tag{4.17}$$

since $\Pi_h p_0$ is linear on K. From (4.13) we obtain that

$$\begin{aligned}\| p - p_{\text{I}} \|_{1,\Omega} \leqslant \| p_0 - p_{\text{I}0} \|_{1,\Omega} &+ \| \varepsilon(p_1 - p_{\text{I}1}) \|_{1,\Omega} \\ &+ \| \varepsilon(\theta_\varepsilon - \theta_{\text{I}\varepsilon}) \|_{1,\Omega} + C\varepsilon \| f \|_{0,\Omega},\end{aligned} \tag{4.18}$$

where we have used the regularity estimate $\| p_0 \|_{2,\Omega} \leqslant C\| f \|_{0,\Omega}$. Now it remains to estimate the terms at the right-hand side of (4.18).

Lemma 4.4. *We have*

$$\| p_0 - p_{\text{I}0} \|_{1,\Omega} \leqslant Ch\| f \|_{0,\Omega}, \tag{4.19}$$
$$\| \varepsilon(p_1 - p_{\text{I}1}) \|_{1,\Omega} \leqslant C(h + \varepsilon)\| f \|_{0,\Omega}. \tag{4.20}$$

Proof. Estimate (4.19) is a direct consequence of the standard finite element interpolation theory since $p_{\text{I}0} = \Pi_h p_0$ by (4.17). Next we note that $\chi^j(x/\varepsilon)$ satisfies

$$\| \chi^j \|_{0,\infty,\Omega} + \varepsilon\| \nabla\chi^j \|_{0,\infty,\Omega} \leqslant C \tag{4.21}$$

for some constant C independent of h and ε. Thus we have, for any $K \in \mathcal{T}_h$,

$$\| \varepsilon(p_{\text{I}1}) \|_{0,K} \leqslant C\varepsilon \left\| \chi^j\frac{\partial}{\partial x_j}(p_0 - \Pi_h p_0) \right\|_{0,K} \leqslant Ch\varepsilon|p_0|_{2,K},$$
$$\begin{aligned}\| \varepsilon\nabla(p_1 - p_{\text{I}1}) \|_{0,K} = \varepsilon\left\| \nabla\left(\chi^j\frac{\partial(p_0 - \Pi_h p_0)}{\partial x_j}\right) \right\|_{0,K} \\ \leqslant C\| \nabla(p_0 - \Pi_h p_0) \|_{0,K} + C\varepsilon|p_0|_{2,K} \\ \leqslant C(h + \varepsilon)|p_0|_{2,K}.\end{aligned}$$

This completes the proof. □

Lemma 4.5. *We have*

$$\| \varepsilon\theta_\varepsilon \|_{1,\Omega} \leqslant C\sqrt{\varepsilon}\| p_0 \|_{1,\infty,\Omega} + C\varepsilon|p_0|_{2,\Omega}. \tag{4.22}$$

Proof. Let $\zeta \in C_0^\infty(\mathbb{R}^2)$ be the cut-off function which satisfies $\zeta \equiv 1$ in $\Omega \backslash \Omega_{\delta/2}$, $\zeta \equiv 0$ in Ω_δ, $0 \leqslant \zeta \leqslant 1$ in $\mathbb{R}^2$, and $|\nabla \zeta| \leqslant C/\delta$ in Ω, where for any $\delta > 0$ sufficiently small, we denote by Ω_δ as

$$\Omega_\delta = \{x \in \Omega : \text{dist}(x, \partial\Omega) \geqslant \delta\}.$$

With this definition, it is clear that $\theta_\varepsilon - \zeta p_1 = \theta_\varepsilon + \zeta(\chi^j \partial p_0/\partial x_j) \in H_0^1(\Omega)$. Multiplying the equation in (4.3) by $\theta_\varepsilon - \zeta p_1$, we get

$$\left(a(\tfrac{x}{\varepsilon}) \nabla \theta_\varepsilon, \nabla \left(\theta_\varepsilon + \zeta \chi^j \frac{\partial p_0}{\partial x_j} \right) \right) = 0,$$

which yields, by using (4.21),

$$\begin{aligned}
\| \nabla \theta_\varepsilon \|_{0,\Omega} &\leqslant C \| \nabla(\zeta \chi^j \partial p_0/\partial x_j) \|_{0,\Omega} \\
&\leqslant C \| \nabla \zeta \cdot \chi^j \partial p_0/\partial x_j \|_{0,\Omega} + C \| \zeta \nabla \chi^j \partial p_0/\partial x_j \|_{0,\Omega} \\
&\quad + C \| \zeta \chi^j \partial^2 p_0/\partial x_j^2 \|_{0,\Omega} \\
&\leqslant C \sqrt{|\partial\Omega| \cdot \delta} \frac{D}{\delta} + C \sqrt{|\partial\Omega| \cdot \delta} \frac{D}{\varepsilon} + C |p_0|_{2,\Omega}, \qquad (4.23)
\end{aligned}$$

where $D = \| p_0 \|_{1,\infty,\Omega}$ and the constant C is independent of the domain Ω. From (4.23) we have

$$\begin{aligned}
\| \varepsilon \theta_\varepsilon \|_{0,\Omega} &\leqslant C \left(\frac{\varepsilon}{\sqrt{\delta}} + \sqrt{\delta} \right) \| p_0 \|_{1,\infty,\Omega} + C \varepsilon |p_0|_{2,\Omega} \\
&\leqslant C \sqrt{\varepsilon} \| p_0 \|_{1,\infty,\Omega} + C \varepsilon |p_0|_{2,\Omega}. \qquad (4.24)
\end{aligned}$$

Moreover, by applying the maximum principle to (4.3), we get

$$\| \theta_\varepsilon \|_{0,\infty,\Omega} \leqslant \| \chi^j \partial p_0/\partial x_j \|_{0,\infty,\partial\Omega} \leqslant C \| p_0 \|_{1,\infty,\Omega}. \qquad (4.25)$$

Combining (4.24) and (4.25) completes the proof. □

Lemma 4.6. *We have*

$$\| \varepsilon \theta_{\mathrm{I}\varepsilon} \|_{1,\Omega} \leqslant C \left(\frac{\varepsilon}{h} \right)^{1/2} \| p_0 \|_{1,\infty,\Omega}. \qquad (4.26)$$

Proof. First we remember that for any $K \in \mathcal{T}_h$, $\theta_{\mathrm{I}\varepsilon} \in H^1(K)$ satisfies

$$L_\varepsilon \theta_{\mathrm{I}\varepsilon} = 0 \text{ in } K, \quad \theta_{\mathrm{I}\varepsilon} = -\chi^j(\tfrac{x}{\varepsilon}) \frac{\partial(\Pi_h p_0)}{\partial x_j} \text{ on } \partial K. \qquad (4.27)$$

By applying maximum principle and (4.21) we get

$$\| \theta_{\mathrm{I}\varepsilon} \|_{0,\infty,K} \leqslant \| \chi^j \partial(\Pi_h p_0)/\partial x_j \|_{0,\infty,\partial K} \leqslant C \| p_0 \|_{1,\infty,K}.$$

Thus we have

$$\| \, \varepsilon \theta_{I\varepsilon} \, \|_{0,\Omega} \leqslant C\varepsilon \| \, p_0 \, \|_{1,\infty,\Omega}. \tag{4.28}$$

On the other hand, since the constant C in (4.23) is independent of Ω, we can apply the same argument leading to (4.23) to obtain

$$\| \, \varepsilon \nabla \theta_{I\varepsilon} \, \|_{0,K} \leqslant C\varepsilon \| \, \Pi_h p_0 \, \|_{1,\infty,K} \left(\frac{\sqrt{|\partial K|}}{\sqrt{\delta}} + \frac{\sqrt{|\partial K| \cdot \delta}}{\varepsilon} \right) + C\varepsilon | \, \Pi_h p_0 \, |_{2,K}$$

$$\leqslant C\sqrt{h} \| \, p_0 \, \|_{1,\infty,K} \left(\frac{\varepsilon}{\sqrt{\delta}} + \sqrt{\delta} \right)$$

$$\leqslant C\sqrt{h\varepsilon} \| \, p_0 \, \|_{1,\infty,K},$$

which implies that

$$\| \, \varepsilon \nabla \theta_{I\varepsilon} \, \|_{0,\Omega} \leqslant C \left(\frac{\varepsilon}{h} \right)^{1/2} \| \, p_0 \, \|_{1,\infty,\Omega}.$$

This completes the proof.   □

Proof of Theorem 4.2. The theorem is now a direct consequence of (4.18) and Lemmas 4.4–4.6 and the regularity estimate $\| \, p_0 \, \|_{2,\Omega} \leqslant C \| \, f \, \|_{0,\Omega}$.
                                                                               □

Remark 4.1. As we pointed out earlier, the multiscale FEM indeed gives correct homogenized result as ε tends to zero. This is in contrast with the traditional FEM which does not give the correct homogenized result as $\varepsilon \to 0$. The error would grow like $O(h^2/\varepsilon^2)$. On the other hand, we also observe that when $h \sim \varepsilon$, the multiscale method attains large error in both H^1 and L^2 norms. This is what we call the *resonance* effect between the grid scale (h) and the small scale (ε) of the problem. This estimate reflects the intrinsic scale interaction between the two scales in the *discrete* problem. Our extensive numerical experiments confirm that this estimate is indeed generic and sharp. From the viewpoint of practical applications, it is important to reduce or completely remove the resonance error for problems with many scales since the chance of hitting a resonance sampling is high. In the next subsection, we propose an over-sampling method to overcome this difficulty.

4.4   The over-sampling technique

As illustrated by our error analysis, large errors result from the "resonance" between the grid scale and the scales of the continuous problem. For the two-scale problem, the error due to the resonance manifests as a ratio between the wavelength of the small scale oscillation and the grid size; the error becomes large when the two scales are close. A deeper analysis shows that the boundary layer in the first order corrector seems to be the main source of the resonance effect. By a judicious choice of boundary conditions for the base function, we can eliminate the boundary layer in the first order corrector. This would give a nice conservative difference structure in the discretization, which in turn leads to *cancellation of resonance errors* and gives an improved rate of convergence.

Motivated by our convergence analysis, we propose an *over-sampling* method to overcome the difficulty due to scale resonance [62]. The idea is quite simple and easy to implement. Since the boundary layer in the first order corrector is thin, $O(\varepsilon)$, we can sample in a domain with size larger than $h+\varepsilon$ and use only the interior sampled information to construct the bases; here, h is the mesh size and ε is the small scale in the solution. By doing this, we can reduce the influence of the boundary layer in the larger sample domain on the base functions significantly. As a consequence, we obtain an improved rate of convergence.

Specifically, let ψ^j be the base functions satisfying the homogeneous elliptic equation in the larger domain $S \supset K$. We then form the actual base ϕ^i by linear combination of ψ^j,

$$\phi^i = \sum_{j=1}^{d} c_{ij} \psi^j.$$

The coefficients c_{ij} are determined by condition $\phi^i(\mathbf{x}_j) = \delta_{ij}$. The corresponding θ_ε^i for ϕ^i are now free of boundary layers. Our extensive numerical experiments have demonstrated that the over-sampling technique does improve the numerical error substantially in many applications. On the other hand, the over-sampling technique results in a *nonconforming* MsFEM method. In [42], we perform a careful estimate of the nonconforming errors in both H^1 norm and the L^2 norm. The analysis shows that the non-conforming error is indeed small, consistent with our numerical results [62, 63]. Our analysis also reveals another source of resonance, which is the mismatch between the mesh size and the "perfect" sample size. In case of a periodic structure, the "perfect" sample size is the length of an integer multiple of the period. We call the new resonance the "cell resonance". In the error expansion, this resonance effect appears as a *higher* order correction. In numerical computations, we found that the cell resonance error is generically small, and is rarely observed in practice. Nonetheless, it is possible to completely eliminate this cell resonance error by using the over-sampling technique to construct the base functions but using piecewise linear functions as test functions. This reduces the nonconforming error and eliminates the resonance error completely (see [60]).

4.5   Performance and implementation issues

The multiscale method given in the previous section is fairly straightforward to implement. Here, we outline the implementation and define some notations that are used in the discussion below. We consider solving problems in a unit square domain. Let N be the number of elements in the x and y directions. The mesh size is thus $h = 1/N$. To compute the base functions, each element is discretized into $M \times M$ subcell elements with mesh size $h_s = h/M$. To implement the over-sampling method, we partition the domain into sampling domains and each of them contains many elements. From the analysis and numerical tests, the size of the sampling domains can be chosen freely as long as the boundary layer is avoided. In practice, though, one wants to maximize the efficiency of over-sampling by choosing the largest possible sample size which reduces the redundant computation of overlapping domains to a minimum.

In general, the multiscale (sampling) base functions are constructed numerically, except for certain special cases. They are solved in each K or S using standard FEM. The linear systems are solved using a robust multigrid method with matrix dependent prolongation and ILLU smoothing (MG-ILLU, see [102]). The global linear system on Ω is solved using the same method. Numerical tests show that the accuracy of the final solution is insensitive to the accuracy of base functions.

Since the base functions are independent of each other, their construction can be carried out in parallel perfectly. In our parallel implementation of over-sampling, the sample domains are chosen such that they can be handled within each processor without communication. The multigrid solver is also modified to better suit the parallelization. In particular, the ILLU smoothing is replaced by Gauss-Seidel iterations. More implementation details can be found in [62].

Cost and performance. In practical computations, a large amount of overhead time comes from constructing the base functions. On a sequential machine, the operation count of our method is about twice that of a conventional FEM for a 2-D problem. However, due to good parallel efficiency, this difference is reduced significantly on a massively parallel computer. For example, using 256 processors on an Intel Paragon, our method with $N = 32$ and $M = 32$ only spends 9% more CPU time than the conventional linear FEM method using 1024×1024 elements [62]. Note that this comparison is made for a single solve of the problem. In practice, multiple solves are often required, then the overhead of base construction is negligible. A detailed study of MsFEM's parallel efficiency has been conducted in [62]. It was also found that MsFEM is helpful for improving multigrid convergence when the coefficient a_ε has very large contrast (i.e., the ratio between the maximum and minimum of a_ε).

Significant computational savings can be obtained for time dependent problems (such as two-phase flows) by constructing the multiscale bases adaptively. Multiscale base functions are updated only for those coarse grid elements where the saturation changes significantly. In practice, the number of such coarse grid elements are small. They are concentrated near the interface separating oil and water. Also, the cost of solving a base function in a small cell is more efficient than solving the fine grid problem globally because the condition number for solving the local base function in each coarse grid element is much smaller than that of the corresponding global fine grid pressure system. Thus, updating a small number of multiscale base functions dynamically is much cheaper than updating the fine grid pressure field globally.

Another advantage of the multiscale finite element method is its ability to scale down the size of a large scale problem. This offers a big saving in computer memory. For example, let N be the number of elements in each spatial direction, and M be the number of subcell elements in each direction for solving the base functions. Then there are total $(MN)^n$ (n is dimension) elements at the fine grid level. For a traditional FEM, the computer memory needed for solving the problem on the fine grid is $O(M^n N^n)$. In contrast, MsFEM requires only $O(M^n + N^n)$ amount of memory. For a typical value of $M = 32$ in a 2-D problem, the traditional FEM needs about 1000 times more memory than MsFEM.

MsFEM for problems with scale separation. If there is a scale separation in representative volumes smaller than the coarse block, then multiscale finite element basis functions can be computed based on the smaller regions. To demonstrate this, we first consider a periodic case. In this case, the basis functions can be approximated by

$$\phi^j(x) = \phi_0^j(x) + \epsilon \chi^i \nabla_i \phi_0^j.$$

Consequently, the approximation of the basis functions can be carried out in a domain of size ϵ via the computation of χ^i. This reduces the computational cost. Moreover, the assembly of stiffness matrix can be also performed in a period, because $a(x/\epsilon)\nabla\phi^i \cdot \nabla\phi^j$ is a periodic function. The results obtained by this approximation give the classical numerical homogenization procedure that is based on the computation of effective coefficients based on periodic problems. We would like to note that this approximation procedure is not limited to periodic problems and can be applied to random homogeneous problems with the strong scale separation, i.e., the size of representative volume is much smaller than the coarse mesh size. In general, this holds for problems where homogenization by periodization (see [75]) is true. Random homogeneous case with ergodicity is one of them. We note that a number of methods used in practice employs this strategy (e.g., [37, 55, 76, 93]).

Convergence and accuracy. Since we need to use an additional grid to compute the base function numerically, it makes sense to compare our MsFEM with a traditional FEM at the subcell grid, $h_s = h/M$. Note that MsFEM only captures the solution at the coarse grid h, while FEM tries to resolve the solution at the fine grid h_s. Our extensive numerical experiments demonstrate that the accuracy of MsFEM on the coarse grid h is comparable to that of FEM on the fine grid. In some cases, MsFEM is even more accurate than the FEM (see below and the next section).

As an example, in Table 4.1 we present the result for

$$a(\mathbf{x}/\varepsilon) = \frac{2 + P\sin(2\pi x/\varepsilon)}{2 + P\cos(2\pi y/\varepsilon)} + \frac{2 + \sin(2\pi y/\varepsilon)}{2 + P\sin(2\pi x/\varepsilon)} \quad (P = 1.8), \quad (4.29)$$

$$f(\mathbf{x}) = -1 \quad \text{and} \quad u|_{\partial\Omega} = 0. \quad (4.30)$$

The convergence of three different methods are compared for fixed $\varepsilon/h = 0.64$, where "-L" indicates that linear boundary condition is imposed on the multiscale base functions, "os" indicates the use of over-sampling, and LFEM stands for standard FEM with linear base functions.

Table 4.1. Convergence for periodic case.

N	ε	MsFEM-L		MsFEM-os-L		LFEM	
		$\|\|E\|\|_{l^2}$	rate	$\|\|E\|\|_{l^2}$	rate	MN	$\|\|E\|\|_{l^2}$
16	0.04	3.54e-4		7.78e-5		256	1.34e-4
32	0.02	3.90e-4	-0.14	3.83e-5	1.02	512	1.34e-4
64	0.01	4.04e-4	-0.05	1.97e-5	0.96	1024	1.34e-4
128	0.005	4.10e-4	-0.02	1.03e-5	0.94	2048	1.34e-4

We see clearly the scale resonance in the results of MsFEM-L and the (almost) first order convergence (i.e., no resonance) in MsFEM-os-L. Evident also is the error of MsFEM-os-L being smaller than those of LFEM obtained on the fine grid. In [62,64], more extensive convergence tests have been presented.

4.6 Applications

Flow in porous media. One of the main application of our multiscale method is the flow and transport through porous media. This is a fundamental problem in hydrology and petroleum engineering. Here, we apply MsFEM to solve the single phase flow, which is a good test problem in practice.

We model the porous media by random distributions of a_ε generated using a spectral method. In fact, $a_\varepsilon = \alpha 10^{\beta p}$, where p is a random field represents porosity, and α and β are scaling constants to give the desired contrast of a_ε. In particular, we have tested the method for a porous medium with a statistically fractal porosity field (see Figure 4.1). The fractal dimension is 2.8. This is a model of flow in an oil reservoir or aquifer with uniform injection in the domain and outflow at the boundaries. We note that the problem has a continuous scale because of the fractal distribution.

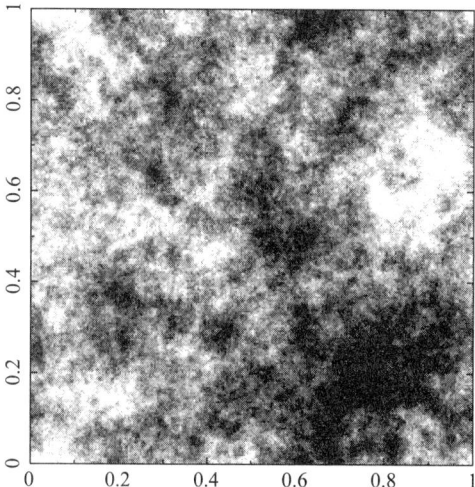

Figure 4.1.   Porosity field with fractal dimension of 2.8 generated using the spectral method.

The pressure field due to uniform injection is solved and the error is shown in Figure 4.2. The horizontal dash line indicates the error of the LFEM solution with $N = 2048$. The coarse-grid solutions are obtained with different number of elements, N, but fixed $NM = 2048$.

We note that error of MsFEM-os-L almost coincide with that of the well-resolved solution obtained using LFEM. However, MsFEM without over-sampling is less accurate. MsFEM-O indicates that oscillatory boundary conditions, obtained from solving some reduced 1-D elliptic equations along ∂K (see [62]), are imposed on the base functions. The decay of error in MsFEM is because of the decay of small scales in a_ε. The next figure shows the results for a log-normally distributed a_ε. In this case, the effect of scale resonance shows clearly for MsFEM-L, i.e., the error increases as h approaches ε. Here $\varepsilon \sim 0.004$ roughly equals the correlation length. Using the oscillatory boundary conditions (MsFEM-

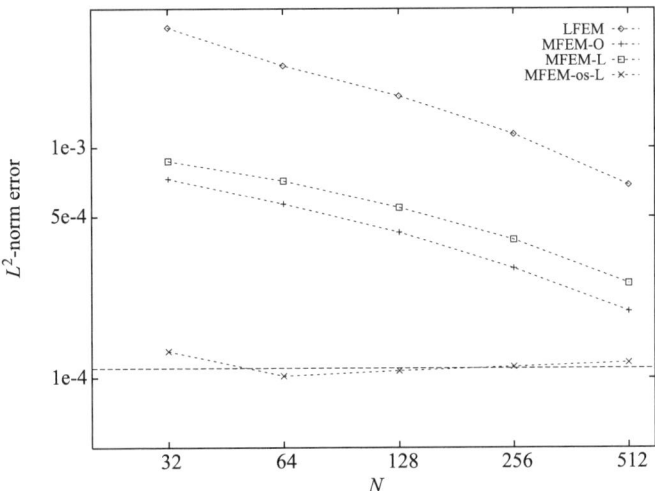

Figure 4.2. The L^2-norm error of the solutions using various schemes for a fractal distributed permeability field.

O) gives better results, but it does not completely eliminate resonance. On the other hand, the multiscale method with over-sampling agrees extremely well with the well-resolved calculation. One may wonder why the errors do not decrease as the number of coarse grid elements increase. This is because we use the same subgrid mesh size, which is the same as the well-resolved grid size, to construct the base functions for various coarse grid sizes ($N = 32, 64, 128$, etc). In some special cases, one can construct multiscale base functions analytically. In this case, the errors for the coarse grid computations will indeed decrease as the number of coarse grid elements increase.

Fine scale recovery. To solve transport problems in the subsurface formations, as in oil reservoir simulations, one needs to compute the velocity field from the elliptic equation for pressure, i.e., $\mathbf{v} = -a_\varepsilon \nabla u$, here u is pressure. In some applications involving isotropic media, the cell-averaged velocity is sufficient, as shown by some computations using the local upscaling methods (cf. [33]). However, for anisotropic media, especially layered ones (Figure 4.3), the velocity in some thin channels can be much higher than the cell average, and these channels often have dominant effects on the transport solutions. In this case, the information about fine scale velocity becomes vitally important. Therefore, an important question for all upscaling methods is how to take those fast-flow channels into account.

For MsFEM, the fine scale velocity can be easily recovered from the multiscale base functions, noting that they provide interpolations from

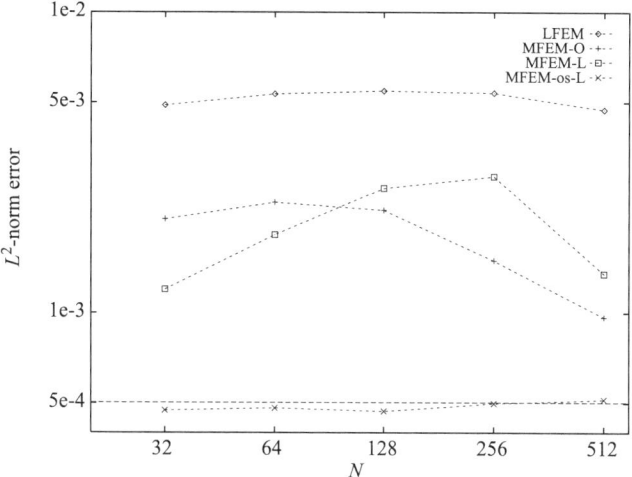

Figure 4.3. The L^2-norm error of the solutions using various schemes for a log-normally distributed permeability field.

the coarse h-grid to the fine h_s-grid. Using the over-sampling technique, the error in velocity is $O(\varepsilon/h)$, as proved in [42]. We remark that the resonance effect seems unavoidable in the velocity. On the other hand, our numerical tests indicate that the error is small when $\varepsilon \approx h$. The cell-averaged velocity can also be obtained and its error is even smaller.

To demonstrate the accuracy of the recovered velocity and effect of small-scale velocity on the transport problem, we show the fractional flow result of a "tracer" test using the layered medium in Figure 4.4:

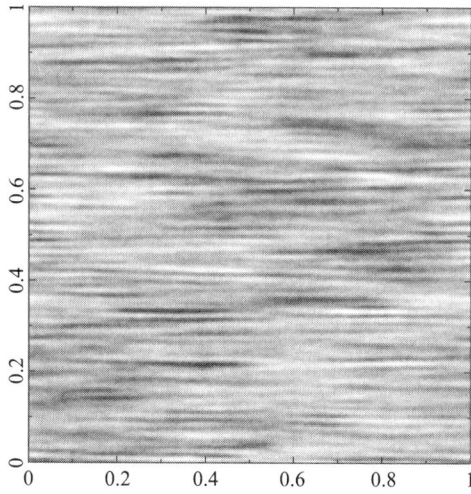

Figure 4.4. A random porosity field with layered structure.

a fluid with red color originally saturating the medium is displaced by
the same fluid with blue color injected by flow in the medium at the left
boundary, where the flow is created by a unit horizontal pressure drop.
The linear convection equation is solved to compute the saturation of the
red fluid (for details, see [34]). To demonstrate that we can recover the
fine grid velocity field from the coarse grid pressure calculation, we plot
the horizontal velocity fields obtained by two methods. In Figure 4.5(a),
we plot the horizontal velocity field obtained by using a fine grid ($N =$
1024) calculation. In Figure 4.5(b), we plot the same horizontal velocity
field obtained by using the coarse grid pressure calculation with $N = 64$
and using the multiscale finite element bases to interpolate the fine grid
velocity field. We can see that the recovered velocity field captures very
well the layer structure in the fine grid velocity field. Further, we use the
recovered fine grid velocity field to compute the saturation in time. In
Figure 4.6(a), we plot the saturation at $t = 0.06$ obtained by the fine grid
calculation. Figure 4.6(b) shows the corresponding saturation obtained
using the recovered velocity field from the coarse grid calculation. The
agreement is striking.

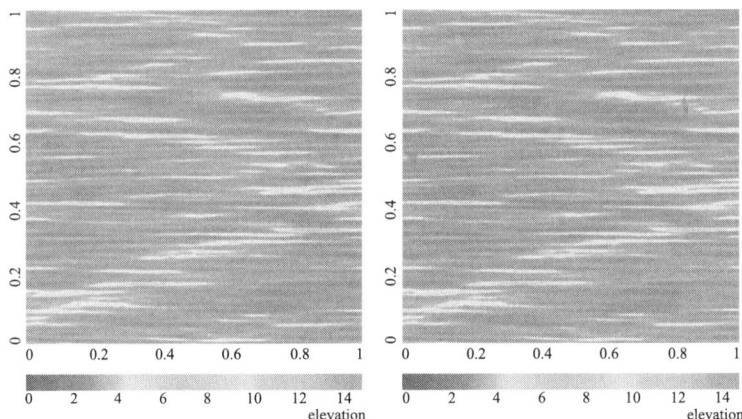

Figure 4.5.   (a) Fine grid horizontal velocity field, $N = 1024$. (b) Recovered
horizontal velocity field from the coarse grid $N = 64$ calculation using multiscale bases.

We also check the fractional flow curves obtained by the two calculations. The fractional flow of the red fluid, defined as

$$F = \int S_{red} v_x \, dy / \int v_x \, dy$$

(S being the saturation), at the right boundary is shown in Figure 4.7.
The top pair of curves are the solutions of the transport problem us-

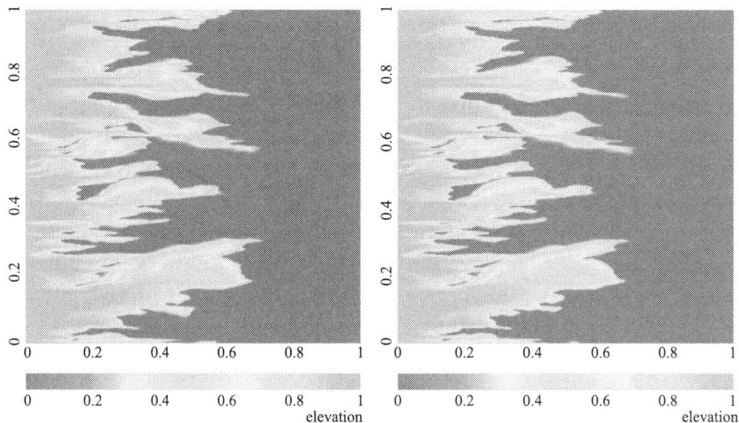

Figure 4.6. (a) Fine grid saturation at $t = 0.06$, $N = 1024$. (b) Saturation computed using the recovered velocity field from the coarse grid calculation.

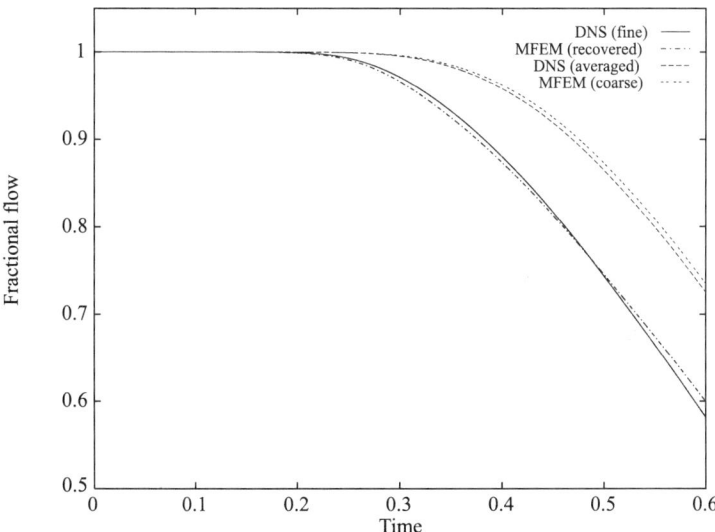

Figure 4.7. Variation of fractional flow with time. DNS: well-resolved direct numerical solution using LFEM ($N = 512$). MsFEM: over-sampling is used ($N = 64$, $M = 8$).

ing the cell-averaged velocity obtained from a well-resolved solution and from MsFEM; the bottom pair are solutions using well-resolved fine scale velocity and the recovered fine scale velocity from the MsFEM calculation. Two conclusions can be made from the comparisons. First, the cell-averaged velocity may lead to a large error in the solution of the transport equation. Second, both recovered fine scale velocity and the cell-averaged velocity obtained from MsFEM give faithful reproductions of respective direct numerical solutions.

Scale-up of one-phase flows. The multiscale finite element method has been used in conjunction with some moment closure models to obtain an upscaled method for one-phase flows, see, e.g. [25, 39, 46]. Note that the multiscale finite element method presented above does not conserve mass. For long time integration, it may lead to significant loss of mass. This is an undesirable feature of the method. In a recent work with Zhiming Chen [25], we have designed and analyzed a mixed multiscale finite element method, and we have applied this mixed method to study the scale up of one-phase flows and found that mass is conserved very well even for long time integration. Below we describe our results in some detail.

In its simplest form, neglecting the effect of gravity, compressibility, capillary pressure, and considering constant porosity and unit mobility, the governing equations for the flow transport in highly heterogeneous porous media can be described by the following partial differential equations [78, 103, 39]:

$$\text{div}(k(x)\nabla p) = 0, \tag{4.31}$$

$$\frac{\partial S}{\partial t} + \mathbf{v} \cdot \nabla S = 0, \tag{4.32}$$

where p is the pressure, S is the water saturation, $k(x) = (k_{ij}(x))$ is the relative permeability tensor, and $\mathbf{v} = -k(x)\nabla p$ is the Darcy velocity. The highly heterogeneous properties of the medium are built into the permeability tensor $k(x)$ which is generated through the use of sophisticated geological and geostatistical modeling tools. The detailed structure of the permeability coefficients makes the direct simulation of the above model infeasible. For example, it is common in real simulations to use millions of grid blocks, with each block having a dimension of tens of meters, whereas the permeability measured from cores is at a scale of centimeters [81]. This gives more than 10^5 degrees of freedom per spatial dimension in the computation. This makes a direct simulation to resolve all small scales prohibitive even with today's most powerful supercomputers. On the other hand, from an engineering perspective, it is often sufficient to predict the macroscopic properties of the solutions. Thus it is highly desirable to derive effective coarse grid models to capture the correct large scale solution without resolving the small scale features. Numerical upscaling is one of the commonly used approaches in practice.

Now we describe how the (mixed) multiscale finite element can be combined with the existing upscaling technique for the saturation equation (4.32) to get a complete coarse grid algorithm for problem (4.31)–

(4.32). The numerical upscaling of the saturation equation has been under intensive study in the literature [34, 46, 58, 78, 104, 106]. Here, we use the upscaling method proposed in [46] and [39] to design an overall coarse grid model for problem (4.31)–(4.32). The work of [46] for upscaling the saturation equation involves a moment closure argument. The velocity and the saturation are separated into a local mean quantity and a small scale perturbation with zero mean. For example, the Darcy velocity is expressed as $\mathbf{v} = \mathbf{v}_0 + \mathbf{v}'$ in (4.32), where $\mathbf{v}_0$ is the average of velocity $\mathbf{v}$ over each coarse element, $\mathbf{v}' = (\mathbf{v}_1', \mathbf{v}_2')$ is the deviation of the fine scale velocity from its coarse scale average. After some manipulations, an average equation for the saturation S can be derived as follows [46]:

$$\frac{\partial S}{\partial t} + \mathbf{v}_0 \cdot \nabla S = \frac{\partial}{\partial x_i}\left(D_{ij}(x,t)\frac{\partial S}{\partial x_j}\right), \tag{4.33}$$

where the diffusion coefficients $D_{ij}(x,t)$ are defined by

$$D_{ii}(x,t) = \langle|\mathbf{v}_i'(x)|\rangle L_i^0(x,t), \quad D_{ij}(x,t) = 0 \ \text{ for } i \neq j,$$

$\langle|\mathbf{v}_i'(x)|\rangle$ stands for the average of $|\mathbf{v}_i'(x)|$ over each coarse element. $L_i^0(x,t)$ is the length of the coarse grid streamline in the x_i direction which starts at time t at point x, i.e.,

$$L_i^0(x,t) = \int_0^t y_i(s)\,ds,$$

where $y(s)$ is the solution of the following system of ODEs:

$$\frac{dy(s)}{ds} = \mathbf{v}_0(y(s)), \quad y(t) = x.$$

Note that the hyperbolic equation (4.32) is now replaced by a convection-diffusion equation. The convection-dominant parabolic equation (4.33) is solved by the characteristic linear finite element method [32], [92] in our simulation. The flow transport model (4.31)–(4.32) is solved in the coarse grid as follows:

1. Solve the pressure equation (4.31) by the over-sampling mixed multiscale finite element method and obtain the fine scale velocity field using the multiscale basis functions.
2. Compute the coarse grid average $\mathbf{v}_0$ and the fine scale deviation $\langle|\mathbf{v}_i'(x)|\rangle$ on the coarse grid.
3. At each time step, solve the convection-diffusion equation (4.33) by the characteristic linear finite element method on the coarse grid in which the lengths $L_i^0(x,t)$ of the streamline are computed for the center of each coarse grid element.

The mixed multiscale finite element method can be readily combined with the above upscaling model for the saturation equation. The local fine grid velocity $\mathbf{v}'$ will be constructed from the multiscale finite element base functions. The main cost in the above algorithm lies in the computation of multiscale bases which can be done a priori and completely in parallel. This algorithm is particularly attractive when multiple simulations must be carried out due to the change of boundary and source distribution as it is often the case in engineering applications. In such a situation, the cost of computing the multiscale base functions is just an over-head. Moreover, once these base functions are computed, they can be used for subsequent time integration of the saturation. Because the evolution equation is now solved on a coarse grid, a larger time step can be used. This also offers additional computational saving. For many oil recovery problems, due to the excessively large fine grid data, upscaling is a necessary step before performing many simulations and realizations on the upscaled coarse grid model. If one can coarsen the fine grid by a factor of 10 in each dimension, the computational saving of the coarse grid model over the original fine model could be as large as a factor 10,000 (three space dimensions plus time).

We perform a coarse grid computation of the above algorithm on the coarse 64×64 mesh. The fractional flow curve using the above algorithm is depicted in Figure 4.8. It gives excellent agreement with the "exact"

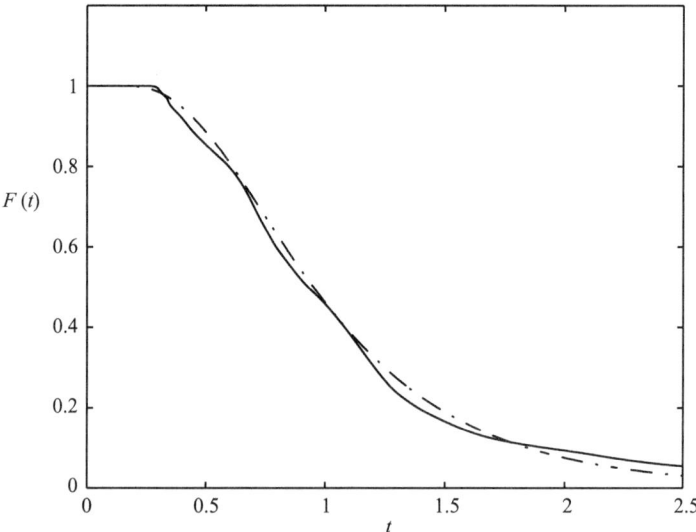

Figure 4.8. The accuracy of the coarse grid algorithm. Solid line is the "exact" fractional flow curve using mixed finite element method solving the pressure equation. The slash-dotted line is the fractional flow curve using above coarse grid algorithm.

fractional flow curve. The contour plots of the saturation S on the fine 1024×1024 mesh at time $t = 0.25$ and $t = 0.5$ computed by using the "exact" velocity field are displayed in Figure 4.10. In Figure 4.9, we show the contour plots of the saturation obtained using the recovered velocity field from the coarse grid pressure calculation $N = 64$. We can see that the contour plots in Figure 4.9 approximate the "exact" ones in Figure 4.10 in certain accuracy but the sharp oil/water interfaces in Figure 4.10 are smeared out. This is due to the parabolic nature of the upscaled equation (4.33). We have also performed many other numerical experiments to test the robustness of this combined coarse grid model. We found that for permeability fields with strong layered structure, the above coarse grid model is very robust. The agreement with the fine grid calculations is very good. We are currently working toward some qualitative and quantitative understanding of this upscaling model.

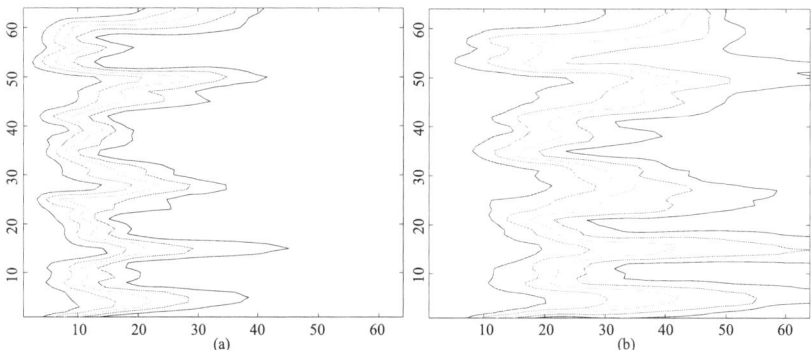

Figure 4.9. The contour plots of the saturation S computed using the upscaled model on a 64×64 mesh at time $t = 0.25$ (left) and $t = 0.5$ (right).

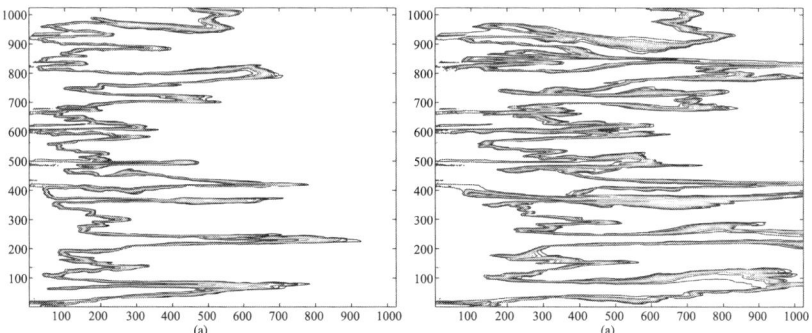

Figure 4.10. The contour plots of the saturation S computed on the fine 1024×1024 mesh using the "exact" velocity field at time $t = 0.25$ (left) and $t = 0.5$ (right).

Finally, we remark that the upscaling equation (4.33) uses small scale information $\mathbf{v}'$ of the velocity field to define the diffusion coefficients. This information can be constructed *locally* through the mixed multi-scale basis functions. This is an important property of our multiscale finite element method. It is clear that solving directly the homogenized pressure equation

$$\mathrm{div}(k^*(x)\nabla p^*) = 0$$

will not provide such small scale information. On the other hand, whenever one can afford to resolve all the small scale feature by a fine grid, one can use fast linear solvers, such as multigrid methods, to solve the pressure equation (4.31) on the fine mesh. From the fine grid computation, one can easily construct the average velocity $\mathbf{v}_0$ and its deviation $\mathbf{v}'$. However, when multiple simulations must be carried out due to the change of boundary conditions, the pressure equation (4.31) will then have to be solved again on the fine mesh. The multiscale finite element method only solves the pressure equation once on a coarse mesh, and the fine grid velocity can be constructed locally through the finite element bases. This is the main advantage of our mixed multiscale finite element method. This process becomes more difficult for the nonlinear two-phase flow due to the dynamic coupling between the pressure and the saturation. We are now investigating the possibility of upscaling the two-phase flow by using multiscale finite element base functions that are constructed from the one-phase flow (time independent). In this case, we need to provide corrections to the pressure equation to account for the scale interaction near the oil/water interface.

It should be noted that some adaptive scale-up strategies have also been developed [34, 106]. The idea is to refine the mesh around the fast-flow channels in order to capture their effect directly. The approach seems to work well when the channels are isolated. For MsFEM, it is also possible to adjust the coarse mesh adaptively based on the recovered velocity. In particular, one does not need to use the fine recovered velocity in the regions with no fast-flow channels; in those regions, the coarse mesh and cell-averaged velocity are sufficient. On the other hand, one can simply keep the fine mesh when the channels are too many. How to develop a consistent upscaling equation for the saturation equation is still open when the capillary pressure effect is neglected, which is the common practice in oil reservoir simulations. One approach is to combine grid adaptivity with multiscale modeling. We use a dynamic adaptive coarse grid [24] to capture the isolated small scale features, such as the flow channels and use the multiscale finite element method to capture the small scale feature within each adaptive coarse grid block. By doing this, we take into account the local flow orientation and anisotropy in upscaling the saturation equation. We are also investigating the possibility to develop a consistent upscaling model for the saturation equation by combining multiscale finite element methods and systematic multiscale modeling for the saturation equation.

4.7 Brief overview of mixed finite element and finite volume element methods

Control volume multiscale finite element method. In this section, we discuss multiscale finite volume element method. Finite volume method is chosen because, by its construction, it satisfies the numerical local conservation which is important in groundwater and reservoir simulations. Let $\mathcal{K}^h$ denote the collection of coarse elements/rectangles K. Consider a coarse element K, and let ξ_K be its center. The element K is divided into four rectangles of equal area by connecting ξ_K to the midpoints of the element's edges. We denote these quadrilaterals by K_ξ, where $\xi \in Z_h(K)$, are the vertices of K. Also, we denote $Z_h = \bigcup_K Z_h(K)$ and $Z_h^0 \subset Z_h$ the vertices which do not lie on the Dirichlet boundary of Ω. The control volume V_ξ is defined as the union of the quadrilaterals K_ξ sharing the vertex ξ.

The key idea of the method is the construction of basis functions on the coarse grids, such that these basis functions capture the small-scale information on each of these coarse grids. As before, the basis functions are constructed from the solution of the leading order homogeneous elliptic equation on each coarse element with some specified boundary conditions. We consider a coarse element K that has d vertices, the local basis functions $\phi_i, i = 1, \cdots, d$, are set to satisfy the following elliptic problem:

$$-\nabla \cdot (k \cdot \nabla \phi_i) = 0 \quad \text{in } K,$$
$$\phi_i = g_i \quad \text{on } \partial K \tag{4.34}$$

for some function g_i defined on the boundary of the coarse element K. As we discussed earlier, Hou et al. [62] have demonstrated that a careful choice of boundary conditions would improve the accuracy of the method. In previous findings, the function g_i for each i is chosen to vary linearly along ∂K or to be the solution of the local one-dimensional problems [74] or the solution of the problem in a slightly larger domain is chosen to define the boundary conditions. For simplicity, we consider linear boundary conditions and also discuss the boundary conditions obtained from a global solution. We will require $\phi_i(x_j) = \delta_{ij}$. Finally, a nodal basis function associated with the vertex x_i in the domain Ω is constructed from the combination of the local basis functions that share this x_i and zero elsewhere. We would like to note that one can use an approximate solution of (4.34) when it is possible. For example, in the case of periodic or random homogeneous cases, the basis functions can be approximated using homogenization expansion $\phi_i = \phi_i^0 + \epsilon \chi_k \nabla_k \phi_i^0$, where χ_k is the solution of the cell problem and ϕ_i^0 is standard finite element basis on the coarse mesh (see [41]).

Next, we denote by V^h the space of our approximate pressure solution, which is spanned by the basis functions $\{\phi_j\}_{x_j \in Z_h^0}$. Then we formulate the finite dimensional problem corresponding to finite volume element formulation of pressure equation. A statement of mass conservation on a coarse-control volume V_x is formed from pressure equation, where the approximate solution is written as a linear combination of the basis functions. Assembly of this conservation statement for all control volumes would give the corresponding linear system of equations that can be solved accordingly. The resulting linear system has incorporated the fine-scale information through the involvement of the nodal basis functions on the approximate solution. To be specific, the problem now is to seek $p^h \in V^h$ with $p^h = \sum_{x_j \in Z_h^0} p_j \phi_j$ such that

$$\int_{\partial V_\xi} k \cdot \nabla p^h \cdot n \, dl = 0 \tag{4.35}$$

for every control volume $V_\xi \subset \Omega$. Here n defines the normal vector on the boundary of the control volume, ∂V_ξ, and S is the fine-scale saturation field at this point. The resulting multiscale method differs from the multiscale finite element method, since it employs the finite volume element method as a global solver, and it is called multiscale finite volume element method (MsFVEM). We would like to note that the coarse-scale velocity field obtained using MsFVEM is conservative in control volume elements V_ξ (not in $\mathcal{K}^h$).

Mixed multiscale finite element methods. For simplicity, we assume Neumann boundary conditions. First, we review the mixed multiscale finite element formulation following [25] (see also [6], [1], and [7]). We can rewrite two-phase flow equation as

$$\begin{aligned} k^{-1}u - \nabla p &= 0 \quad \text{in } \Omega, \\ \mathrm{div}\, u &= 0 \quad \text{in } \Omega, \\ k(x)\nabla p \cdot n &= g(x) \quad \text{on } \partial\Omega. \end{aligned} \tag{4.36}$$

The variational problem associated with (4.36) is to seek $(u, p) \in H(\mathrm{div}, \Omega) \times L^2(\Omega)/R$ such that $u \cdot n = g$ on $\partial\Omega$ and

$$\begin{aligned} (k^{-1}u, v) + (\mathrm{div}\, v, p) &= 0 \quad \forall v \in H_0(\mathrm{div}, \Omega), \\ (\mathrm{div}\, u, q) &= 0 \quad \forall q \in L^2(\Omega)/R, \end{aligned} \tag{4.37}$$

where $H_0(\mathrm{div}, \Omega)$ is $H(\mathrm{div}, \Omega)$ with homogeneous Neumann boundary conditions. By defining

$$a(u, v) = (k^{-1}u, v), \quad b(v, q) = (\mathrm{div}\, v, q), \tag{4.38}$$

we can rewrite the weak formulation as

$$a(u, v) + b(v, p) = 0 \quad \forall v \in H_0(\text{div}, \Omega),$$
$$b(u, q) = 0 \quad \forall q \in L^2(\Omega)/R.$$

Let $V_h \subset H(\text{div}, \Omega)$ and $Q_h \subset L^2(\Omega)/R$ be finite dimensional spaces and $V_h^0 = V_h \cap H_0(\text{div}, \Omega)$. The numerical approximation problem associated with (4.37) is to find $(u_h, p_h) \in V_h \times Q_h$ such that $u_h \cdot n = g_h$ on $\partial\Omega$, where $g_h = g_{0,h} n$ on $\partial\Omega$ and $g_{0,h} = \sum_{e \in \{\partial K \cap \partial\Omega, K \in \mathcal{T}_h\}} (\int_e g \, ds) N_e$, $N_e \in V_h$, is corresponding basis function to edge e,

$$(k^{-1} u_h, v_h) + (\text{div} v_h, p_h) = 0 \quad \forall v_h \in V_h^0$$
$$(\text{div} u_h, q_h) = 0 \quad \forall q_h \in Q_h. \tag{4.39}$$

One can define a linear operator $B_h : V_h^0 \to Q_h'$ by $b(u_h, q_h) = (B_h u_h, q_h)$. Suppose that the following conditions are satisfied:

$$a(u_h, u_h) \quad \text{is} \quad \ker B_h\text{-coercive}, \tag{4.40}$$

$$\inf_{q_h \in Q_h} \sup_{v_h \in V_h} \frac{b(v_h, q_h)}{\|v_h\|_{H(\text{div}, \Omega)} \|q_h\|_{L^2(\Omega)}} \geq C. \tag{4.41}$$

Then the following approximation property follows (see e.g., [19]).

Lemma 4.7. *If (u, p) and (u_h, p_h) respectively solve problems (4.37) and (4.39) and conditions (4.40) and (4.41) hold, then*

$$\|u - u_h\|_{H(\text{div}, \Omega)} + \|p - p_h\|_{0, \Omega}$$
$$\leq \inf_{\substack{v_h \in V_h \\ v_h - g_{0,h} \in V_h^0}} \|u - v_h\|_{H(\text{div}, \Omega)} + \inf_{q_h \in Q_h} \|p - q_h\|_{0, \Omega}. \tag{4.42}$$

Following Chen and Hou [25] (see also [6]), one can construct multiscale basis functions for velocity in each coarse block K :

$$\text{div}(k(x)\nabla w_i^K) = \frac{1}{|K|} \quad \text{in } K,$$
$$k(x)\nabla w_i^K n^K = \begin{cases} g_i^K \text{ on } e_i^K, \\ 0 \quad \text{else,} \end{cases} \tag{4.43}$$

where $g_i^K = \frac{1}{|e_i^K|}$ and e_i^K are the edges of K. Then, we can define the finite dimensional space for velocity by

$$V_h = \bigoplus_K \{\Psi_i^K\},$$
$$V_h^0 = V_h \cap H_0(\text{div}, \Omega),$$

where $\Psi_i^K = k(x)\nabla w_i^K$.

4.8   MsFEM using limited global information

Motivation. Multiscale finite element methods and their modifications are used in two-phase flow simulations through heterogeneous porous media. First, we briefly describe the underlying fine-scale equations. We present two-phase flow equations neglecting the effects of gravity, compressibility, capillary pressure and dispersion on the fine scale. Porosity, defined as the volume fraction of the void space, will be taken to be constant and therefore serves only to rescale time. The two phases will be referred to as water and oil and designated by the subscripts w and o, respectively. We can then write Darcy's law, with all quantities dimensionless, for each phase j as follows:

$$\mathbf{v}_j = -\lambda_j(S)k\nabla p, \qquad (4.44)$$

where $\mathbf{v}_j$ is phase velocity, S is water saturation (volume fraction), p is pressure, $\lambda_j = k_{rj}(S)/\mu_j$ is phase mobility, where k_{rj} and μ_j are the relative permeability and viscosity of phase j respectively, and $\mathbf{k}$ is the permeability tensor, which is here taken to be diagonal, $\mathbf{k} = kI$, where I is the identity matrix.

Combining Darcy's law with conservation of mass, div $(\mathbf{v}_w + \mathbf{v}_o) = 0$, allows us to write the flow equation in the following form:

$$\mathrm{div}(\lambda(S)k\nabla p) = f, \qquad (4.45)$$

where the total mobility $\lambda(S)$ is given by $\lambda(S) = \lambda_w(S) + \lambda_o(S)$ and f is a source term. The saturation dynamics affects the flow equations. One can derive the equation describing the dynamics of the saturation

$$\frac{\partial S}{\partial t} + \mathrm{div}\,(\mathbf{F}) = 0, \qquad (4.46)$$

where $\mathbf{F} = \mathbf{v}f_w(S)$, with $f_w(S)$, the fractional flow of water, given by $f_w = \lambda_w/(\lambda_w + \lambda_o)$, and the total velocity $\mathbf{v}$ by

$$\mathbf{v} = \mathbf{v}_w + \mathbf{v}_o = -\lambda(S)k\nabla p. \qquad (4.47)$$

In the presence of capillary effects, an additional diffusion term is present in (4.46).

If $k_{rw} = S$, $k_{ro} = 1 - S$ and $\mu_w = \mu_o$, then the flow equation reduces to

$$\mathrm{div}\,(k\nabla p^{sp}) = f.$$

This equation, the linear advection pollutant transport equation, will be referred to as the single-phase flow equation associated with (4.45), and p^{sp} will be referred to as the single-phase flow solution.

As we see from (4.45) and (4.46), the pressure equation is solved many times for different saturation profiles. Thus, computing the basis functions once at time zero is very beneficial and the basis functions are only updated near sharp interfaces. In fact, our numerical results show that only slight improvement can be achieved by updating the basis functions near sharp fronts. However, we have found that for heterogeneous permeability fields with very strong non-local effects, the use of some type of global information can improve multiscale finite element results significantly, which will be discussed next.

We present a representative numerical example for a permeability field generated using two-point geostatistics. To generate this permeability field, we have used GSLIB algorithm [30]. The permeability is log-normally distributed with prescribed variance $\sigma^2 = 1.5$ (σ^2 here refers to the variance of $\log k$) and some correlation structure. The correlation structure is specified in terms of dimensionless correlation lengths in the x and z-directions, $l_x = 0.4$ and $l_z = 0.04$, nondimensionalized by the system length. Linear boundary conditions are used for constructing multiscale basis function in (4.34). Spherical variogram is used [30]. In this numerical example, the fine-scale field is 120×120, while the coarse-scale field is 12×12 defined in the rectangle with the length 5 and the width 1. For the two-phase flow simulations, the system is considered to initially contain only oil ($S = 0$) and water is injected at inflow boundaries ($S = 1$ is prescribed), i.e., we specify $p = 1$, $S = 1$ along the $x = 0$ edge and $p = 0$ along the $x = 5$ edge, and no flow boundary conditions on the lateral boundaries. Relative permeability functions are specified as $k_{rw} = S^2$, $k_{ro} = (1 - S)^2$; water and oil viscosities are set to $\mu_w = 1$ and $\mu_o = 5$. Porosity is constant and serves only to nondimensionalize time. Results are presented in terms of the fraction of oil in the produced fluid (i.e., oil cut, designated F) against pore volume injected (PVI). PVI represents dimensionless time and is computed via $\int Q\,dt/V_p$ where V_p is the total pore volume of the system and Q is the total flow rate.

In our first numerical test, Figure 4.11, we compare the fractional flows. The dashed line corresponds to the calculations performed using a simple saturation upscaling (no subgrid treatment), while dotted line corresponds to the calculations performed by solving the saturation equation on the fine grid using the reconstructed fine-scale velocity field. We observe from this figure that the second approach is very accurate, while the first approach over-predicts the breakthrough time. The saturation snapshots are compared in Figure 4.12. One can observe that there is a very good agreement.

In the next set of numerical results, we consider strongly channelized permeability fields. These permeability fields are proposed in some recent benchmark tests, such as the SPE comparative solution project [27]. In

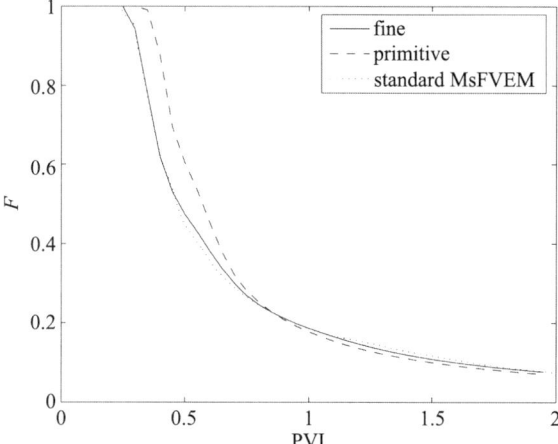

Figure 4.11. Fractional flow comparison for a permeability field generated using two-point geostatistics.

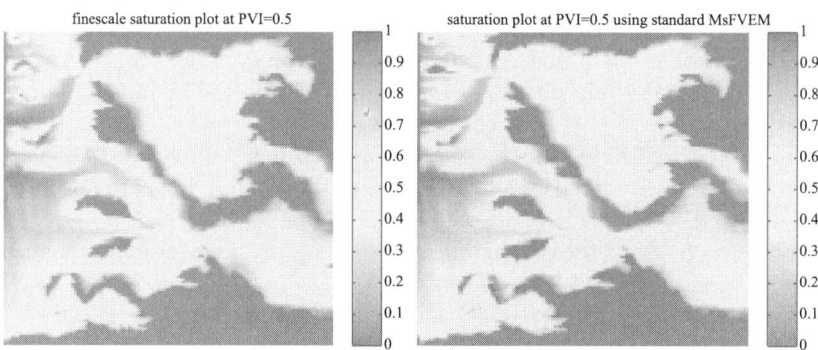

Figure 4.12. Saturation maps at PVI=0.5 for fine-scale solution (left figure) and standard MsFVEM (right figure).

Figure 4.13, one of the layers of this 3-D permeability field is depicted. All the layers have 220×60 fine-scale resolution, and we take the coarse grid to be 22×6. As it can be observed, the permeability field contains a high permeability channel, where most flow will occur in our simulation. In Figure 4.14, the fractional flows are compared. The boundary conditions are taken to be $p = 1$, $S = 1$ along the $x = 0$ edge and $p = 0$ along the $x = 5$ edge, and no flow boundary conditions on the lateral boundaries. Again, the dashed line corresponds to the calculations performed using a simple saturation upscaling (no subgrid treatment), while dotted line corresponds to the calculations performed by solving

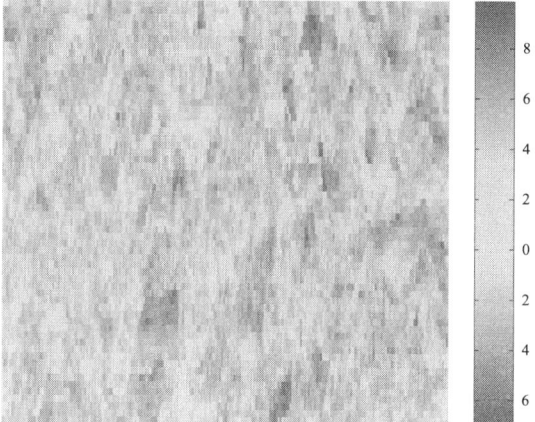

Figure 4.13. Log-permeability for one of the layers of upper Ness.

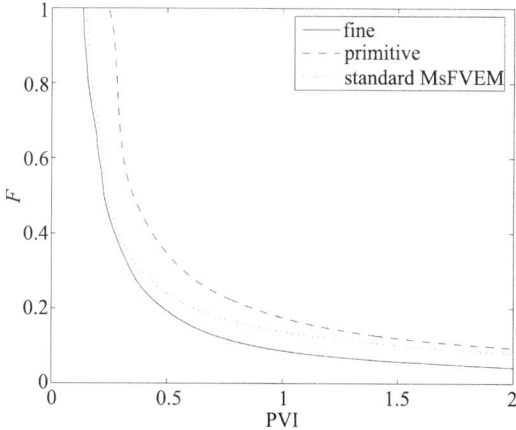

Figure 4.14. Fractional flow comparison for a channelized permeability field.

the saturation equation on the fine grid using the reconstructed fine-scale velocity field. We observe from this figure that the second approach is not very accurate in contrast to the permeability field generated using two-point geostatistics [30]. This is because the local basis functions can not account accurately the global connectivity of the media. Indeed, in the next figure, Figure 4.15, the saturation fields at time PVI = 0.5 are compared. We see that multiscale finite element methods with local basis functions introduce some errors. In the bottom left corner, there is a saturation pocket which is not in the reference solution computed using a fine grid. The reason for this is that the local basis functions in the lower left corner contains high permeability region. However, this high permeability region does not have global connectivity, and the local ba-

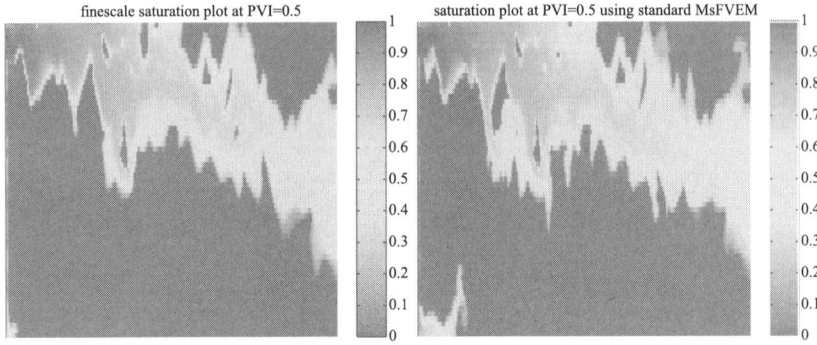

Figure 4.15. Saturation maps at PVI=0.5 for fine-scale solution (left figure) and standard MsFVEM (right figure).

sis functions can not take this effect into account. Next, we discuss how global information can be incorporated into multiscale basis functions to improve the accuracy of the computations.

Multiscale finite volume element method. The main idea of the modified multiscale finite volume element method (MsFVEM) is to use the solution of the fine-scale problem at time zero to determine the boundary conditions for the basis functions. This approach is proposed in [40] to handle the permeability fields which are strongly channelized. For this type of permeability fields, some type of global information is needed. Next, we describe the method. We denote the solution of pressure equation at time zero by $p^{sp}(x)$. In defining $p^{sp}(x)$, we use the actual boundary conditions of the global problem. $p^{sp}(x)$ depends on global boundary conditions, and, generally, is updated each time when global boundary conditions are changed. The boundary conditions in (4.34) for modified basis functions are defined in the following way. For each rectangular element K with vertices x_i ($i = 1, 2, 3, 4$) denote by $\phi_i(x)$ a restriction of the nodal basis on K, such that $\phi_i(x_j) = \delta_{ij}$. At the edges where $\phi_i(x) = 0$ at both vertices, we take boundary condition for $\phi_i(x)$ to be zero. Consequently, the basis functions are localized. We only need to determine the boundary condition at two edges which have the common vertex x_i ($\phi_i(x_i) = 1$). Denote these two edges by $[x_{i-1}, x_i]$ and $[x_i, x_{i+1}]$ (see Figure 4.16). We only need to describe the boundary condition, $g_i(x)$, for the basis function $\phi_i(x)$, along the edges $[x_i, x_{i+1}]$ and $[x_i, x_{i-1}]$. If $p^{sp}(x_i) \neq p^{sp}(x_{i+1})$, then

$$g_i(x)|_{[x_i, x_{i+1}]} = \frac{p^{sp}(x) - p^{sp}(x_{i+1})}{p^{sp}(x_i) - p^{sp}(x_{i+1})},$$

$$g_i(x)|_{[x_i, x_{i-1}]} = \frac{p^{sp}(x) - p^{sp}(x_{i-1})}{p^{sp}(x_i) - p^{sp}(x_{i-1})}.$$

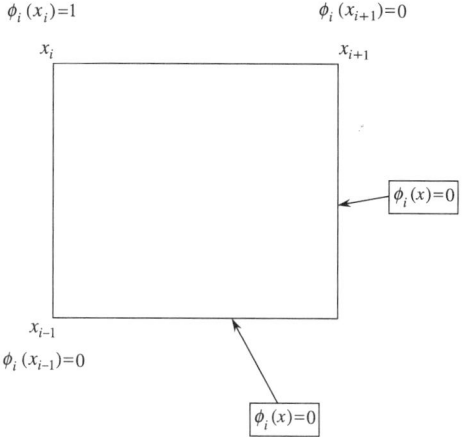

Figure 4.16. Schematic description of nodal points.

If $p^{sp}(x_i) = p^{sp}(x_{i+1}) \neq 0$, then

$$g_i(x)|_{[x_i,x_{i+1}]} = \phi_i^0(x) + \frac{1}{2p^{sp}(x_i)}(p^{sp}(x) - p^{sp}(x_{i+1})),$$

where $\phi_i^0(x)$ is a linear function on $[x_i, x_{i+1}]$ such that $\phi_i^0(x_i) = 1$ and $\phi_i^0(x_{i+1}) = 0$. Similarly,

$$g_{i+1}(x)|_{[x_i,x_{i+1}]} = \phi_{i+1}^0(x) + \frac{1}{2p^{sp}(x_{i+1})}(p^{sp}(x) - p^{sp}(x_{i+1})), \quad (4.48)$$

where $\phi_{i+1}^0(x)$ is a linear function on $[x_i, x_{i+1}]$ such that $\phi_{i+1}^0(x_{i+1}) = 1$ and $\phi_{i+1}^0(x_i) = 0$. If $p^{sp}(x_i) = p^{sp}(x_{i+1}) \neq 0$, then one can also use simply linear boundary conditions. If $p^{sp}(x_i) = p^{sp}(x_{i+1}) = 0$, then linear boundary conditions are used. In the applications considered in this paper, the initial pressure is always positive. Finally, the basis function $\phi_i(x)$ is constructed by solving (4.34). The choice of the boundary conditions for the basis functions is motivated by the analysis. In particular, we would like to recover the exact fine-scale solution along each edge if the nodal values of the pressure are equal to the values of exact fine-scale pressure. This is the underlying idea for the choice of boundary conditions. Using this property and Cea's lemma one can show that the pressure obtained from the numerical solution is equal to the underlying fine-scale pressure.

Mixed multiscale finite element methods. Next, following [1], we present a mixed multiscale finite element method that employs single-phase flow information. Suppose that p^{sp} solves the single-phase flow equation. We set $b_i^K = (k\nabla p^{sp}|_{e_i^K}) \cdot n^K$ and assume that b_i^K is uniformly bounded. Then the new basis functions for velocity is constructed by solving the following local problems (4.43) with $g_i^K = b_i^K / \beta_i^K$, where $\beta_i^K = \int_{e_i^K} k\nabla p^{sp} \cdot n^K ds$. For further analysis, we assume that $\beta_i^K \neq 0$. In general, if $\beta_i^K = 0$ one can use standard mixed multiscale finite element basis functions. Let $N_i^K = k(x)\nabla w_i^K$ and the multiscale finite dimensional space V_h^0 for velocity be defined by

$$V_h := \bigoplus_K \{N_i^K\} \subset H(\mathrm{div}, \Omega),$$

$$V_h^0 := V_h \cap H_0(\mathrm{div}, \Omega).$$

First, we will show that the resulting multiscale finite element solution for velocity is exact for single-phase flow (i.e., $\lambda(x) = 1$). Let $v_h|_K = \beta_i^K N_i^K$. Then β_i^K is the interpolation value of the fine scale solution. Furthermore, a direct calculation yields $(v_h|_{e_i^K}) \cdot n^K = k\nabla p^{sp} \cdot n^K$. Because

$$\mathrm{div}\, v_h = \beta_i^K \mathrm{div}\, N_i^K = \frac{1}{|K|} \int_{\partial K} k\nabla p^{sp} \cdot n^K ds = \frac{1}{|K|} \int_K \mathrm{div}(k\nabla p^{sp} d) = 0,$$

the following equation is obtained immediately:

$$\mathrm{div}\, v_h = 0 \quad \text{in } K, \tag{4.49}$$

$$v_h \cdot n^K = k\nabla p^{sp} \cdot n^K \quad \text{on } \partial K. \tag{4.50}$$

Because $\mathrm{div}(k\nabla p^{sp}) = 0$, we get $v_h = k\nabla p^{sp}$ and the following proposition.

Proposition Let $\beta_i^K = \int_{e_i^K} k\nabla p^{sp} \cdot n^K ds$. Then on each coarse block K,

$$k\nabla p^{sp} = \beta_i^K N_i^K. \tag{4.51}$$

Lemma 4.8. *If $|\beta_i^K| \geqslant Ch$ with C is independent of h, then*
(1) $a(u_h, u_h)$ *is $\ker B_h$-coercive;*
(2) $\inf_{q_h \in Q_h} \sup_{v_h \in V_h^0} \frac{b(v_h, q_h)}{\|v_h\|_{H(div, \Omega)} \|q_h\|_{L^2(\Omega)}} \geqslant C.$

Numerical results. Next, we show the numerical results obtained using modified multiscale finite element type methods for the permeability layer depicted in Figure 4.13 and two-phase flow parameters presented earlier. We consider two types of boundary conditions in a rectangular region $[0,5] \times [0,1]$. For the first type of boundary conditions, we specify $p = 1$, $S = 1$ along the $x = 0$ edge and $p = 0$ along the $x = 5$ edge. On the rest of the boundaries, we assume no flow boundary condition. We call this type of the boundary condition as side-to-side. The other type of boundary conditions is obtained by specifying $p = 1$, $S = 1$ along the $x = 0$ edge for $0.5 \leqslant z \leqslant 1$ and $p = 0$ along the $x = 5$ edge for

$0 \leqslant z \leqslant 0.5$. On the rest of the boundaries, we assume no flow boundary condition.

In Figure 4.17, the fractional flows are plotted for standard and modified MsFVEM. We observe from this figure that modified MsFVEM is more accurate and provides nearly the same fractional flow response as the direct fine-scale calculations. In Figure 4.18, we compare the saturation fields at PVI=0.5. As we see, the saturation field obtained using modified MsFVEM is very accurate and there is no longer the saturation pocket at the left bottom corner. Thus, the modified MsFVEM captures the connectivity of the media accurately.

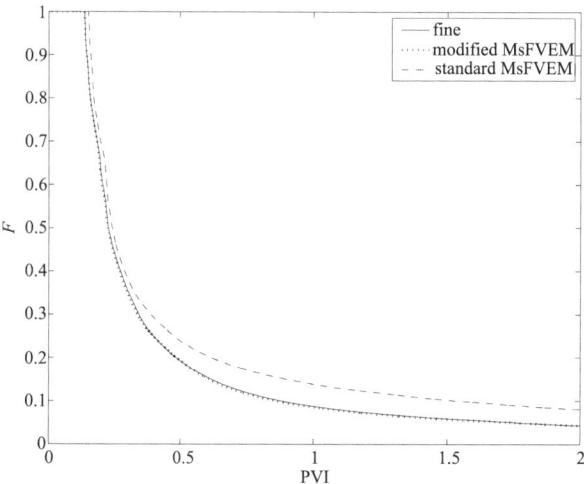

Figure 4.17. Fractional flow comparison for standard MsFVEM and modified MsFVEM for side-to-side flow.

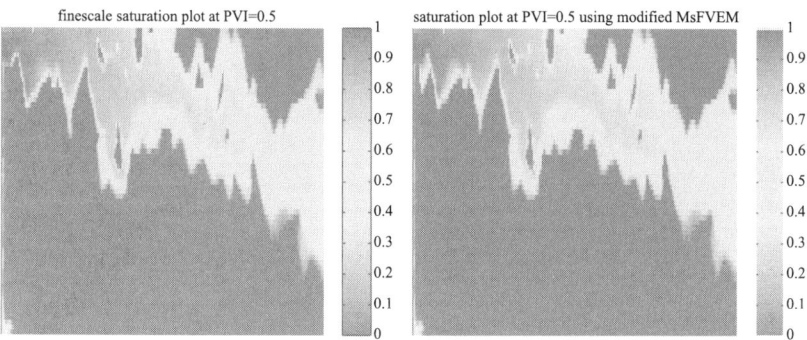

Figure 4.18. Saturation maps at PVI=0.5 for fine-scale solution (left figure) and modified MsFVEM (right figure). Side-to-side boundary condition is used.

In the next set of numerical results, we test the modified multiscale finite element methods for a different layer (layer 40) of SPE comparative solution project. In Figures 4.19 and 4.20, the fractional flows and total flow rates (Q) are compared for two different boundary conditions. One can see clearly that the modified MsFVEM method gives nearly exact results for these integrated responses. The standard MsFVEM tends to over-predict the total flow rate at time zero. This initial error persists at later times. This phenomena is often observed in upscaling of two-phase flows. More numerical results and discussions can be found in [40]. These numerical results demonstrate that modified multiscale finite element methods which use a limited global information are more accurate. Moreover, modified multiscale finite element methods are capable of capturing the long-range flow features accurately for channelized permeability fields.

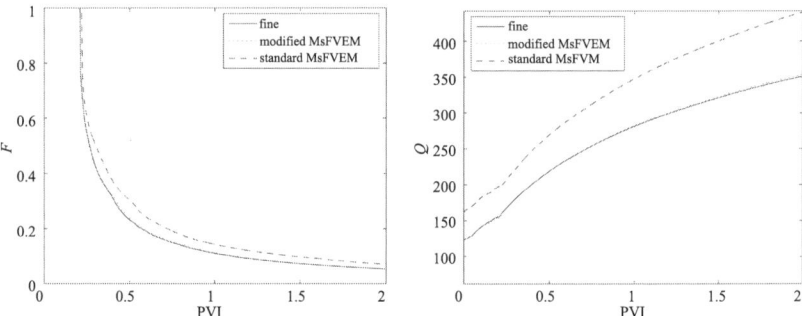

Figure 4.19. Fractional flow (left figure) and total production (right figure) comparison for standard MsFVEM and modified MsFVEM for side-to-side flow (layer 40).

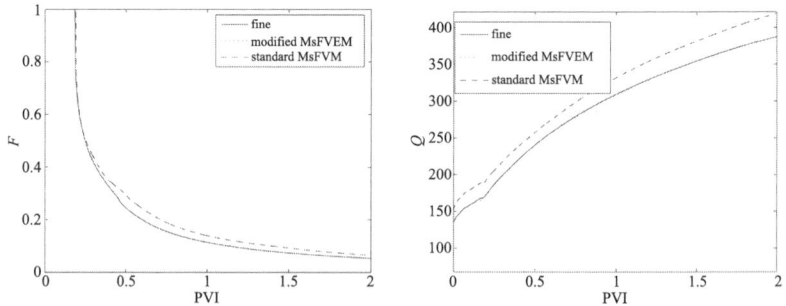

Figure 4.20. Fractional flow (left figure) and total production (right figure) comparison for standard MsFVEM and modified MsFVEM for corner-to-corner (layer 40).

For the next set of results, we consider another layer of the upper Ness (layer 59). In Figure 4.21, both fractional flow (left figure) and total flow (right figure) are plotted. We observe that the modified MsFVEM gives almost the exact results for these quantities, while the standard MsFVEM overpredicts the total flow rate, and there are deviations in the fractional flow curve around PVI $\approx$ 0.6. Note that unlike the previous case, fractional flow for standard MsFVEM is nearly exact at later times (PVI $\approx$ 2). In Figure 4.22, the saturation maps are plotted at PVI = 0.5. The left figure represents the fine-scale, the middle figure represents the results obtained using standard MsFVEM, and the right figure represents the results obtained using the modified MsFVEM. We observe from this figure that the saturation map obtained using standard MsFVEM has some errors. These errors are more evident near the lower left corner. The results of the saturation map obtained using the modified MsFVEM is nearly the same as the fine-scale saturation field. It is evident from these figures that the modified MsFVEM performs better than the standard MsFVEM.

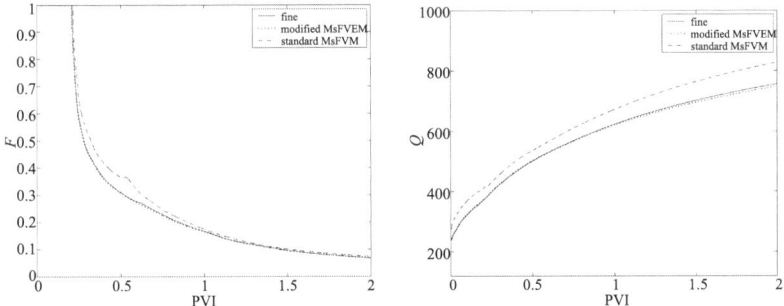

Figure 4.21. Fractional flow (left figure) and total production (right figure) comparison for standard MsFVEM and modified MsFVEM for corner-to-corner flow.

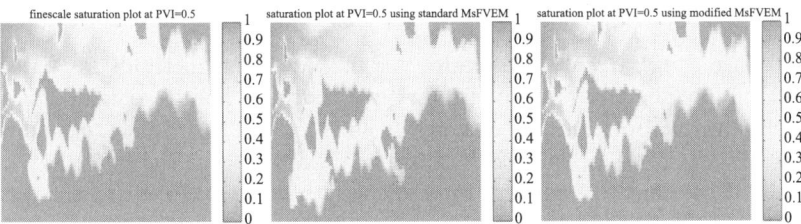

Figure 4.22. Saturation maps at PVI = 0.5 for fine-scale solution (left figure), standard MsFVEM (middle figure), and modified MsFVEM (right figure). Corner-to-corner boundary condition is used.

4.9 Analysis

Galerkin finite element methods with limited global information. We have proposed some analysis for modified multiscale finite element method in [40] and [2]. The main idea is to show that the pressure evolution in two-phase flow simulations is strongly influenced by the initial pressure. To demonstrate this, we consider a channelized permeability field, where the value of the permeability in the channel is large. We assume the permeability has the form kI, where I is an identity matrix. In a channelized medium, the dominant flow is within the channels. Our analysis assumes a single channel and restricted to 2-D. Here, we briefly mention the main findings. Denote the initial stream function and pressure by $\eta = \psi(x, t = 0)$ and $\zeta = p(x, t = 0)$ (ζ is also denoted by p^{sp} previously). The stream function is defined

$$\partial\psi/\partial x_1 = -v_2, \quad \partial\psi/\partial x_2 = v_1. \tag{4.52}$$

Then the equation for the pressure can be written as

$$\frac{\partial}{\partial\eta}\left(|k|^2\lambda(S)\frac{\partial p}{\partial\eta}\right) + \frac{\partial}{\partial\zeta}\left(\lambda(S)\frac{\partial p}{\partial\zeta}\right) = 0. \tag{4.53}$$

For simplicity, $S = 0$ at time zero is assumed. We consider a typical boundary condition that gives high flow within the channel, such that the high flow channel will be mapped into a large slab in (η, ζ) coordinate system. If the heterogeneities within the channel in η direction is not strong (e.g., narrow channel in Cartesian coordinates), the saturation within the channel will depend on ζ. In this case, the leading order pressure will depend only on ζ, and it can be shown that

$$p(\eta, \zeta, t) = p_0(\zeta, t) + \text{high order terms}, \tag{4.54}$$

where $p_0(\zeta, t)$ is the dominant pressure. This asymptotic expansion shows that the time-varying pressure strongly depends on the initial pressure (i.e., the leading order term in the asymptotic expansion is a function of initial pressure and time only). In our analysis, we will assume that $|p(x,t) - \hat{p}(p^{sp}, t)|_{H^1}$ is small.

Since the analysis of the multiscale finite element methods is carried out only for the pressure equation, we will assume t (time) is fixed. Then, assuming the function $\hat{p}$ is sufficiently smooth, one can state the following. There exists A_K in each K, such that $\|\nabla p(x) - A_K \nabla p^{sp}(x)\|_{L^2(\Omega)}$ is small. Note that this assumption indicates that the fine-scale features of pressure solutions of two-phase equations does not change significantly during a simulation (e.g., streamlines do not vary significantly in each coarse block). This phenomena can be observed in numerical simulations of two-phase flows when $\mu_o/\mu_w > 1$.

The assumption for the case with scale separation indicates that the coarse-scale features of two-phase flow and single-phase flow are similar (e.g., coarse-scale streamlines do not vary significantly). We will use the following assumption.

Assumption G. There exists a sufficiently smooth scalar valued function $G(\eta)$ $(G \in C^3)$, such that

$$|p - G(p^{sp})|_{1,\Omega} \leqslant C\delta, \tag{4.55}$$

where p^{sp} is single-phase flow pressure and δ is sufficiently small.

We note that G is $p_0(\zeta, t)$ at fixed t in (4.54). Moreover, one does not need to know the function G for computing the multiscale approximation of the solution. It is only necessary that G has certain smoothness properties, however, it is important that the basis functions span p^{sp} in each coarse block.

Theorem 4.3. *Under Assumption G and $p^{sp} \in W^{1,s}(\Omega)$ $(s > 2)$, multiscale finite element method converges with the rate given by*

$$|p - p_h|_{1,\Omega} \leqslant C\delta + Ch^{1-2/s}|p^{sp}|_{W^{1,s}(\Omega)} + Ch^{1-2/s}|p^{sp}|_{1,\Omega} + Ch\|f\|_{0,\Omega}$$
$$\leqslant C\delta + Ch^{1-2/s}. \tag{4.56}$$

The proof of this theorem is given in [2]. Note that Theorem (4.3) shows that MsFEM converges for problems without any scale separation and the proof of this theorem does not use homogenization techniques. Next, we present the proof.

Proof. Following standard practice of finite element estimation, we seek $q_h = c_i \phi_i^K$, where ϕ_i^K are single-phase flow based multiscale finite element basis functions. Then from Cea's lemma, we have

$$|p - p_h|_{1,\Omega} \leqslant |p - G(p^{sp})|_{1,\Omega} + |G(p^{sp}) - c_i \phi_i^K|_{1,\Omega}. \tag{4.57}$$

Next, we present an estimate for the second term. We choose $c_i = G(p^{sp}(x_i))$, where x_i are vertices of K. Furthermore, using Taylor expansion of G around $\overline{p}_K$, which is the average of p^{sp} over K,

$$G(p^{sp}(x_i)) = G(\overline{p}_K) + G'(\overline{p}_K)(p^{sp}(x_i) - \overline{p}_K) + \frac{1}{2}G''(\xi_{x_i})(p^{sp}(x_i) - \overline{p}_K)^2,$$

where $\xi_{x_i} = \overline{p}_K + \theta(p^{sp}(x_i) - \overline{p}_K)$, $0 < \theta < 1$, we have in each K,

$$\begin{aligned}
c_i \phi_i^K &= G(\overline{p}_K)\phi_i^K + G'(\overline{p}_K)(p^{sp}(x_i) - \overline{p}_K)\phi_i^K \\
&\quad + \frac{1}{2}G''(\xi_{x_i})(p^{sp}(x_i) - \overline{p}_K)^2 \phi_i^K \\
&= G(\overline{p}_K) + G'(\overline{p}_K)(p^{sp}(x_i)\phi_i^K - \overline{p}_K) \\
&\quad + \frac{1}{2}G''(\xi_{x_i})(p^{sp}(x_i) - \overline{p}_K)^2 \phi_i^K. \tag{4.58}
\end{aligned}$$

In the last step, we have used $\sum_i \phi_i^K = 1$. Similarly, in each K,

$$G(p^{sp}(x)) = G(\bar{p}_K) + G'(\bar{p}_K)(p^{sp}(x) - \bar{p}_K) + \frac{1}{2}G''(\xi_x)(p^{sp}(x) - \bar{p}_K)^2, \tag{4.59}$$

where $\xi_x = \bar{p}_K + \theta(p^{sp}(x) - \bar{p}_K)$, $0 < \theta < 1$. Using (4.58) and (4.59), we get

$$
\begin{aligned}
|G(p^{sp}) - c_i\phi_i^K|_{1,K} &\leqslant |G'(\bar{p}_K)(p^{sp}(x) - p^{sp}(x_i)\phi_i^K)|_{1,K} \\
&\quad + |\frac{1}{2}G''(\xi_{x_i})(p^{sp}(x_i) - \bar{p}_K)^2\phi_i^K|_{1,K} \\
&\quad + |\frac{1}{2}G''(\xi_x)(p^{sp}(x) - \bar{p}_K)^2|_{1,K}.
\end{aligned}
\tag{4.60}
$$

Because of $|p^{sp}(x) - p^{sp}(x_i)|_{1,K} \leqslant Ch\|f\|_{0,K}$, the estimate of the first term is the following:

$$|G'(\bar{p}_K)(p^{sp}(x) - p^{sp}(x_i)\phi_i^K)|_{1,K} \leqslant Ch\|f\|_{0,K}.$$

For the second term on the right-hand side of (4.60), assuming $p^{sp}(x) \in W^{1,s}(\Omega)$, we have

$$
\begin{aligned}
|G''(\xi_{x_i})(p^{sp}(x_i) - \bar{p}_K)^2\phi_i^K|_{1,K} &\leqslant Ch^{2-4/s}|p^{sp}|_{W^{1,s}(K)}^2 \\
&\leqslant Ch^{1-2/s}|p^{sp}|_{W^{1,s}(K)},
\end{aligned}
$$

where $s > 2$. Here, we have used the inequality (e.g., [4])

$$|u(x) - u(y)| \leqslant C|x - y|^{1-2/s}|u|_{W^{1,s}}$$

for $s > 2$, where C depends only on s.

For the third term, since G'' and G''' are bounded, we have the following estimate:

$$
\begin{aligned}
&|G''(\xi_x)(p^{sp}(x) - \bar{p}_K)^2|_{1,K} \\
&\quad \leqslant C\|(p^{sp}(x) - \bar{p}_K)^2\nabla p^{sp}(x)\|_{0,K} + C\|(p^{sp}(x) - \bar{p}_K)\nabla p^{sp}(x)\|_{0,K} \\
&\quad \leqslant Ch^{2-4/s}|p^{sp}|_{W^{1,s}(K)}^2|p^{sp}|_{1,K} + Ch^{1-2/s}|p^{sp}|_{1,K} \\
&\quad \leqslant Ch^{2-4/s}|p^{sp}|_{W^{1,s}(\Omega)}^2|p^{sp}|_{1,K} + Ch^{1-2/s}|p^{sp}|_{1,K} \\
&\quad \leqslant Ch^{1-2/s}|p^{sp}|_{1,K}.
\end{aligned}
\tag{4.61}
$$

Combining the above estimates, we have for (4.60),

$$|G(p^{sp}) - c_i\phi_i^K|_{1,K} \leqslant Ch^{1-2/s}|p^{sp}|_{W^{1,s}(K)} + Ch^{1-2/s}|p^{sp}|_{1,K} + Ch\|f\|_{0,K}. \tag{4.62}$$

Summing (4.62) over all K and taking into account Assumption G, we have

$$|p - p_h|_{1,\Omega} \leqslant C\delta + Ch^{1-2/s}|p^{sp}|_{W^{1,s}(\Omega)} + Ch^{1-2/s}|p^{sp}|_{1,\Omega} + Ch\|f\|_{0,\Omega}$$
$$\leqslant C\delta + Ch^{1-2/s}. \tag{4.63}$$

Consequently, if $s > 2$ (see e.g., [9]), single-phase flow based multiscale finite element method converges.

Extensions of Galerkin finite element methods with limited global information. The multiscale finite element methods considered above employ information from only one single-phase flow solution. In general, depending on the source term, boundary data, and mobility $\lambda(S)$ (if it contains sharp variations), it might be necessary to use information from multiple global solutions for the computation of accurate two-phase flow solution. The previous multiscale finite element methods can be extended to take into account additional global information. Next, we present an extension of the Galerkin multiscale finite element method that uses the partition of unity method [12] (also see e.g., [97], [56], [70]).

Assume that $u_1, u_2,..., u_N$ are the global functions such that $|p - G(u_1, u_2, ..., u_N)|_{1,\Omega}$ is sufficiently small. Here, $u_1, ..., u_N$ can be possible pressure snapshots for different mobility $\lambda(S)$ or pressure fields corresponding to different source terms and/or boundary conditions. We would like to note that in a very interesting paper [89], the authors prove under certain conditions on f (source term) and $\lambda = 1$ that p is a smooth function of single-phase flow solutions (elliptic pressure equations) with boundary conditions x_1 and x_2 (it is also extended to multi-dimensional case). In this case, u_1 and u_2 are the solutions of single-phase flow equations with boundary conditions $u_i = x_i$ $(i = 1, 2)$, and it was shown that $p(u_1, u_2) \in H^2$. Next, we will formulate the method.

Let ω_i be a patch (see Figure 4.23), and define ϕ_i^0 to be piecewise linear basis function in patch ω_i, such that $\phi_i^0(x_j) = \delta_{ij}$. For simplicity

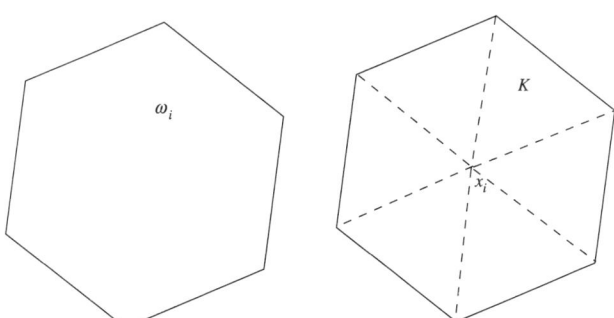

Figure 4.23. Schematic description of patch.

of notation, denote $u_1 = 1$. Then, the multiscale finite element method for each patch ω_i is constructed by

$$\psi_{ij} = \phi_i^0 u_j, \qquad (4.64)$$

where $j = 1, \cdots, N$ and i is the index of nodes (see Figure 4.23). First, we note that in each K, $\sum_{i=1}^{n} \psi_{ij} = u_j$ is the desired single-phase flow solution.

We will use the following assumption. *There exists a sufficiently smooth scalar valued function $G(\eta)$, $\eta \in R^N$ ($G \in C^3$), such that*

$$|p - G(u_1, ..., u_N)|_{1,\Omega} \leqslant C\delta, \qquad (4.65)$$

where δ is sufficiently small.

As before the form of the function G is not important for the computations, however, it is crucial that the basis functions span $u_1,..., u_N$ in each coarse block. The next theorem shows that MsFEM converges for problems without scale separation in this case.

Theorem 4.4. *Assume* (4.65) *and* $u_i \in W^{1,s}(\Omega)$, $s > 2$, $i = 1, ..., N$. *Then*

$$|p - p_h|_{1,\Omega} \leqslant C\delta + Ch^{1-2/s}.$$

The proof of this theorem is given in [2].

Mixed finite element methods with limited global information. One can carry out the analysis of mixed multiscale finite element method with limited global information. First, we re-formulate our assumption for the analysis of mixed multiscale finite element methods. From (4.55), it follows that

$$\|\nabla p - G'(p^{sp})\nabla p^{sp}\|_{0,\Omega} \leqslant C\delta.$$

Using the fact that k and $\lambda(x)$ are bounded, we have

$$\|\lambda(x)k\nabla p - G'(p^{sp})\lambda(x)k\nabla p^{sp}\|_{0,\Omega} \leqslant C\delta.$$

Noting that $u = \lambda(x)k\nabla p$ and $u^{sp} = k\nabla p^{sp}$, it follows that there exists a coarse-scale function scalar $A(x)$ such that

$$\|u - A(x)u^{sp}\|_{0,\Omega} \leqslant \delta. \qquad (4.66)$$

Since $A(x)u^{sp}$ approximates u, we assume that it has small divergence,

$$\left| \int_K \text{div}(A(x)u^{sp})dx \right| \leqslant C\delta_1 h^2 \qquad (4.67)$$

in each K, where δ_1 is a small number. For our analysis, we note that (4.67) gives

$$\left| \int_{\partial K} A(x)u^{sp}n^K ds \right| \leqslant C\delta_1 h^2. \qquad (4.68)$$

We will assume that $A(x) \in C^\gamma$ $(0 < \gamma \leqslant 1)$. (4.68) can be written as

$$\left| \sum_i A_i \int_{\partial e_i^K} u^{sp} n^K ds \right| \leqslant C\delta_1 h^2. \tag{4.69}$$

Here A_i's are defined as

$$A_i = \int_{\partial e_i^K} A(x) u^{sp} n^K ds / \int_{\partial e_i^K} u^{sp} n^K ds,$$

since

$$\int_{\partial e_i^K} u^{sp} n^K ds = \beta_i^K \neq 0.$$

Note that *not* for any $A(x)$, A_i is necessarily a value of $A(x)$ along the edge e_i^K because $u^{sp} n^K$ can change sign. However, we only need to define $A(x)$ for each edge by its value A_i (e.g., the value of $A(x)$ at the center of edge). Then, for any such $A(x)$, (4.66) is satisfied provided $\delta < h^\gamma$. This can be directly verified. Thus, our main assumption will be (4.66) and (4.69), where $A(x)$ is defined, for example, at the center of each edge e_i^K. We would like to note that from the fact that $\mathrm{div}(A(x)u^{sp})$ is small in each K, it follows that $A(x)$, for example, can be taken as an approximation of stream function corresponding to u^{sp}. As before, the form of $A(x)$ is not important for the computations of multiscale solutions.

The following theorem about the convergence of mixed multiscale finite element methods for problems without scale separation is proven in [2].

Theorem 4.5. *Assume (4.66) and (4.69) and $A(x) \in C^\gamma$, $0 < \gamma \leqslant 1$. Let (u, p) and (u_h, p_h) respectively solve problems (4.37) and (4.39) with single-phase flow based mixed multiscale finite element. Then*

$$\|u - u_h\|_{H(div,\Omega)} + \|p - p_h\|_{0,\Omega} \leqslant C\delta + C\delta_1 + Ch^\gamma. \tag{4.70}$$

5 Multiscale finite element methods for nonlinear partial differential equations

Next, we show that MsFEM can be naturally generalized to nonlinear partial differential equations. The goal of MsFEM is to find a numerical approximation of a homogenized solution *without* solving auxiliary problems (e.g., periodic cell problems) that arise in homogenization. The homogenized solutions are sought on a coarse grid space S^h, where $h \gg \epsilon$.

Let K^h be a partition of Ω. We denote by S^h standard family of finite dimensional space, which possesses approximation properties, e.g., piecewise linear functions over triangular elements,

$$S^h = \{v_h \in C^0(\overline{\Omega}) : \text{the restriction } v_h \text{ is linear for each element } K$$
$$\text{and } v_h = 0 \text{ on } \partial\Omega\}. \tag{5.1}$$

In further presentation, K is a triangular element that belongs to K^h. To formulate MsFEM for general nonlinear problems, we will need (1) a *multiscale mapping* that gives us the desired approximation containing the small scale information and (2) a *multiscale numerical formulation* of the equation.

We consider the formulation and analysis of MsFEM for general nonlinear elliptic equations, $u_\epsilon \in W_0^{1,p}(\Omega)$,

$$-\text{div}\, a_\epsilon(x, u_\epsilon, \nabla u_\epsilon) + a_{0,\epsilon}(x, u_\epsilon, \nabla u_\epsilon) = f, \tag{5.2}$$

where $a_\epsilon(x, \eta, \xi)$ and $a_{0,\epsilon}(x, \eta, \xi)$, $\eta \in \mathbb{R}$, $\xi \in \mathbb{R}^d$, satisfy the following assumptions:

$$|a_\epsilon(x, \eta, \xi)| + |a_{0,\epsilon}(x, \eta, \xi)| \leqslant C\,(1 + |\eta|^{p-1} + |\xi|^{p-1}), \tag{5.3}$$

$$(a_\epsilon(x, \eta, \xi_1) - a_\epsilon(x, \eta, \xi_2), \xi_1 - \xi_2) \geqslant C\,|\xi_1 - \xi_2|^p, \tag{5.4}$$

$$(a_\epsilon(x, \eta, \xi), \xi) + a_{0,\epsilon}(x, \eta, \xi)\eta \geqslant C|\xi|^p. \tag{5.5}$$

Denote

$$H(\eta_1, \xi_1, \eta_2, \xi_2, r) = (1 + |\eta_1|^r + |\eta_2|^r + |\xi_1|^r + |\xi_2|^r) \tag{5.6}$$

for arbitrary η_1, $\eta_2 \in \mathbb{R}$, ξ_1, $\xi_2 \in \mathbb{R}^d$, and $r > 0$. We further assume that

$$|a_\epsilon(x, \eta_1, \xi_1) - a_\epsilon(x, \eta_2, \xi_2)| + |a_{0,\epsilon}(x, \eta_1, \xi_1) - a_{0,\epsilon}(x, \eta_2, \xi_2)|$$
$$\leqslant C\,H(\eta_1, \xi_1, \eta_2, \xi_2, p-1)\,\nu(|\eta_1 - \eta_2|)$$
$$+ C\,H(\eta_1, \xi_1, \eta_2, \xi_2, p-1-s)\,|\xi_1 - \xi_2|^s, \tag{5.7}$$

where $\eta \in \mathbb{R}$ and $\xi \in \mathbb{R}^d$, $s > 0$, $p > 1$, $s \in (0, \min(p-1,1))$ and ν is the modulus of continuity, a bounded, concave, and continuous function in $\mathbb{R}_+$ such that $\nu(0) = 0$, $\nu(t) = 1$ for $t \geqslant 1$ and $\nu(t) > 0$ for $t > 0$. These assumptions guarantee the well-posedness of the nonlinear elliptic problem (5.2). Here $\Omega \subset \mathbb{R}^d$ is a Lipschitz domain and ϵ denotes the small scale of the problem. The homogenization of nonlinear partial differential equations has been studied previously (see, e.g., [90]). It can be shown that a solution u_ϵ converges (up to a sub-sequence) to u in an appropriate norm, where $u \in W_0^{1,p}(\Omega)$ is a solution of a homogenized equation

$$-\text{div}\, a^*(x, u, Du) + a_0^*(x, u, Du) = f. \tag{5.8}$$

Multiscale mapping. Introduce the mapping $E^{\text{MsFEM}} : S^h \to V_\epsilon^h$ in the following way. For each element $v_h \in S^h$, $v_{\epsilon,h} = E^{\text{MsFEM}} v_h$ is defined as the solution of

$$-\operatorname{div} a_\epsilon(x, \eta^{v_h}, \nabla v_{\epsilon,h}) = 0 \quad \text{in } K, \tag{5.9}$$

$v_{\epsilon,h} = v_h$ on ∂K and $\eta^{v_h} = \frac{1}{|K|} \int_K v_h dx$ for each K. We would like to point out that different boundary conditions can be chosen to obtain more accurate solutions and this will be discussed later. Note that for linear problems, E^{MsFEM} is a linear operator, where for each $v_h \in S^h$, $v_{\epsilon,h}$ is the solution of the linear problem. Consequently, V_ϵ^h is a linear space that can be obtained by mapping a basis of S^h. This is precisely the construction presented in [62] for linear elliptic equations.

Multiscale numerical formulation. Multiscale finite element formulation of the problem is the following. Find $u_h \in S^h$ (consequently, $u_{\epsilon,h}(= E^{\text{MsFEM}} u_h) \in V_\epsilon^h$) such that

$$\langle A_{\epsilon,h} u_h, v_h \rangle = \int_\Omega f v_h dx \quad \forall v_h \in S^h, \tag{5.10}$$

where

$$\langle A_{\epsilon,h} u_h, v_h \rangle = \sum_{K \in K^h} \int_K ((a_\epsilon(x, \eta^{u_h}, \nabla u_{\epsilon,h}), \nabla v_h)$$
$$+ a_{0,\epsilon}(x, \eta^{u_h}, \nabla u_{\epsilon,h}) v_h) dx. \tag{5.11}$$

Note that the above formulation of MsFEM is a generalization of the Petrov-Galerkin MsFEM introduced in [60] for linear problems. MsFEM, introduced above, can be generalized to different kinds of nonlinear problems and this will be discussed later.

5.1 Multiscale finite volume element method (MsFVEM)

The formulation of multiscale finite element (MsFEM) can be extended to a finite volume method. By its construction, the finite volume method has local conservative properties [53] and it is derived from a local relation, namely the balance equation/conservation expression on a number of subdomains which are called control volumes. Finite volume element method can be considered as a Petrov-Galerkin finite element method, where the test functions are constants defined in a dual grid. Consider a triangle K, and let z_K be its barycenter. The triangle K is divided into three quadrilaterals of equal area by connecting z_K to the midpoints of its three edges. We denote these quadrilaterals by K_z, where $z \in Z_h(K)$

are the vertices of K. Also we denote $Z_h = \bigcup_K Z_h(K)$, and Z_h^0 are all vertices that do not lie on Γ_D, where Γ_D is Dirichlet boundaries. The control volume V_z is defined as the union of the quadrilaterals K_z sharing the vertex z (see Figure 5.1). The multiscale finite volume element method (MsFVEM) is to find $u_h \in S^h$ (consequently, $u_{\epsilon,h} = E^{\text{MsFVEM}} u_h$) such that

$$-\int_{\partial V_z} a_\epsilon\left(x, \eta^{u_h}, \nabla u_{\epsilon,h}\right) \cdot n\, dS + \int_{V_z} a_{0,\epsilon}\left(x, \eta^{u_h}, \nabla u_{\epsilon,h}\right)\, dx$$
$$= \int_{V_z} f\, dx, \quad \forall z \in Z_h^0, \qquad (5.12)$$

where n is the unit normal vector pointing outward on ∂V_z. Note that the number of control volumes that satisfies (5.12) is the same as the dimension of S^h. We will present numerical results for both multiscale finite element and multiscale finite volume element methods.

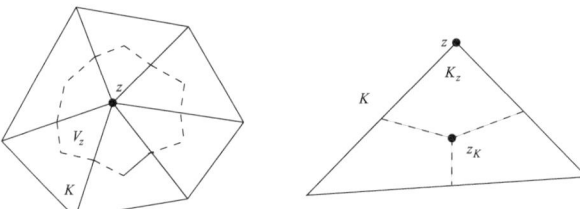

Figure 5.1. *Left*: Portion of triangulation sharing a common vertex z and its control volume. *Right*: Partition of a triangle K into three quadrilaterals.

5.2   Examples of V_ϵ^h

Linear case. For linear operators, V_ϵ^h can be obtained by mapping a basis of S^h. Define a basis of S^h, $S^h = \text{span}(\phi_0^i)$, where ϕ_0^i are standard linear basis functions. In each element $K \in K^h$, we define a set of nodal basis $\{\phi_\epsilon^i\}$, $i = 1, \cdots, n_d$, with $n_d\,(= 3)$ being the number of nodes of the element, satisfying

$$-\text{div}\, a_\epsilon(x) \nabla \phi_\epsilon^i = 0 \quad \text{in } K \in K^h \qquad (5.13)$$

and $\phi_\epsilon^i = \phi_0^i$ on ∂K. Thus, we have

$$V_\epsilon^h = \text{span}\{\phi_\epsilon^i;\ i = 1, \cdots, n_d,\ K \subset K^h\} \subset H_0^1(\Omega).$$

Oversampling technique can be used to improve the method [62].

Special nonlinear case. For the special case,

$$a_\epsilon(x, u_\epsilon, \nabla u_\epsilon) = a_\epsilon(x) b(u_\epsilon) \nabla u_\epsilon,$$

V_ϵ^h can be related to the linear case. Indeed, for this case, the local problems associated with the multiscale mapping E^{MsFEM} (see (5.9)) have the form

$$-\operatorname{div} a_\epsilon(x) b(\eta^{v_h}) \nabla v_{\epsilon,h} = 0 \quad \text{in } K.$$

Because η^{v_h} are constants over K, the local problems satisfy the linear equations,

$$-\operatorname{div} a_\epsilon(x) \nabla \phi_\epsilon^i = 0 \quad \text{in } K,$$

and V_ϵ^h can be obtained by mapping a basis of S^h as it is done for the first example. Thus, for this case one can construct the base functions in the beginning of the computations.

V_ϵ^h **using subdomain problems.** One can use the solutions of smaller (than $K \in \mathbf{K}^h$) subdomain problems to approximate the solutions of the local problems (5.9). This can be done in various ways based on a homogenization expansion. For example, instead of solving (5.9) we can solve (5.9) in a subdomain S with boundary conditions v_h restricted onto the subdomain boundaries, ∂S. Then the gradient of the solution in a subdomain can be extended periodically to K to approximate $\nabla v_{\epsilon,h}$ in (5.11). $v_{\epsilon,h}$ can be easily reconstructed based on $\nabla v_{\epsilon,h}$. When the multiscale coefficient has a periodic structure, the multiscale mapping can be constructed over one periodic cell with a specified average.

5.3 Convergence of MsFEM for nonlinear partial differential equations

In [44] it was shown using G-convergence theory that

$$\lim_{h \to 0} \lim_{\epsilon \to 0} \|u_h - u\|_{W_0^{1,p}(\Omega)} = 0, \tag{5.14}$$

(up to a subsequence) where u is a solution of (5.8) and u_h is an Ms-FEM solution given by (5.10). This result can be obtained without any assumption on the nature of the heterogeneities and cannot be improved because there could be infinitely many scales, $\alpha(\epsilon)$, present such that $\alpha(\epsilon) \to 0$ as $\epsilon \to 0$.

For the periodic case, it can be shown that the convergence of MsFEM in the limit as $\epsilon/h \to 0$. To show the convergence for $\epsilon/h \to 0$, we consider $h = h(\epsilon)$, such that $h(\epsilon) \gg \epsilon$ and $h(\epsilon) \to 0$ as $\epsilon \to 0$. We would like to note that this limit as well as the proof of the periodic case is different from (5.14), where the double-limit is taken. In contrast to the proof of (5.14), the proof of the periodic case requires the correctors for the solutions of the local problems.

Next we will present the convergence results for MsFEM solutions. For general nonlinear elliptic equations under assumptions (5.3)–(5.7)

the strong convergence of MsFEM solutions can be shown. In the proof
of this theorem we show the form of the truncation error (in a weak sense)
in terms of the resonance errors between the mesh size and small scale
ϵ. The resonance errors are derived explicitly. To obtain the convergence
rate from the truncation error, one needs some lower bounds. Under the
general conditions, such as (5.3)–(5.7), one can prove strong convergence
of MsFEM solutions without an explicit convergence rate (cf. [95]). To
convert the obtained convergence rates for the truncation errors into the
convergence rate of MsFEM solutions, additional assumptions, such as
monotonicity, are needed.

Next, we formulate convergence theorems. The proofs can be found
in [41].

Theorem 5.1. *Assume that $a_\epsilon(x, \eta, \xi)$ and $a_{0,\epsilon}(x, \eta, \xi)$ are periodic functions with respect to x and let u be a solution of (5.8) and u_h is an
MsFEM solution given by (5.10). Moreover, we assume that ∇u_h is uniformly bounded in $L^{p+\alpha}(\Omega)$ for some $\alpha > 0$. Then*

$$\lim_{\epsilon \to 0} \|u_h - u\|_{W_0^{1,p}(\Omega)} = 0, \tag{5.15}$$

where $h = h(\epsilon) \gg \epsilon$ and $h \to 0$ as $\epsilon \to 0$ (up to a subsequence).

Theorem 5.2. *Let u and u_h be the solutions of the homogenized problem
(5.8) and MsFEM (5.10), respectively, with the coefficient $a_\epsilon(x, \eta, \xi) =
a(x/\epsilon, \xi)$ and $a_{0,\epsilon} = 0$. Then*

$$\|u_h - u\|_{W_0^{1,p}(\Omega)}^p \leq c \left(\frac{\epsilon}{h}\right)^{\frac{s}{(p-1)(p-s)}} + c \left(\frac{\epsilon}{h}\right)^{\frac{p}{p-1}} + ch^{\frac{p}{p-1}}. \tag{5.16}$$

5.4 Multiscale finite element methods for nonlinear parabolic equations

We consider

$$\frac{\partial}{\partial t} u_\epsilon - \operatorname{div}\left(a_\epsilon(x, t, u_\epsilon, \nabla u_\epsilon)\right) + a_{0,\epsilon}(x, t, u_\epsilon, \nabla u_\epsilon) = f, \tag{5.17}$$

where ϵ is a small scale. Our motivation in considering (5.17) mostly
stems from the applications of flow in porous media (multi-phase flow
in saturated porous media, flow in unsaturated porous media) though
many applications of nonlinear parabolic equations of these kinds occur
in transport problems. Many problems in subsurface modeling have multiscale nature where the heterogeneities associated with the media is no
longer periodic. It was shown that a solution u_ϵ converges to u (up to a
subsequence) in an appropriate sense where u is a solution of

$$\frac{\partial}{\partial t} u - \operatorname{div}\left(a^*(x, t, u, \nabla u)\right) + a_0^*(x, t, u, \nabla u) = f. \tag{5.18}$$

In [43] the homogenized fluxes a^* and a_0^* are computed under the assumption that the heterogeneities are strictly stationary random fields with respect to both space and time.

The numerical homogenization procedure presented in the previous section can be extended to parabolic equations. To do this we will first formulate MsFEM in a slightly different manner from that presented in [62] for *the linear problem*. Consider a standard finite dimensional S^h space over a coarse triangulation of Ω, (5.1) and define $E^{\text{MsFEM}} : S^h \to V_\epsilon^h$ in the following way. For each $u_h \in S^h$, there is a corresponding element $u_{h,\epsilon}$ in V_ϵ^h that is defined by

$$\frac{\partial}{\partial t} u_{h,\epsilon} - \text{div}\,(a_\epsilon(x,t)\nabla u_{h,\epsilon}) = 0 \quad \text{in } K \times [t_n, t_{n+1}], \qquad (5.19)$$

with boundary condition $u_{h,\epsilon} = u_h$ on ∂K, and $u_{h,\epsilon}(t = t_n) = u_h$. For the linear equations E^{MsFEM} is a linear operator and the obtained multiscale space, V_ϵ^h, is a linear space on $\Omega \times [t_n, t_{n+1}]$. Moreover, the basis in the space V_ϵ^h can be obtained by mapping the basis functions of S^h. For *the nonlinear parabolic equations* considered in this paper the operator E^{MsFEM} is constructed similar to (5.19) using the local problems, i.e., for each $u_h \in S^h$, there is a corresponding element $u_{h,\epsilon}$ in V_ϵ^h that is defined by

$$\frac{\partial}{\partial t} u_{h,\epsilon} - \text{div}\,(a_\epsilon(x, t, \eta, \nabla u_{h,\epsilon})) = 0 \quad \text{in } K \times [t_n, t_{n+1}], \qquad (5.20)$$

with boundary condition $u_{h,\epsilon} = u_h$ on ∂K, and $u_{h,\epsilon}(t = t_n) = u_h$. Here $\eta = \frac{1}{|K|} \int_K u_h dx$. Note that E^{MsFEM} is a nonlinear operator and V_ϵ^h is no longer a linear space.

The following method that can be derived from general multiscale finite element framework is equivalent to our numerical homogenization procedure. Find $u_h \in V_\epsilon^h$ such that

$$\int_{t_n}^{t_{n+1}} \int_\Omega \frac{\partial}{\partial t} u_h v_h dx dt + A(u_h, v_h) = \int_{t_n}^{t_{n+1}} \int_\Omega f v_h dx dt, \quad \forall v_h \in S^h,$$

where

$$A(u_h, w_h) = \sum_K \int_{t_n}^{t_{n+1}} \int_K ((a_\epsilon(x, t, \eta^{u_h}, \nabla v_\epsilon), \nabla w_h)$$
$$+ a_{0,\epsilon}(x, t, \eta^{u_h}, \nabla v_\epsilon) w_h) dx dt,$$

where v_ϵ is the solution of the local problem (5.20), $u_h = l^{u_h}$ in each K, $\eta^{u_h} = \frac{1}{|K|} \int_K l^{u_h} dx$, and u_h is known at $t = t_n$.

We would like to note that the operator E^{MsFEM} can be constructed using larger domains as it is done in MsFEM with oversampling [62]. This way one reduces the effects of the boundary conditions and initial conditions. In particular, for the temporal oversampling it is only sufficient to start the computations before t_n and end them at t_{n+1}. Consequently, the oversampling domain for $K \times [t_n, t_{n+1}]$ consists of $[\tilde{t}_n, t_{n+1}] \times S$, where $\tilde{t}_n < t_n$ and $K \subset S$. More precise formulation and detail numerical studies of oversampling technique for nonlinear equations are currently under investigation. Further we would like to note that oscillatory initial conditions can be imposed (without using oversampling techniques) based on the solution of the elliptic part of the local problems (5.20). These initial conditions at $t = t_n$ are the solutions of

$$-\mathrm{div}\,(a_\epsilon(x, t, \eta, \nabla u_{h,\epsilon})) = 0 \quad \text{in } K, \tag{5.21}$$

or

$$-\mathrm{div}\,(\bar{a}_\epsilon(x, \eta, \nabla u_{h,\epsilon})) = 0 \quad \text{in } K, \tag{5.22}$$

where

$$\bar{a}_\epsilon(x, \eta, \xi) = \frac{1}{t_{n+1} - t_n} \int_{t_n}^{t_{n+1}} a(T(x/\epsilon^\beta, \tau/\epsilon^\alpha)\omega, \eta, \xi)d\tau$$

and $u_{h,\epsilon} = u_h$ on ∂K. The latter can become efficient depending on the inter-play between the temporal and spatial scales. This issue is discussed below.

Note that in the case of periodic media the local problems can be solved in a single period in order to construct $A(u_h, v_h)$. In general, one can solve the local problems in a domain different from K (an element) to calculate $A(u_h, v_h)$, and our analysis is applicable to these cases. Note that the numerical advantages of our approach over the fine scale simulation is similar to that of MsFEM. In particular, for each Newton's iteration a linear system of equations on *a coarse grid* is solved.

For some special cases the operator E^{MsFEM} introduced in the previous section can be simplified (see [44]). In general one can avoid solving the local parabolic problems if the ratio between temporal and spatial scales is known, and solve instead a simplified equation. For example, assuming that $a_\epsilon(x, t, \eta, \xi) = a(x/\epsilon^\beta, t/\epsilon^\alpha, \eta, \xi)$, we have the following. If $\alpha < 2\beta$ one can solve instead of (5.20) the local problem $-\mathrm{div}\,(a_\epsilon(x, t, \eta^{u_h}, \nabla v_\epsilon)) = 0$, if $\alpha > 2\beta$ one can solve instead of (5.20) the local problem $-\mathrm{div}\,(\bar{a}_\epsilon(x, \eta^{u_h}, \nabla v_\epsilon)) = 0$, where $\bar{a}_\epsilon(x, \eta, \xi)$ is an average over time of $a_\epsilon(x, t, \eta, \xi)$, while if $\alpha = 2\beta$ we need to solve the parabolic equation in $K \times [t_n, t_{n+1}]$, (5.20).

We would like to note that, in general, one can use (5.21) or (5.22) as oscillatory initial conditions, and these initial conditions can be efficient for some cases. For example, for $\alpha > 2\beta$ with initial conditions given by (5.22) the solutions of the local problems (5.20) can be computed easily since they are approximated by (5.22). Moreover, one can expect better accuracy with (5.22) for the case $\alpha > 2\beta$ because this initial condition is more compatible with the local heterogeneities compare to the artificial linear initial conditions (cf. (5.20)). The comparison of various oscillatory initial conditions including the ones obtained by oversampling method is a subject of future studies.

Finally, we would like to mention that one can prove the following theorem.

Theorem 5.3. $u_h = \sum_i \theta_i(t)\phi_i^0(x)$ *converges to u, a solution of the homogenized equation in* $V_0 = L^p(0, T, W_0^{1,p}(\Omega))$ *as* $\lim_{h \to 0} \lim_{\epsilon \to 0}$ *under additional not restrictive assumptions (see* [44]*).*

Remark 5.1. The proof of the theorem uses the convergence of the solutions and the fluxes, and consequently it is applicable for the case of general heterogeneities that uses G-convergence theory. Since the G-convergence of the operators occurs up to a subsequence, the numerical solution converges to a solution of a homogenized equation (up to a subsequence of ϵ).

5.5   Numerical results

In this section we present several ingredients pertaining to the implementation of multiscale finite element method for nonlinear elliptic equations. More numerical examples relevant to subsurface applications can be found in [41]. We will present numerical results for both MsFEM and multiscale finite volume element method (MsFVEM). We use an inexact-Newton algorithm as an iterative technique to tackle the nonlinearity. For the numerical examples below, we use $a_\epsilon(x, u_\epsilon, \nabla u_\epsilon) = a_\epsilon(x, u_\epsilon)\nabla u_\epsilon$. Let $\{\phi_0^i\}_{i=1}^{N_{dof}}$ be the standard piecewise linear basis functions of S^h. Then MsFEM solution may be written as

$$u_h = \sum_{i=1}^{N_{dof}} \alpha_i\, \phi_0^i \tag{5.23}$$

for some $\alpha = (\alpha_1, \alpha_2, \cdots, \alpha_{N_{dof}})^T$, where α_i depends on ϵ. Hence, we need to find α such that

$$F(\alpha) = 0, \tag{5.24}$$

where $F : \mathbb{R}^{N_{dof}} \to \mathbb{R}^{N_{dof}}$ is a nonlinear operator such that

$$F_i(\alpha) = \sum_{K \in K^h} \int_K (a_\epsilon(x, \eta^{u_h})\nabla u_{\epsilon,h}, \nabla \phi_0^i)\, dx - \int_\Omega f\, \phi_0^i\, dx. \tag{5.25}$$

We note that in (5.25) α is implicitly buried in η^{u_h} and $u_{\epsilon,h}$. An inexact-Newton algorithm is a variation of Newton's iteration for nonlinear system of equations, where the Jacobian system is only approximately solved. To be specific, given an initial iterate α^0, for $k = 0, 1, 2, \cdots$, until convergence do the following:

- Solve $F'(\alpha^k)\delta^k = -F(\alpha^k)$ by some iterative technique until $\|F(\alpha^k) + F'(\alpha^k)\delta^k\| \leqslant \beta_k \|F(\alpha^k)\|$.
- Update $\alpha^{k+1} = \alpha^k + \delta^k$.

In this algorithm $F'(\alpha^k)$ is the Jacobian matrix evaluated at iteration k. We note that when $\beta_k = 0$ then we have recovered the classical Newton iteration. Here we have used

$$\beta_k = 0.001 \left(\frac{\|F(\alpha^k)\|}{\|F(\alpha^{k-1})\|} \right)^2, \tag{5.26}$$

with $\beta_0 = 0.001$. Choosing β_k this way, we avoid over-solving the Jacobian system when α^k is still considerably far from the exact solution.

Next we present the entries of the Jacobian matrix. For this purpose, we use the following notations. Let $K_i^h = \{K \in K^h : z_i \text{ is a vertex of } K\}$, $I^i = \{j : z_j \text{ is a vertex of } K \in K_i^h\}$, and $K_{ij}^h = \{K \in K_i^h : K \text{ shares } \overline{z_i z_j}\}$. We note that we may write $F_i(\alpha)$ as follows:

$$F_i(\alpha) = \sum_{K \in K_i^h} \left(\int_K (a_\epsilon(x, \eta^{u_h}) \nabla u_{\epsilon,h}, \nabla \phi_0^i) \, dx - \int_K f \, \phi_0^i \, dx \right), \tag{5.27}$$

with

$$-\operatorname{div} a_\epsilon(x, \eta^{u_h}) \nabla u_{\epsilon,h} = 0 \quad \text{in } K \quad \text{and} \quad u_{\epsilon,h} = \sum_{z_m \in Z_K} \alpha_m \phi_0^m \quad \text{on } \partial K,$$
$$\tag{5.28}$$

where Z_K is all the vertices of element K. It is apparent that $F_i(\alpha)$ is not fully dependent on all $\alpha_1, \alpha_2, \cdots, \alpha_d$. Consequently, $\frac{\partial F_i(\alpha)}{\partial \alpha_j} = 0$ for $j \notin I^i$. To this end, we denote $\psi_\epsilon^j = \frac{\partial u_{\epsilon,h}}{\partial \alpha_j}$. By applying chain rule of differentiation to (5.28) we have the following local problem for ψ_ϵ^j:

$$-\operatorname{div} a_\epsilon(x, \eta^{u_h}) \nabla \psi_\epsilon^j = \frac{1}{3} \operatorname{div} \frac{\partial a_\epsilon(x, \eta^{u_h})}{\partial u} \nabla u_{\epsilon,h} \text{ in } K \quad \text{and} \quad \psi_\epsilon^j = \phi_\epsilon^j \text{ on } \partial K.$$
$$\tag{5.29}$$

The fraction $1/3$ comes from taking the derivative in the chain rule of differentiation. In the formulation of the local problem, we have replaced the nonlinearity in the coefficient by η^{v_h}, where for each triangle K, $\eta^{v_h} = 1/3 \sum_{i=1}^3 \alpha_i^K$, which gives $\partial \eta^{v_h}/\partial \alpha_i = 1/3$. Moreover, for a rectangular element the fraction $1/3$ should be replaced by $1/4$.

Thus, provided that $v_{\epsilon,h}$ has been computed, then we may compute ψ_ϵ^j using (5.29). Using the above descriptions we have the expressions for the entries of the Jacobian matrix:

$$\frac{\partial F_i}{\partial \alpha_i} = \sum_{K \in K_i^h} \left(\frac{1}{3} \int_K \left(\frac{\partial a_\epsilon(x, \eta^{u_h})}{\partial u} \nabla u_{\epsilon,h}, \nabla \phi_0^i \right) dx \right.$$
$$\left. + \int_K (a_\epsilon(x, \eta^{u_h}) \nabla \psi_i, \nabla \phi_0^i) \, dx \right), \qquad (5.30)$$

$$\frac{\partial F_i}{\partial \alpha_j} = \sum_{K \in K_{ij}^h} \left(\frac{1}{3} \int_K \left(\frac{\partial a_\epsilon(x, \eta^{u_h})}{\partial u} \nabla u_{\epsilon,h}, \nabla \phi_\epsilon^i \right) dx \right.$$
$$\left. + \int_K (a_\epsilon(x, \eta^{u_h}) \nabla \psi_\epsilon^j, \nabla \phi_0^i) \, dx \right) \qquad (5.31)$$

for $j \neq i$, $j \in I^i$.

The implementation of the oversampling technique is similar to the procedure presented earlier, except the local problems in larger domains are used. As in the non-oversampling case, we denote $\psi_\epsilon^j = \frac{\partial v_{\epsilon,h}}{\partial \alpha_j}$, such that after applying chain rule of differentiation to the local problem we have

$$-\mathrm{div}\, a_\epsilon(x, \eta^{u_h}) \nabla \psi_\epsilon^j = \frac{1}{3} \mathrm{div}\, \frac{\partial a_\epsilon(x, \eta^{u_h})}{\partial u} \nabla v_{\epsilon,h} \text{ in } S \text{ and } \psi_\epsilon^j = \phi_0^j \text{ on } \partial S,$$
$$(5.32)$$

where η^{u_h} is computed over the corresponding element K and ϕ_0^j is understood as the nodal basis functions on oversampled domain S. Then all the rest of the inexact-Newton algorithms are the same as in the non-oversampling case. Specifically, we also use (5.30) and (5.31) to construct the Jacobian matrix of the system. We note that we will only use ψ_ϵ^j from (5.32) pertaining to the element K.

From the derivation (both for oversampling and non-oversampling) it is obvious that the Jacobian matrix is not symmetric but sparse. Computation of this Jacobian matrix is similar to computing the stiffness matrix resulting from standard finite element, where each entry is formed by accumulation of element by element contribution. Once we have the matrix stored in memory, then its action to a vector is straightforward. Because it is a sparse matrix, devoting some amount of memory for entries storage is inexpensive. The resulting linear system is solved using preconditioned bi-conjugate gradient stabilized method.

We want to solve the following problem:

$$-\mathrm{div}\, a(x/\epsilon, u_\epsilon) \nabla u_\epsilon = -1 \quad \text{in } \Omega \subset \mathbb{R}^2,$$
$$u_\epsilon = 0 \quad \text{on } \partial \Omega, \qquad (5.33)$$

where $\Omega = [0,1] \times [0,1]$, $a(x/\epsilon, u_\epsilon) = k(x/\epsilon)/(1 + u_\epsilon)^{l(x/\epsilon)}$, with

$$k(x/\epsilon) = \frac{2 + 1.8\sin(2\pi x_1/\epsilon)}{2 + 1.8\cos(2\pi x_2/\epsilon)} + \frac{2 + \sin(2\pi x_2/\epsilon)}{2 + 1.8\cos(2\pi x_1/\epsilon)} \qquad (5.34)$$

and $l(x/\epsilon)$ is generated from $k(x/\epsilon)$ such that the average of $l(x/\epsilon)$ over Ω is 2. Here we use $\epsilon = 0.01$. Because the exact solution for this problem is not available, we use a well resolved numerical solution using standard finite element method as a reference solution. The resulting nonlinear system is solved using inexact-Newton algorithm. The reference solution is solved on 512×512 mesh. Tables 5.2 and 5.4 present the relative errors of the solution with and without oversampling, respectively. In Tables 5.3 and 5.5, the relative errors for multiscale finite volume element method are presented. The relative errors are computed as the corresponding error divided by the norm of the solution. In each table, the second, third, and fourth columns list the relative error in L^2, H^1, and L^∞ norm, respectively. As we can see from these two tables, the oversampling significantly improves the accuracy of the multiscale method.

Table 5.2. Relative MsFEM errors without oversampling.

N	L^2-norm		H^1-norm		L^∞-norm	
	Error	Rate	Error	Rate	Error	Rate
32	0.029		0.115		0.03	
64	0.053	-0.85	0.156	-0.44	0.0534	-0.94
128	0.10	-0.94	0.234	-0.59	0.10	-0.94

Table 5.3. Relative MsFVEM errors without oversampling.

N	L^2-norm		H^1-norm		L^∞-norm	
	Error	Rate	Error	Rate	Error	Rate
32	0.03		0.13		0.04	
64	0.05	-0.65	0.19	-0.60	0.05	-0.24
128	0.058	-0.19	0.25	-0.35	0.057	-0.19

Table 5.4. Relative MsFEM errors with oversampling.

N	L^2-norm		H^1-norm		L^∞-norm	
	Error	Rate	Error	Rate	Error	Rate
32	0.0016		0.036		0.0029	
64	0.0012	0.38	0.019	0.93	0.0016	0.92
128	0.0024	-0.96	0.0087	1.14	0.0026	-0.71

Table 5.5. Relative MsFVEM errors with oversampling.

N	L^2-norm		H^1-norm		L^∞-norm	
	Error	Rate	Error	Rate	Error	Rate
32	0.002		0.038		0.005	
64	0.003	-0.43	0.021	0.87	0.003	0.72
128	0.001	1.10	0.009	1.09	0.001	1.08

For our next example, we consider the problem with non-periodic coefficients, where $a_\epsilon(x, \eta) = k_\epsilon(x)/(1 + \eta)^{\alpha_\epsilon(x)}$. $k_\epsilon(x) = \exp(\beta_\epsilon(x))$ is chosen such that $\beta_\epsilon(x)$ is a realization of a random field with the spherical variogram [30] and with the correlation lengths $l_x = 0.2$, $l_y = 0.02$ and with the variance $\sigma = 1$. $\alpha_\epsilon(x)$ is chosen such that $\alpha_\epsilon(x) = k_\epsilon(x) + const$ with the spatial average of 2. As for the boundary conditions we use "left-to-right flow" in $\Omega = [0, 5] \times [0, 1]$ domain, $u_\epsilon = 1$ at the inlet $(x_1 = 0)$, $u_\epsilon = 0$ at the outlet $(x_1 = 5)$, and no flow boundary conditions on the lateral sides $x_2 = 0$ and $x_2 = 1$. In Table 5.6 we present the relative error for multiscale method with oversampling. Similarly, in

Table 5.6. Relative MsFEM errors for random heterogeneities, spherical variogram, $l_x = 0.20$, $l_z = 0.02$, $\sigma = 1.0$.

N	L^2-norm		H^1-norm		L^∞-norm		hor. flux	
	Error	Rate	Error	Rate	Error	Rate	Error	Rate
32	0.0006		0.0505		0.0025		0.025	
64	0.0002	1.58	0.029	0.8	0.001	1.32	0.017	0.57
128	0.0001	1	0.016	0.85	0.0005	1	0.011	0.62

Table 5.7 we present the relative error for multiscale finite volume method with oversampling. Clearly, the oversampling method captures the effects induced by the large correlation features. Both H^1 and horizontal flux errors are under five percent. Similar results have been observed for various kinds of non-periodic heterogeneities. In the next set of numerical examples, we test MsFEM for problems with fluxes $a_\epsilon(x, \eta)$ that are discontinuous in space. The discontinuity in the fluxes is introduced by multiplying the underlying permeability function, $k_\epsilon(x)$, by a constant in certain regions, while leaving it unchanged in the rest of the domain. As an underlying permeability field, $k_\epsilon(x)$, we choose the random field used for the results in Table 5.6. In the first set of examples, the discontinuities are introduced along the boundaries of the coarse elements. In particular, $k_\epsilon(x)$ on the left half of the domain is multiplied by a constant J, where $J = \exp(1)$, or $\exp(2)$, or $\exp(4)$. The results in Tables 5.8–5.10 show that MsFEM converges and the error falls below five percent for relatively large coarsening. For the second set of examples (Tables 5.11–5.13), the discontinuities are not aligned with the boundaries of the coarse elements. In particular, the discontinuity boundary is given by $y = x\sqrt{2} + 0.5$, i.e., the discontinuity line intersects the coarse

Table 5.7. Relative MsFVEM errors for random heterogeneities, spherical variogram, $l_x = 0.20$, $l_z = 0.02$, $\sigma = 1.0$.

N	L^2-norm		H^1-norm		L^∞-norm		hor. flux	
	Error	Rate	Error	Rate	Error	Rate	Error	Rate
32	0.0006		0.0515		0.0025		0.027	
64	0.0002	1.58	0.029	0.81	0.0013	0.94	0.018	0.58
128	0.0001	1	0.016	0.85	0.0005	1.38	0.012	0.58

Table 5.8. Relative MsFEM errors for random heterogeneities, spherical variogram, $l_x = 0.20$, $l_z = 0.02$, $\sigma = 1.0$, aligned discontinuity, jump = $\exp(1)$.

N	L^2-norm		H^1-norm		L^∞-norm		hor. flux	
	Error	Rate	Error	Rate	Error	Rate	Error	Rate
32	0.0006		0.0641		0.0020		0.039	
64	0.0002	1.58	0.0382	0.75	0.0010	1.00	0.027	0.53
128	0.0001	1.00	0.0210	0.86	0.0005	1.00	0.018	0.59

Table 5.9. Relative MsFEM errors for random heterogeneities, spherical variogram, $l_x = 0.20$, $l_z = 0.02$, $\sigma = 1.0$, aligned discontinuity, jump = $\exp(2)$.

N	L^2-norm		H^1-norm		L^∞-norm		hor. flux	
	Error	Rate	Error	Rate	Error	Rate	Error	Rate
32	0.0008		0.0817		0.0040		0.061	
64	0.0004	1.00	0.0493	0.73	0.0023	0.80	0.041	0.57
128	0.0002	1.00	0.0256	0.95	0.0011	1.06	0.025	0.71

Table 5.10. Relative MsFEM errors for random heterogeneities, spherical variogram, $l_x = 0.20$, $l_z = 0.02$, $\sigma = 1.0$, aligned discontinuity, jump = $\exp(4)$.

N	L^2-norm		H^1-norm		L^∞-norm		hor. flux	
	Error	Rate	Error	Rate	Error	Rate	Error	Rate
32	0.0011		0.1010		0.0068		0.195	
64	0.0006	0.87	0.0638	0.66	0.0045	0.59	0.109	0.84
128	0.0003	1.00	0.0349	0.87	0.0024	0.91	0.063	0.79

Table 5.11. Relative MsFEM errors for random heterogeneities, spherical variogram, $l_x = 0.20$, $l_z = 0.02$, $\sigma = 1.0$, nonaligned discontinuity, jump = $\exp(1)$.

N	L^2-norm		H^1-norm		L^∞-norm		hor. flux	
	Error	Rate	Error	Rate	Error	Rate	Error	Rate
32	0.0006		0.0623		0.0023		0.035	
64	0.0002	1.58	0.0366	0.77	0.0014	0.72	0.024	0.54
128	0.0001	1.00	0.0203	0.85	0.0006	1.22	0.016	0.59

Table 5.12. Relative MsFEM errors for random heterogeneities, spherical variogram, $l_x = 0.20$, $l_z = 0.02$, $\sigma = 1.0$, nonaligned discontinuity, jump = $\exp(2)$.

N	L^2-norm		H^1-norm		L^∞-norm		hor. flux	
	Error	Rate	Error	Rate	Error	Rate	Error	Rate
32	0.0010		0.0785		0.0088		0.052	
64	0.0003	1.74	0.0440	0.84	0.0052	0.76	0.031	0.75
128	0.0001	1.59	0.0239	0.88	0.0022	1.24	0.017	0.87

grid blocks. Similar to the aligned case, various jump magnitudes are considered. These results demonstrate the robustness of our approach

Table 5.13. Relative MsFEM errors for random heterogeneities, spherical variogram, $l_x = 0.20$, $l_z = 0.02$, $\sigma = 1.0$, nonaligned discontinuity, jump = $\exp(4)$.

N	L^2-norm		H^1-norm		L^∞-norm		hor. flux	
	Error	Rate	Error	Rate	Error	Rate	Error	Rate
32	0.0067		0.1775		0.1000		0.164	
64	0.0016	2.07	0.0758	1.23	0.0288	1.80	0.077	1.09
128	0.0009	0.83	0.0687	0.14	0.0423	-0.55	0.039	0.98

for anisotropic fields where h and ϵ are nearly the same, and the fluxes that are discontinuous spatial functions.

As for CPU comparisons, we have observed more than 92 percent CPU savings when using MsFEM without oversampling. With the oversampling approach, the CPU savings depend on the size of the oversampled domain. For example, if the oversampled domain size is two times larger than the target coarse block (half coarse block extension on each side) we have observed 70 percent CPU savings for 64×64 and 80 percent CPU savings for 128×128 coarse grid. In general, the computational cost will decrease if the oversampled domain size is close to the target coarse block size, and this cost will be close to the cost of MsFEM without oversampling. Conversely, the error decreases if the size of the oversampled domains increases. In the numerical examples studied in our paper, we have observed the same errors for the oversampling methods using either one coarse block extension or half coarse block extensions. The latter indicates that the leading resonance error is eliminated by using a smaller oversampled domain. Oversampled domains with one coarse block extension are previously used in simulations of flow through heterogeneous porous media. As it is indicated in [62], one can use large oversampled domains for simultaneous computations of the several local solutions. Moreover, parallel computations will improve the speed of the method because MsFEM is well suited for parallel computation [62]. For the problems where $a_\epsilon(x, \eta, \xi) = a_\epsilon(x)b(\eta)\xi$ (see section 5.2 and the next section for applications) our multiscale computations are very fast because the base functions are built in the beginning of the computations. In this case, we have observed more than 95 percent CPU savings.

Applications of MsFEM to Richards' equation are presented in [41].

5.6   Generalizations of MsFEM and some remarks

Next, we present the framework of MsFEM for general equations. Consider

$$L_\epsilon u_\epsilon = f, \qquad (5.35)$$

where ϵ is a small scale and $L_\epsilon : X \to Y$ is an operator. Moreover, we assume that L_ϵ G-converges to L^* (up to a sub-sequence), where u is a solution of

$$L^* u = f, \tag{5.36}$$

(we refer to [90], page 14 for the definition of G-convergence for operators). The objective of MsFEM is to approximate u in S^h. Denote S^h a family of finite dimensional space such that it possesses an approximation property (see [107], [91]) as before. Here h is a scale of computation and $h \gg \epsilon$. For (5.35) *multiscale mapping*, $E^{\mathrm{MsFEM}} : S^h \to V_\epsilon^h$, will be defined as follows. For each element $v_h \in S^h$, $v_{\epsilon,h} = E^{\mathrm{MsFEM}} v_h$ is defined as

$$L_\epsilon^{\mathrm{map}} v_{\epsilon,h} = 0 \quad \text{in } K, \tag{5.37}$$

where $L_\epsilon^{\mathrm{map}}$ can be, in general, different from L_ϵ and allows us to capture the effects of the small scales. Moreover, the domains different from the target coarse block K can be used in the computations of the local solutions. To solve (5.37) one needs to impose boundary and initial conditions. This issue needs to be resolved on a case by case basis, and the main idea is to interpolate v_h onto the underlying fine grid. Further, we seek a solution of (5.35) in V_ϵ^h as follows. Find $u_h \in S^h$ (consequently $u_{\epsilon,h} \in V_\epsilon^h$) such that

$$\langle L_\epsilon^{\mathrm{global}} u_{\epsilon,h}, v_h \rangle = \langle f, v_h \rangle, \quad \forall v_h \in S^h, \tag{5.38}$$

where $\langle u, v \rangle$ denotes the duality between X and Y, and $L_\epsilon^{\mathrm{global}}$ can be, in general, different from L_ϵ. For example, for nonlinear elliptic equations we have $L_\epsilon u = -\mathrm{div}\, a_\epsilon(x, u, \nabla u) + a_{0,\epsilon}(x, u, \nabla u)$, $L_\epsilon^{\mathrm{map}} u = \mathrm{div}\, a_\epsilon(x, \eta^u, \nabla u)$ in K, and $L_\epsilon^{\mathrm{global}} = \mathrm{div}\, a_\epsilon(x, \eta^u, \nabla u) + a_{0,\epsilon}(x, \eta^u, \nabla u)$ in K. The convergence of MsFEM is to show that $u_h \to u$ and $u_{\epsilon,h} \to u_\epsilon$, where $u_{\epsilon,h} = E^{\mathrm{MsFEM}} u_h$ in appropriate space. The correct choices of $L_\epsilon^{\mathrm{map}}$ and $L_\epsilon^{\mathrm{global}}$ are the essential part of MsFEM and guarantees the convergence of the method.

In conclusion, we have presented a natural extension of MsFEM to nonlinear problems. This is accomplished by considering a multiscale map instead of the base functions that are considered in linear Ms-FEM [62]. Our approaches share some common elements with recently introduced HMM [37], where macroscopic and microscopic solvers are also needed. In general, the finding of "correct" macroscopic and microscopic solvers is the main difficulty of the multiscale methods. Our approaches follow MsFEM and, consequently, finite element methods constitute its main ingredient. The resonance errors, that arise in linear problems also arise in nonlinear problems. Note that the resonance errors are the common feature of multiscale methods unless periodic problems are considered and the solutions of the local problems in an exact period are used. To reduce the resonance errors we use oversampling technique and show that the error can be greatly reduced by sampling from the larger domains. The multiscale map for MsFEM uses the solutions of the local problems in the target coarse block. This way one can sample the heterogeneities of the coarse block. If there is a scale separation and, in addition, some kind of periodicity, one can use the solutions of the smaller size problems to approximate the multiscale map. Note that a potential disadvantage of periodicity assumption is that the periodicity can act to disrupt large-scale connectivity features of the flow. For the examples similar to the non-periodic ones considered in this paper, with the use of the smaller size problems for approximating the solutions of the local problems, we have found very large errors (of order 50 percent).

6 Multiscale simulations of two-phase immiscible flow in adaptive coordinate system

Previously, we discussed some applications of MsFEM to two-phase flows. In this section, we explore the use of adaptive coordinate system in multiscale simulations of two-phase porous media flows. In particular, we would like to present upscaling of transport equations and its coupling to MsFEM.

As we discussed earlier, the use of global information can improve the multiscale finite element method. In particular, the solution of the pressure equation at initial time is used to construct the boundary conditions for the basis functions. It is interesting to note that the multiscale finite element methods that employ a limited global information reduces to standard multiscale finite element method in flow-based coordinate system. This can be verified directly and the reason behind it is that we have already employed a limited global information in flow-based coordinate system. To achieve high degree of speed-up in two-phase flow computations, we also consider the upscaling of transport equation and its coupling to pressure equation.

We would like to derive an upscaled model for the transport equation. We will assume the velocity is independent of time, $\lambda(S) = 1$, and restrict ourselves to the two-dimensional case. Then using the pressure-streamline framework, one obtains

$$S_t^\epsilon + v_0^\epsilon f(S^\epsilon)_p = 0, \qquad (6.1)$$
$$S(p, \psi, t = 0) = S_0,$$

where ϵ denotes the small scale, v_0^ϵ denotes the Jacobian of the transformation and is positive, and p denotes the initial pressure. For simplicity, we assume $\mathbf{k}(x) = k(x)\mathbf{I}$ and we have $\nabla\psi\cdot\nabla p = 0$. For deriving upscaled equations, we will first homogenize (6.1) along the streamlines, and then to homogenize across the streamlines. The homogenization along the streamlines can be done following Bourgeat and Mikelic [17] or following Hou and Xin [69] and E [36]. The latter uses two-scale convergence theory and we refer to [96] for the results on homogenization of (6.1) using two-scale convergence theory. We note that the homogenization results of Bourgeat and Mikelic is for general heterogeneities without an assumption on periodicity, and thus, is more appropriate for problems considered in the paper. Following [17], the homogenization of (6.1) can be easily derived (see Proposition 3.4 in [17]). Here, we briefly sketch the proof.

For ease of notations, we ignore the ψ dependence of v_0^ϵ and S^ϵ, and treat ψ as a parameter. We consider

$$v_0^\epsilon(p) = v_0\left(p, \frac{p}{\epsilon}\right).$$

Moreover, we assume that the domain is a unit interval. Then, for each ψ, it can be shown that $S^\epsilon(p, \psi, t) \to \tilde{S}(p, \psi, t)$ in $L^1((0,1) \times (0,T))$, where $\tilde{S}$ satisfies

$$\tilde{S}_t + \tilde{v}_0 f(\tilde{S})_p = 0, \tag{6.2}$$

and where $\tilde{v}_0$ is harmonic average of v_0^ϵ, i.e.,

$$\frac{1}{v_0^\epsilon} \to \frac{1}{\tilde{v}_0} \quad \text{weak}^* \text{ in } L^\infty(0,1).$$

The proof of this fact follows from Proposition 3.4 of [17].

Following [17] and assuming for simplicity $\int_0^1 \frac{d\eta}{v_0^\epsilon(\eta)} = \int_0^1 \frac{d\eta}{\tilde{v}_0(\eta)} = 1$, we introduce

$$\frac{dX^\epsilon(p)}{dp} = v_0^\epsilon(X^\epsilon(p)), \quad \frac{dX^0(p)}{dp} = \tilde{v}_0(X^0(p)).$$

Then (Lemma 3.1 of [17]):

$$X^\epsilon \to X^0 \quad \text{in } C[0,1] \text{ as } \epsilon \to 0. \tag{6.3}$$

Consequently,

$$\int_0^T \int_0^1 |S^\epsilon(p,\tau) - \tilde{S}(p,\tau)| \, dp \, d\tau$$

$$= \int_0^T \int_0^1 |S^\epsilon(X^\epsilon(p), \tau) - \tilde{S}(X^\epsilon(p), \tau)| v_0^\epsilon(X^\epsilon(p)) \, dp \, d\tau$$

$$\leqslant \int_0^T \int_0^1 |S^\epsilon(X^\epsilon(p), \tau) - \tilde{S}(X^0(p), \tau)| v_0^\epsilon(X^\epsilon(p)) \, dp \, d\tau$$

$$+ \int_0^T \int_0^1 |\tilde{S}(X^\epsilon(p), \tau) - \tilde{S}(X^0(p), \tau)| v_0^\epsilon(X^\epsilon(p)) \, dp \, d\tau$$

$$\leqslant \int_0^T \int_0^1 |S^\epsilon(X^\epsilon(p), \tau) - \tilde{S}(X^0(p), \tau)| dp d\tau$$

$$+ \int_0^T \int_0^1 |\tilde{S}(X^\epsilon(p), \tau) - \tilde{S}(X^0(p), \tau)| dp d\tau. \qquad (6.4)$$

The first term on the right-hand side of (6.4) converges to zero because $S^\epsilon(X^\epsilon(p), \tau)$ and $\tilde{S}(X^0(p), \tau)$ satisfy the same equation $u_t + f(u)_p = 0$, however, with the following initial conditions $S^\epsilon(X^\epsilon(p), t = 0) = S_0 \circ X^\epsilon$, $\tilde{S}(X^0(p), \tau) = S_0 \circ X^0$. Because of (6.3) and comparison principle,

$$\int_0^T \int_0^1 |S^\epsilon(X^\epsilon(p), \tau) - \tilde{S}(X^0(p), \tau)| dp d\tau \leqslant C \int_0^1 |S_0 \circ X^\epsilon - S_0 \circ X^0| dp,$$

the first term converges to zero. The convergence of the second term for each ψ follows from the argument in [17] (page 368) using Lebesgue's dominated convergence theorem.

Next, we provide a convergence rate (see also [96]) of the fine saturation S^ϵ to the homogenized limit $\tilde{S}$ as $\epsilon \to 0$.

Theorem 6.1. *Assume that $v_0^\epsilon(p)$ is bounded uniformly:*

$$C^{-1} \leqslant v_0^\epsilon \left(p, \frac{p}{\epsilon} \right) \leqslant D.$$

Denote by $F(t, T)$ the solution to $S_t + f(S)_T = 0$. The solution $\tilde{S}$ of (6.2) converges to S^ϵ (assuming initial conditions that do not depend on the fast scale) at a rate given by

$$\|S^\epsilon - \tilde{S}\|_\infty \leqslant G\epsilon$$

when F remains Lipschitz for all time, and

$$\|S^\epsilon - \tilde{S}\|_n \leqslant G\epsilon^{1/n}$$

when F develops at most a finite number of discontinuities.

Proof. First, we note that the velocity bound implies that $\tilde{C}^{-1} \leqslant \tilde{v}_0(p) \leqslant \tilde{D}$, uniformly in ψ, ζ. We transform the equations for S^ϵ (6.1) and $\tilde{S}$ (6.2) to the time of flight variable defined by

$$\begin{array}{ll} \dfrac{dT^\epsilon}{dp} = \dfrac{1}{v_0^\epsilon(p, \psi)} & \text{for } S^\epsilon \text{ and} \\ T^\epsilon(0) = \quad 0 \end{array} \qquad \begin{array}{ll} \dfrac{d\tilde{T}}{dp} = \dfrac{1}{\tilde{v}(p, \psi, \frac{\psi}{\epsilon})} & \text{for } \tilde{S}. \\ \tilde{T}(0) = \quad 0 \end{array}$$

Both equations reduce to

$$S_t + f(S)_T = 0.$$

The solution to this equation is $F(t,T)$. Since the initial condition does not depend on ϵ neither does F, $S = F(t,T^\epsilon(P,\Psi))$, $\tilde{S} = F(t,\tilde{T}(P,\Psi))$. Using these expressions for the saturation we can obtain the desired estimates by following the same steps as in the linear case. When F remains Lipschitz for all times we can easily obtain a pointwise estimate in terms of the Lipschitz constant M,

$$\|S^\epsilon - \tilde{S}\|_\infty = \|F(t,T^\epsilon) - F(t,\tilde{T})\|_\infty \leqslant M\|T^\epsilon - \tilde{T}\|_\infty \leqslant G\epsilon.$$

Otherwise we will need the time of flight bound that we derived for the linear flux that reduces here to

$$|T^\epsilon(P) - \tilde{T}(P)| \leqslant 2C\epsilon. \tag{6.5}$$

We will divide the domain in regions where F is Lipschitz with constant M in the second variable, denoted by A_2, and shock regions, denoted by A_1, and estimate the difference of S^ϵ and $\tilde{S}$ in each region separately. To fix the notation, let that there be n discontinuities in $F(t,\cdot)$ of magnitude less than ΔF, which does not have to be small, at $\{T = T_i\}_{i=1,\cdots,n}$. We will denote the thin strips of width $2C\epsilon$ around the discontinuities with A_1,

$$A_1 = \{T \text{ such that } |T - T_i| \leqslant 2C\epsilon \text{ for some } i = 1, \cdots, n\}$$

and with A_2 its complement. We selected the width of the strip based on (6.5), so that for any point P, if $T^\epsilon(P) \notin A_1$, then $T^\epsilon(P)$ and $\tilde{T}(P)$ are on the same side of any jump T_i. When $T^\epsilon(P) \in A_2$, F is Lipschitz in the region between T^ϵ and $\tilde{T}$, and we can show

$$\int_{A_2} (S^\epsilon - \tilde{S})^2 dpd\psi = \int_{A_2} (F(t,T^\epsilon) - F(t,\tilde{T}))^2 dpd\psi$$

$$\leqslant M^2\|T^\epsilon - \tilde{T}\|_\infty^2 |T^\epsilon(A_2)^{-1}|$$

$$\leqslant N^2\epsilon^2 |T^\epsilon(A_2)^{-1}|,$$

where we used the time of flight bound (6.5). By $|T^\epsilon(A_2)^{-1}|$ we denoted the image of A_2 under the inverse of $T^\epsilon(P)$. Inside the strip A_1, even though S^ϵ and $\tilde{S}$ differ by an $O(1)$ quantity we can use the smallness of the area of the strip to make the L_2 norm of their difference small:

$$\int_{A_1} (S^\epsilon - \tilde{S})^2 dpd\psi = \int_{A_2} (F(t,T^\epsilon) - F(t,\tilde{T}))^2 dpd\psi$$

$$\leqslant (\Delta S + N\epsilon)^2 |T^\epsilon(A_1)^{-1}|$$

$$\leqslant (\Delta S + N\epsilon)^2 4CDn\epsilon.$$

We estimated the area $|T^\epsilon(A_1)^{-1}|$ by using the definition of A_1 and the fact that the Jacobian of the transformation $T^\epsilon(P)^{-1}$ is v_0^ϵ and is bounded uniformly in p, ψ. Putting together the two estimates for regions A_1 and A_2 we obtain $\|S^\epsilon - \tilde{S}\|_2 \leqslant G\epsilon^{1/2}$. Estimates in terms of the other L_p norms follow similarly.

The homogenized operator given by (6.2) still contains variation of order ϵ through the fast variable $\frac{\psi}{\epsilon}$, however there it does not contain any derivatives in that variable. Its dependence on $\frac{\psi}{\epsilon}$ is only parametric. We can homogenize the dependence of the partially homogenized operator on $\frac{\psi}{\epsilon}$ and arrive at a homogenized operator that is independent of the small scale. In the latter case, we will only obtain weak convergence of the partially homogenized solution. When we homogenized along the streamlines, the resulting equation was of hyperbolic type like the original equation. In a seminal and celebrated paper, Tartar [100] showed that homogenization across streamlines leads to transport with the average velocity plus a time-dependent diffusion term, referred to as macrodispersion, a physical phenomenon that was not present in the original fine equation. In particular, if the velocity field does not depend on p inside the cells, that is, $\tilde{v}(\psi, \frac{\psi}{\epsilon})$, then the homogenized solution, $\overline{\tilde{S}}$, (weak* limit of $\tilde{S}$, which will be denoted by $\overline{S}$), satisfies

$$\overline{S}_t + \overline{\tilde{v}}_0 \overline{S}_p = \int_0^t \int \overline{S}_{pp}(p - \lambda(t-\tau), \psi, \tau) d\mu_{\frac{\psi}{\epsilon}}(\lambda) d\tau. \qquad (6.6)$$

Here, $d\nu_{\frac{\psi}{\epsilon}}$ is the Young measure associated with the sequence $\tilde{v}_0(\psi, \cdot)$ and $d\mu_{\frac{\psi}{\epsilon}}$ is a Young measure that satisfies

$$\left(\int \frac{d\nu_{\frac{\psi}{\epsilon}}(\lambda)}{\frac{s}{2\pi i q} + \lambda} \right)^{-1} = \frac{s}{2\pi i q} + \overline{\tilde{v}}_0 - \int \frac{d\mu_{\frac{\psi}{\epsilon}}(\lambda)}{\frac{s}{2\pi i q} + \lambda}.$$

We have denoted by $\overline{\tilde{v}}_0$ the weak limit of the velocity. This equation has no dependence on the small scale and we consider it to be the full homogenization of the fine saturation equation. Efendiev and Popov [47] have extended this method for the Riemann problem in the case of nonlinear flux. Note that the homogenization across streamlines provides a weak limit of partially homogenized solution. Because the original solution S^ϵ strongly converges to partially homogenized solution for each ψ, it can be easily shown that $S^\epsilon \to \overline{S}$ weakly. We omit this proof here.

In numerical simulations, it is difficult to use (6.6) as a homogenized operator, and often a second order approximation of this equation is used. These approximate equations can be also derived using perturbation analysis. In particular, using the higher moments of the saturation and the velocity, one can model the macrodispersion. In the context of two-phase flow this idea was introduced by Efendiev, Durlofsky, and Lee [46], [45], Chen and Hou [25] and Hou et al., [61]. In our case, the computation of the macrodispersion is much simpler because the transport equations have been already averaged along the streamlines, and thus we will be applying perturbation technique to one dimensional problem.

We expand $\tilde{S}$, $\tilde{v}_0$ (following [46]) as an average over the cells in the pressure-streamline frame and the corresponding fluctuations:

$$\begin{aligned} \tilde{S} &= \overline{S}(p, \psi, t) + S'(p, \psi, \zeta, t), \\ \tilde{v}_0 &= \overline{\tilde{v}_0}(p, \psi, t) + \tilde{v}_0'(p, \psi, \zeta, t). \end{aligned} \tag{6.7}$$

We will derive the homogenized equation for $f(S) = S$. Averaging equations (6.2) with respect to ψ we find an equation for the mean of the saturation:

$$\overline{S}_t + \overline{\tilde{v}_0}\overline{S}_p + \overline{\tilde{v}_0' S'_p} = 0.$$

An equation for the fluctuations is obtained by subtracting the above equation from (6.2),

$$S'_t + (\tilde{v}_0 - \overline{\tilde{v}_0})\overline{S}_p + \tilde{v}_0 S'_p - \overline{\tilde{v}_0' S'_p} = 0.$$

Together, the equations for the saturation are

$$\begin{aligned} \overline{S}_t + \overline{\tilde{v}_0}\overline{S}_p + \overline{\tilde{v}_0' S'_p} &= 0, \\ S'_t + \tilde{v}_0'\overline{S}_p + \tilde{v}_0 S'_p - \overline{\tilde{v}_0' S'_p} &= 0. \end{aligned} \tag{6.8}$$

We can consider the second equation to be the auxiliary (cell) problem and the first equation to be the upscaled equation. We note that the cell problem for a hyperbolic equation is $O(1)$ whereas for an elliptic it is $O(\epsilon)$. We can obtain an approximate numerical method by solving the cell problem only near the shock region in space time, where the macrodispersion term is largest. In that case it is best to diagonalize these equations by adding the first to the second one:

$$\begin{aligned} \overline{S}_t + \overline{\tilde{v}_0}\overline{S}_p &= -\overline{\tilde{v}_0'(\tilde{S}_p - \overline{S}_p)}, \\ \tilde{S}_t + \tilde{v}_0\tilde{S}_p &= 0. \end{aligned}$$

Compared to (6.8), it has fewer forcing terms and no cross fluxes, which leads to a numerical method with less numerical diffusion that is easier to implement.

6.1   Numerical averaging across streamlines

The derivation in the previous sections contained no approximation. In this section, we follow the same idea as in the derivation to solve the equation for the fluctuations along the characteristics, but with the purpose of deriving an equation on the coarse grid. To achieve this, we will not perform analytical upscaling in the sense of deriving a continuous upscaled equation as in the previous section. We will first discretize the equation with a finite volume method in space and then upscaled the resulting equation. Our upscaled equation will therefore be dependent on the numerical scheme.

We use the same definition for the average saturation and the fluctuations as in (6.7) and follow the same steps until equation (6.8). We discretize the macrodispersion term in the equation for the average saturation

$$\overline{\tilde{v}_0' S_p'} = \frac{\overline{\tilde{v}_0' S'}^{i+1} - \overline{\tilde{v}_0' S'}^{i}}{\Delta p} + O(\Delta p).$$

A superscript $\cdot^i$ refers to a discrete quantity defined at the center of the conservation cell. Instead of solving the equation for the fluctuations on the fine characteristics as before, which would lead to a fine grid algorithm, we solve it on the coarse characteristics defined by

$$\frac{dP}{dt} = \overline{\tilde{v}_0}, \quad \text{with } P(p,0) = p.$$

Compared to the equation that we obtained in the previous section for S', this equation for S' has an extra term, which appears second:

$$S' = - \int_0^t (\tilde{v}_0'(P(p,\tau),\psi)\overline{S}_p(P(p,\tau),\psi,\tau)$$
$$+ \tilde{v}_0'(P(p,\tau),\psi)S_p'(P(p,\tau),\psi,\tau) + \overline{\tilde{v}_0' S_p'}) d\tau.$$

The second term is second-order in fluctuating quantities, and we expect it to be smaller than the first term so we neglect it. As before, we multiply by $\tilde{v}_0'$ and average over ψ to find

$$\overline{\tilde{v}_0' S'} = - \int_0^t \overline{\tilde{v}_0' \tilde{v}_0(P(p,\tau),\psi)\overline{S}_p(P(p,\tau),\psi,\tau)} d\tau.$$

In this form at time t it is necessary to know information about the past saturation in $(0,t)$ to compute the future saturation. Following [46], it can be easily shown that $\overline{S}_p(P(p,\tau))$ depends weakly on time, in the sense that the difference between $\overline{S}_p(P(p,\tau))$ and $\overline{S}_p(P(p,t))$ is of third-order in fluctuating quantities. Therefore we can take $\overline{S}_p(P(p,\tau))$ out of the time integral to find

$$\overline{\tilde{v}_0' S'} = - \int_0^t \overline{\tilde{v}_0' \tilde{v}_0'(P(p,\tau),\psi)} d\tau \overline{S}_p.$$

The term inside the time integral is the covariance of the velocity field along each streamline. The macrodispersion in this form can be computed independent of the past saturation.

The nonlinearity of the flux function introduces an extra source of error in the approximation. We expand $f(\tilde{S})$ near $\overline{S}$ (cf. [45]) and keep only the first term:

$$
\begin{aligned}
\tilde{S} &= \overline{S}(p,\psi,t) + S'(p,\psi,\zeta,t), \\
\tilde{v}_0 &= \overline{v}_0(p,\psi,t) + \tilde{v}_0'(p,\psi,\eta,t), \\
\overline{f}(\tilde{S}) &= f(\overline{S}) + f_S(\overline{S})S' + O(S'^2), \\
f(S)_p &= f_S(\overline{S})\overline{S}_p + f(\overline{S})S' + \cdots .
\end{aligned}
\tag{6.9}
$$

This approximation is not accurate near the shock because S' is not small near sharp fronts. The region near the shock is important because the macrodispersion is large. Due to the dependence of the jump in the saturation on the mobility we expect this approximation to be better for lower mobilities. Nevertheless, this approximation works well in practice. For more accuracy, it is also possible to retain more terms in the Taylor expansion. We will show that in realistic examples these higher-order terms are not important in our setting.

Using these definitions we derive the following equations for the average saturation and the fluctuations (see [96] for more details):

$$
\begin{aligned}
\overline{S}_t + \overline{v}_0 f(\overline{S})_p + \overline{\tilde{v}_0'(f_S(\overline{S})S')}_p &= 0, \\
S_t' + \tilde{v}_0' f_S(\overline{S})\overline{S}_p + \tilde{v}_0 f_S(\overline{S})S_p' - \overline{\tilde{v}_0' S_p'} &= 0.
\end{aligned}
\tag{6.10}
$$

The macrodispersion is discretized as

$$
\overline{\tilde{v}_0'(f_S(\overline{S})S')}_p = \frac{\overline{\tilde{v}_0' f_S(\overline{S})S'}^{i+1} - \overline{\tilde{v}_0' f_S(\overline{S})S'}^{i}}{\Delta p} + O(\Delta p).
$$

We solve the second equation on the coarse characteristics defined by

$$
\frac{dP}{dt} = \overline{v}_0 f_S(\overline{S}), \quad \text{with} P(p,0) = p
$$

and form the terms that appear in the macrodispersion

$$
\overline{\tilde{v}_0' f_S(\overline{S})S'}
$$
$$
= -\int_0^t \overline{\tilde{v}_0' f_S(\overline{S})\tilde{v}_0'(P(p,\tau),\psi) f_S(\overline{S}(P(p,\tau),\psi,\tau))\overline{S}_p(P(p,\tau),\psi,\tau)} d\tau.
$$

As before we have dropped terms that are second-order in fluctuating quantities. It can be shown (see [96]) that $f_S(\overline{S}(P(p,\tau),\psi,\tau))\overline{S}_p(P(p,\tau),$

$\psi, \tau)$ does not vary significantly along the streamlines and it can be taken out of the integration in time:

$$\overline{\tilde{v}'_0 f_S(\overline{S})S'} = -\int_0^t \overline{\tilde{v}'_0 \tilde{v}'_0 (P(p,\tau),\psi)} d\tau f_S(\overline{S})^2 \overline{S}_p. \tag{6.11}$$

This expression is similar to the one obtained in the linear case, however the macrodispersion depends on the past saturation through the equation for the coarse characteristics.

Even though the macrodispersion depends on the past saturation it is possible to compute it incrementally as it is done in [45]. Given its value $D(t)$ at time t we compute the values at $t+\Delta t$ using the macrodispersion at the previous time:

$$D(t + \Delta t) = \int_0^{t+\Delta t} \cdots d\tau = \int_0^t \cdots d\tau + \int_t^{t+\Delta t} \cdots d\tau.$$

This is possible because in the derivation for the approximate expression for the macrodispersion we took the terms that depend on $S(\tau)$ outside the time integration. The integrand, the average covariance of the velocity field along the streamlines, needs to be computed only once at the beginning. Then updating the macrodispersion takes $O(n^2)$ computations, as many as it takes to update $\overline{S}$.

6.2 Numerical results

In this section, we first show representative simulation results for $\lambda(S) = 1$ for flux functions $f(S) = S$ and nonlinear $f(S)$ with viscosity ratio $\mu_o/\mu_w = 5$. For such setting, the pressure and saturation equations are decoupled and we can investigate the accuracy of saturation upscaling independently from the pressure upscaling. At the end of the section we will present numerical results for two-phase flow. We consider two type of permeability fields. The first type includes a permeability field generated using two-point geostatistics with correlation lengths $l_x = 0.3$, $l_z = 0.03$ and $\sigma^2 = 1.5$ (see Figure 6.1, left). The second type of permeability fields corresponds to a channelized system, and we consider two examples. The first example (middle figure of Figure 6.1) is a synthetic channelized reservoir generated using both multi-point geostatistics (for the channels) and two-point geostatistics (for permeability distribution within each facies). The second channelized system is one of the layers of the benchmark test (representing the North Sea reservoir), the SPE comparative project [27] (upper Ness layers). These permeability fields are highly heterogeneous, channelized, and difficult to upscale. Because the permeability fields are highly heterogeneous, they are refined to 400×400 in order to obtain accurate comparisons.

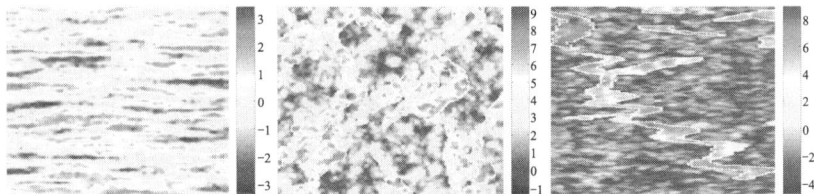

Figure 6.1. Permeability fields used in the simulations. Left - permeability field with exponential variogram, middle - synthetic channelized permeability field, right - layer 36 of SPE comparative project [27].

Simulation results will be presented for saturation snapshots as well as the oil cut as a function of pore volume injected (PVI). Note that the oil cut is also referred to as the fractional flow of oil. The oil cut (or fractional flow) is defined as the fraction of oil in the produced fluid and is given by q_o/q_t, where $q_t = q_o + q_w$, with q_o and q_w being the flow rates of oil and water at the production edge of the model. In particular, $q_w = \int_{\partial\Omega^{\text{out}}} f(S)\mathbf{v} \cdot \mathbf{n} dl$, $q_t = \int_{\partial\Omega^{\text{out}}} \mathbf{v} \cdot \mathbf{n} dl$, and $q_o = q_t - q_w$, where $\partial\Omega^{\text{out}}$ is the outer flow boundary. We will use the notation Ω for total flow q_t and F for fractional flow q_o/q_t in numerical results. Pore volume injected, defined as $\text{PVI} = \frac{1}{V_p} \int_0^t q_t(\tau) d\tau$, with V_p being the total pore volume of the system, provide a dimensionless time for the displacement.

When using multiscale finite element methods for two-phase flow, one can update the basis functions near the sharp fronts. Indeed, sharp fronts modify the local heterogeneities and this can be taken into account by re-solving the local equations, (4.34), for basis functions. If the saturation is smooth in the coarse block, it can be approximated by its average in (4.34), and consequently, the basis functions do not needed to be updated. It can be shown that this approximation yields first-order errors (in terms of coarse mesh size). In our simulations, we have found only a slight improvement when the basis functions are updated, thus the numerical results for the MsFVEM presented in this paper do not include the basis function update near the sharp fronts. Since a pressure-streamline coordinate system is used the boundary conditions are given by $P = 1$, $S = 1$ along the $p = 1$ edge and $P = 0$ along the $p = 0$ edge, and no flow boundary condition on the rest of the boundaries.

For the upscaled saturation equation, which is a convection-diffusion equation, we need to observe an extra CFL-like condition to obtain a stable numerical scheme $\Delta t \leqslant \frac{\Delta p^2}{2\nu}$, where ν is the diffusivity. In our case the diffusivity is $\int_{\text{cell}} \int_0^t \tilde{v}_0'(p(\tau), \psi)\tilde{v}_0'(p, \psi) d\tau d\psi$. If the macrodispersion is large this can be a very restrictive condition. To remedy this, we used an implicit discretization for the macrodispersion. This is straightforward since the problem is one-dimensional. The resulting system was solved by a tridiagonal solver very fast. Since the order of the highest derivative in the equation has increased, we require extra boundary conditions. For the computation of the macrodispersion term, we impose no flux on both boundaries of the domain.

In the upscaled algorithm, a moving mesh is used to concentrate the points of computation near the sharp front. Since the saturation equation is one dimensional in the pressure-streamline coordinates, the implementation of the moving mesh is straightforward and efficient. For the details we refer to [96]. We compare the saturation right before the breakthrough time so that the shock front is largest. For this comparison we also average the fine saturation over the coarse blocks, since the upscaled model is defined on a coarser grid. In Figures 6.2 and 6.3, we plot the saturation for linear and nonlinear (with $\mu_o/\mu_w = 5$) $f(S)$. As we see in both cases, we have very accurate representation of the saturation profile.

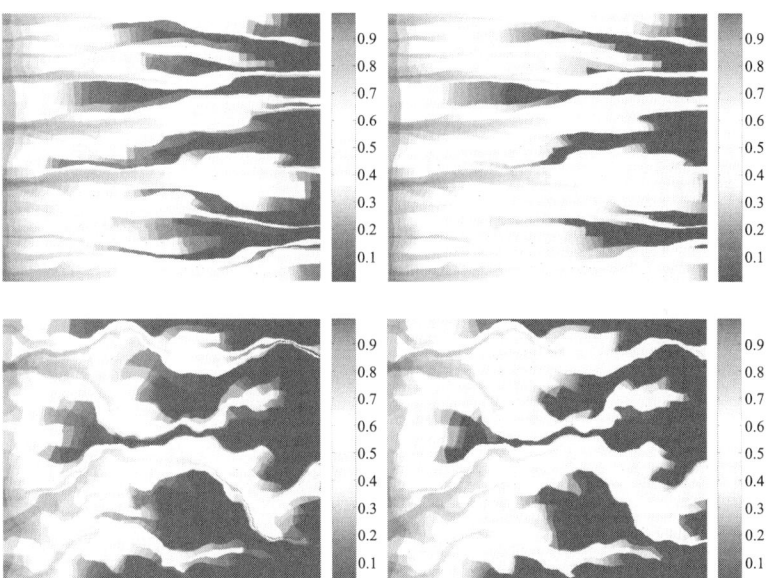

Figure 6.2. Saturation snapshots for variogram based permeability field (top) and synthetic channelized permeability field (bottom). Linear flux is used. Left figures represent the upscaled saturation plots and the right figures represent the fine-scale saturation plots.

We proceed with a quantitative description of the error. We will distinguish between two sources of errors. We will refer to the difference between the upscaled and the exact equation as the upscaling or modeling error and to the difference between the solution of continuous upscaled

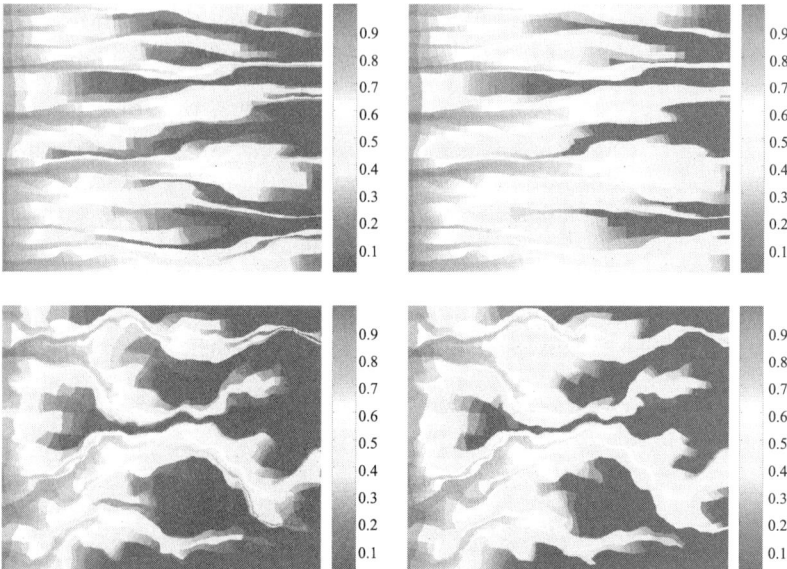

Figure 6.3. Saturation snapshots for variogram based permeability field (top) and synthetic channelized permeability field (bottom). Nonlinear flux is used. Left figures represent the upscaled saturation plots and the right figures represent the fine-scale saturation plots.

equations and the solution to the numerical scheme as the discretization error. We will refer to the difference between the solutions of the continuous fine equations and the numerical scheme of the upscaled equations as the total error. To separate the upscaling error from the total error we will solve the upscaled equations on the fine grid, which is the grid on which we solve to the fine equation. We will also solve them on the coarse grid to compute the total error. The errors are computed in the p, ψ frame and are relative errors. We display the upscaling error against the number of coarse cells for the computations of the previous section in Tables 6.14–6.16 As we see from these tables that upscaling using macrodispersion decreases the upscaling errors. We also see that the effects of macrodispersion are more significant in the case of linear flux when the jump discontinuity in the saturation profile is larger.

In Tables 6.17–6.19, we show the total error, that is, the modeling and discretization error when we use a moving mesh to solve the saturation equation. It is interesting that the convergence of $\tilde{S}$ to S is observed even though the upscaling error is larger than the numerical error of the fine solution, which is 0.02 for the linear flux and 0.002 for the nonlinear flux in the L_1 norm, as mentioned before. The reason is that the location of the moving mesh points was selected so that the points are as

Table 6.14. Upscaling error for permeability generated using two-point geostatistics.

Linear flux	25×25	50×50	100×100	200×200
L_1 error of $\tilde{S}$	0.0021	6.57×10^{-4}	2.15×10^{-4}	8.75×10^{-5}
L_1 error of $\overline{S}$ with macrodispersion	0.115	0.0696	0.0364	0.0135
L_1 error of $\overline{S}$ fine without macrodispersion	0.1843	0.0997	0.0505	0.0191
Nonlinear flux	25×25	50×50	100×100	200×200
L_1 error of $\tilde{S}$	0.0023	8.05×10^{-4}	2.89×10^{-4}	1.29×10^{-4}
L_1 error of $\overline{S}$ with macrodispersion	0.116	0.0665	0.0433	0.0177
L_1 error of $\overline{S}$ fine without macrodispersion	0.151	0.0805	0.0432	0.0186

Table 6.15. Upscaling error for synthetic channelized permeability field.

Linear flux	25×25	50×50	100×100	200×200
L_1 error of $\tilde{S}$	0.0222	0.0171	0.0122	0.0053
L_1 error of $\overline{S}$ with macrodispersion	0.0819	0.0534	0.0333	0.0178
L_1 error of $\overline{S}$ fine without macrodispersion	0.123	0.0834	0.0486	0.0209
Nonlinear flux	25×25	50×50	100×100	200×200
L_1 error of $\tilde{S}$	0.0147	0.0105	0.0075	0.0040
L_1 error of $\overline{S}$ with macrodispersion	0.0842	0.0658	0.0371	0.0207
L_1 error of $\overline{S}$ fine without macrodispersion	0.119	0.0744	0.0424	0.0214

Table 6.16. Upscaling error for SPE 10, layer 36.

Linear flux	25×25	50×50	100×100	200×200
L_1 error of $\tilde{S}$	0.0128	0.0093	0.0072	0.0042
L_1 error of $\overline{S}$ with macrodispersion	0.0554	0.0435	0.0307	0.0176
L_1 error of $\overline{S}$ fine without macrodispersion	0.123	0.0798	0.0484	0.0258
Nonlinear flux	25×25	50×50	100×100	200×200
L_1 error of $\tilde{S}$	0.0089	0.0064	0.0054	0.0033
L_1 error of $\overline{S}$ with macrodispersion	0.0743	0.0538	0.0348	0.0189
L_1 error of $\overline{S}$ fine without macrodispersion	0.0924	0.0602	0.0395	0.0202

dense near the shock as the fine solution using the parameter $h_{\min}$. This was done to observe the upscaling error clearly and also to have similar

Table 6.17. Total error for permeability field generated using two-point geostatistics.

Linear flux	25×25	50×50	100×100	200×200
L_1 upscaling error of $\tilde{S}$	0.0021	6.57×10^{-4}	2.15×10^{-4}	8.75×10^{-5}
L_1 error of $\tilde{S}$ computed on coarse grid	0.0185	0.0062	0.0019	0.0015
L_1 upscaling error of $\overline{S}$	0.115	0.0696	0.0364	0.0135
L_1 error of computed on coarse grid	0.139	0.0779	0.0390	0.0144
Nonlinear flux	25×25	50×50	100×100	200×200
L_1 upscaling error of $\tilde{S}$	0.0023	8.05×10^{-4}	2.89×10^{-4}	1.29×10^{-4}
L_1 error of $\tilde{S}$ computed on coarse grid	0.0268	0.0099	0.0027	9.38×10^{-4}
L_1 upscaling error of $\overline{S}$	0.116	0.0665	0.0433	0.0177
L_1 error of $\overline{S}$ computed on coarse grid	0.146	0.0797	0.0461	0.0184

Table 6.18. Total error for synthetic channelized permeability field.

Linear flux	25×25	50×50	100×100	200×200
L_1 upscaling error of $\tilde{S}$	0.0222	0.0171	0.0122	0.0053
L_1 error of $\tilde{S}$ computed on coarse grid	0.0326	0.0161	0.0107	0.0113
L_1 upscaling error of $\overline{S}$	0.0819	0.0534	0.0333	0.0178
L_1 error of $\overline{S}$ computed on coarse grid	0.135	0.0849	0.0477	0.0274
Nonlinear flux	25×25	50×50	100×100	200×200
L_1 upscaling error of $\tilde{S}$	0.0147	0.0105	0.0075	0.0040
L_1 error of $\tilde{S}$ computed on coarse grid	0.0494	0.0295	0.0150	0.0130
L_1 upscaling error of $\overline{S}$	0.0842	0.0658	0.0371	0.0207
L_1 error of $\overline{S}$ computed on coarse grid	0.17	0.11	0.0541	0.0303

Table 6.19. Total error for SPE10, layer 36.

Linear flux	25×25	50×50	100×100	200×200
L_1 upscaling error of $\tilde{S}$	0.0128	0.0093	0.0072	0.0042
L_1 error of $\tilde{S}$ computed on coarse grid	0.023	0.0095	0.0069	0.0052
L_1 upscaling error of $\overline{S}$	0.0554	0.0435	0.0307	0.0176
L_1 error of $\overline{S}$ computed on coarse grid	0.0683	0.052	0.0361	0.0205
Nonlinear flux	25×25	50×50	100×100	200×200
L_1 upscaling error of $\tilde{S}$	0.0089	0.0064	0.0054	0.0033
L_1 error of $\tilde{S}$ computed on coarse grid	0.0338	0.0148	0.0074	0.0037
L_1 upscaling error of $\overline{S}$	0.0743	0.0538	0.0348	0.0189
L_1 error of $\overline{S}$ computed on coarse grid	0.115	0.0720	0.0406	0.0204

CFL constraints on the time step, which allows a clean comparison of computational times. We compare the require CPU times in Table 6.20. We note that it took 26 units of time to interpolate one quantity from the Cartesian to the pressure-streamline frame. The upscaled solutions were computed on a 25×25 grid and the fine solution was computed on a 400×400 grid so we expect the $\overline{S}$ computations to be 256 times or more faster. The extra gain comes from a less restrictive CFL condition since we use an averaged velocity. The computations in the Cartesian frame are much slower.

Table 6.20. Computational cost.

	fine x, y	fine p, ψ	$\tilde{S}$	$\overline{S}$
layered, linear flux	5648	257	9	1
layered, nonlinear flux	14543	945	28	4
percolation, linear flux	8812	552	12	1
percolation, nonlinear flux	23466	579	12	1
SPE10 36, linear flux	40586	1835	34	2
SPE10 36, nonlinear flux	118364	7644	25	2

The application of the proposed method to two-phase immiscible flow can be performed using the implicit pressure and explicit saturation (IM-PES) framework. This procedure consists of computing the velocity and then using the velocity field in updating the saturation field. When updating the saturation field, we consider the velocity field to be time independent and we can use our upscaling procedure at each IMPES time step. First, we note that in the proposed method, the mapping is done between the current pressure-streamline and initial pressure-streamline. This mapping is nearly the identity for the cases when $\mu_o > \mu_w$. In Figure 6.4, we plot the level sets of the pressure and stream function at time $t = 0.4$ in a Cartesian coordinate system (left plot) and in the coordinate system of the initial pressure and streamline (right plot). Clearly, the level sets are much smoother in initial pressure-streamline frame compared to Cartesian frame. This also explains the observed convergence of upscaling methods as we refined the coarse grid. In Figure 6.5, we plot the saturation snapshots right before the breakthrough. In Figure 6.6, the fractional flow is plotted. Again, the moving mesh algorithm is used to track the front separately. The convergence table is presented in Table 6.21. We see from this table that the errors decrease as first order which indicates that the pressure and saturation is smooth functions of initial pressure and streamline.

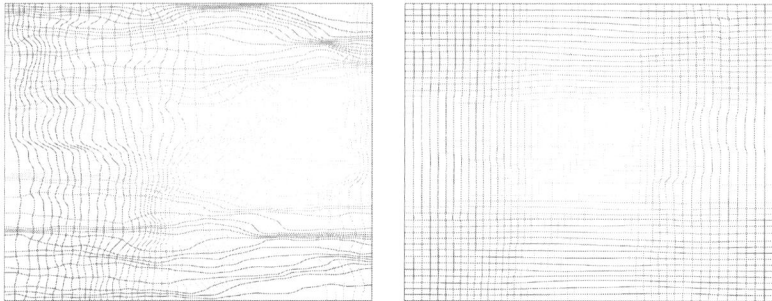

Figure 6.4. Left: Pressure and streamline function at time $t = 0.4$ in Cartesian frame. Right: Pressure and streamline function at time $t = 0.4$ in initial pressure-streamline frame.

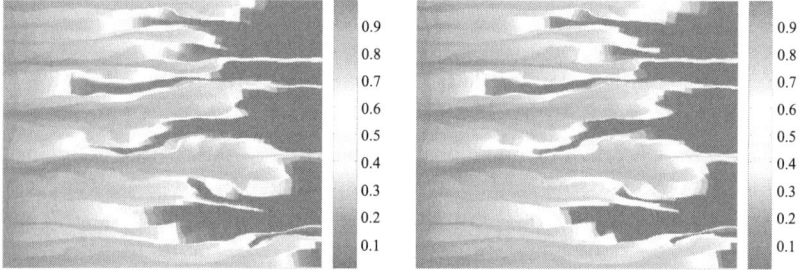

Figure 6.5. Left: Saturation plot obtained using coarse-scale model. Right: The fine-scale saturation plot. Both plots are on coarse grid. Variogram based permeability field is used. $\mu_o/\mu_w = 5$.

Table 6.21. Convergence of the upscaling method for two-phase flow for variogram based permeability field.

with $\tilde{S}$	50×50	100×100	200×200
L_2 pressure error at $t = \frac{3T_{\text{final}}}{4}$	0.0014	0.007	0.004
L_2 velocity error at $t = \frac{3T_{\text{final}}}{4}$	0.0235	0.0137	0.0072
L_1 saturation error $t = T_{\text{final}}$	0.0105	0.0052	0.0027
with $\overline{S}$	50×50	100×100	200×200
L_2 pressure error at $t = \frac{3T_{\text{final}}}{4}$	0.0046	0.0021	0.0008
L_2 velocity error at $t = \frac{3T_{\text{final}}}{4}$	0.0530	0.0335	0.0246
L_1 saturation error $t = T_{\text{final}}$	0.0546	0.0294	0.0134

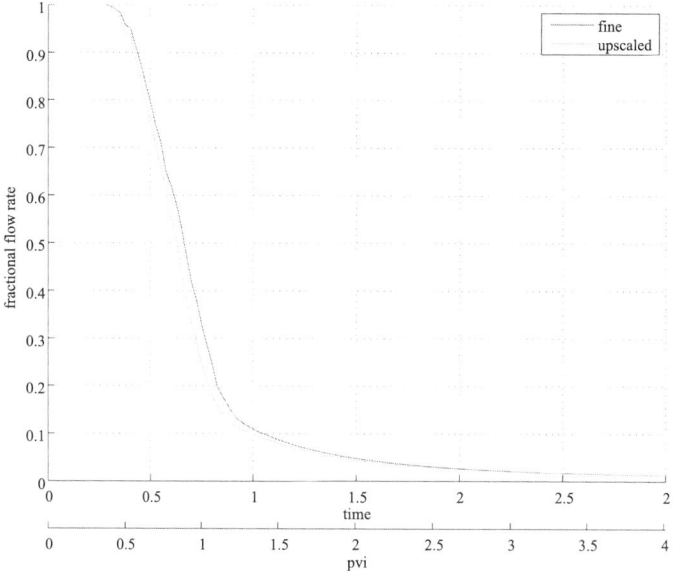

Figure 6.6. Comparison of fractional flow for coarse- and fine-scale models. Variogram based permeability field is used. $\mu_o/\mu_w = 5$.

7 Conclusions

In these lecture notes, we reviewed some of the recent advances in developing systematic multiscale methods with particular emphasis on multiscale finite element methods and their applications to fluid flows in heterogeneous porous media. In particular, the local approaches and their convergence properties for various flow problems are discussed. Moreover, improved subgrid capturing techniques through a judicious choice of local boundary conditions or through oversampling techniques or through the use of limited global information are reviewed. Other topics, such as homogenization, the sampling techniques in numerical homogenization, and multiscale simulations of two-phase flows in heterogeneous porous media are also presented. Although the results presented in this paper are encouraging, there is scope for further exploration. These include the development and mathematical analysis of efficient numerical homogenization techniques for nonlinear convection-diffusion equations with various Peclet numbers (e.g., convection dominated), inexpensive approximations of multiscale basis functions, further exploration of accurate boundary conditions based on local multiscale solutions, the use of limited global information for nonlinear problems, development of adaptive criteria for multiscale basis functions (selection of coarse grid), applications of MsFEM to more multi-phase/multi-component porous media flows, and etc.

References

[1] J. Aarnes, *On the use of a mixed multiscale finite element method for greater flexibility and increased speed or improved accuracy in reservoir simulation*, SIAM MMS, 2 (2004), 421-439.

[2] J. Aarnes, Y. R. Efendiev and L. Jiang, *Analysis of multiscale finite element methods using global information for two-phase flow simulations* (submitted).

[3] J. Aarnes and T. Y. Hou, *An efficient domain decomposition preconditioner for multiscale elliptic problems with high aspect ratios*, Acta Mathematicae Applicatae Sinica, **18** (2002), 63-76.

[4] R. A. Adams, *Sobolev Spaces*, Pure and Applied Mathematics, Vol. 65, Academic Press, New York-London, 1975.

[5] T. Arbogast, *Numerical subgrid upscaling of two-phase flow in porous media,* in: Numerical Treatment of Multiphase Flows in Porous Media, Z. Chen et al., eds., Lecture Notes in Physics 552, Springer, Berlin, 2000, 35-49.

[6] T. Arbogast, *Implementation of a locally conservative numerical subgrid upscaling scheme for two-phase Darcy flow*, Comput. Geosci., 6 (2002), 453-481.

[7] T. Arbogast and K. Boyd, *Subgrid upscaling and mixed multiscale finite elements.* SIAM Num. Anal. (to appear).

[8] M. Avellaneda, T. Y. Hou and G. Papanicolaou, *Finite difference approximations for partial differential equations with rapidly oscillating coefficients*, Mathematical Modelling and Numerical Analysis, 25 (1991), 693-710.

[9] M. Avellaneda and F. -H. Lin, *Compactness methods in the theory of homogenization*, Comm. Pure Appl. Math., 40 (1987), 803-847.

[10] I. Babuska, U. Banerjee and J. E. Osborn, *Survey of meshless and generalized finite element methods: A unified approach*, Acta Numerica, 2003, 1-125.

[11] I. Babuska, G. Caloz and E. Osborn, *Special finite element methods for a class of second order elliptic problems with rough coefficients*, SIAM J. Numer. Anal., 31 (1994), 945-981.

[12] I. Babuška and J. M. Melenk, *The partition of unity method*, Internat. J. Numer. Methods Engrg., 40 (1997), 727-758.

[13] I. Babuska and E. Osborn, *Generalized finite element methods: Their performance and their relation to mixed methods*, SIAM J. Numer. Anal., 20 (1983), 510-536.

[14] I. Babuska and W. G. Szymczak, *An error analysis for the finite element method applied to convection-diffusion problems,* Comput. Methods Appl. Math. Engrg, 31 (1982), 19-42.

[15] A. Bensoussan, J. L. Lions and G. Papanicolaou, *Asymptotic Analysis for Periodic Structures,* Studies in Mathematics and Its Applications, Vol. 5, North-Holland Publ., 1978.

[16] A. Bourgeat, *Homogenized behavior of two-phase flows in naturally fractured reservoirs with uniform fractures distribution,* Comp. Meth. Appl. Mech. Engrg, 47 (1984), 205-216.

[17] A. Bourgeat and A. Mikelić, *Homogenization of two-phase immiscible flows in a one-dimensional porous medium,* Asymptotic Anal., 9 (1994), 359-380.

[18] M. Brewster and G. Beylkin, *A multiresolution strategy for numerical homogenization,* ACHA, 2(1995), 327-349.

[19] F. Brezzi and M. Fortin, *Mixed and Hybrid Finite Element Methods,* Springer-Verlag, Berlin-Heidelberg-New York, 1991.

[20] F. Brezzi and A. Russo, *Choosing bubbles for advection-diffusion problems,* Math. Models Methods Appl. Sci, 4 (1994), 571-587.

[21] F. Brezzi, L. P. Franca, T. J. R. Hughes and A. Russo, $b = \int g$, Comput. Methods in Appl. Mech. and Engrg., 145 (1997), 329-339.

[22] J. E. Broadwell, *Shock structure in a simple discrete velocity gas,* Phys. Fluids, 7 (1964), 1243-1247.

[23] T. Carleman, *Problèms Mathématiques dans la Théorie Cinétique de Gaz,* Publ. Sc. Inst. Mittag-Leffler, Uppsala, 1957.

[24] H. Ceniceros and T. Y. Hou, *An efficient dynamically adaptive mesh for potentially singular solutions.* J. Comput. Phys., 172 (2001), 609-639.

[25] Z. Chen and T. Y. Hou, *A mixed multiscale finite element method for elliptic problems with oscillating coefficients,* Math. Comp., 72 (2002), 541-576 (electronic).

[26] A. J. Chorin, *Vortex models and boundary layer instabilities,* SIAM J. Sci. Statist. Comput., 1 (1980), 1-21.

[27] M. Christie and M. Blunt, *Tenth SPE comparative solution project: A comparison of upscaling techniques,* SPE Reser. Eval. Eng., 4 (2001), 308-317.

[28] M. E. Cruz and A. Petera, *A parallel Monte Carlo finite element procedure for the analysis of multicomponent random media,* Int. J. Numer. Methods Engrg, 38 (1995), 1087-1121.

[29] J. E. Dendy, J. M. Hyman and J. D. Moulton, *The black box multigrid numerical homogenization algorithm,* J. Comput. Phys., 142 (1998), 80-108.

[30] C. V. Deutsch and A. G. Journel, *GSLIB: Geostatistical Software Library and User's Guide,* 2nd ed, Oxford University Press, New York, 1998.

[31] M. Dorobantu and B. Engquist, *Wavelet-based numerical homogenization*, SIAM J. Numer. Anal., 35 (1998), 540-559.

[32] J. Douglas, Jr. and T. F. Russell, *Numerical methods for convection-dominated diffusion problem based on combining the method of characteristics with finite element or finite difference procedures*, SIAM J. Numer. Anal., 19 (1982), 871-885.

[33] L. J. Durlofsky, *Numerical calculation of equivalent grid block permeability tensors for heterogeneous porous media*, Water Resour. Res., 27 (1991), 699-708.

[34] L. J. Durlofsky, R.C. Jones and W.J. Milliken, *A nonuniform coarsening approach for the scale-up of displacement processes in heterogeneous porous media*, Adv. Water Resources, 20 (1997), 335-347.

[35] B. B. Dykaar and P. K. Kitanidis, *Determination of the effective hydraulic conductivity for heterogeneous porous media using a numerical spectral approach: 1. Method*, Water Resour. Res., 28 (1992), 1155-1166.

[36] W. E, *Homogenization of linear and nonlinear transport equations*, Comm. Pure Appl. Math., XLV (1992), 301-326.

[37] W. E and B. Engquist, *The heterogeneous multi-scale methods*, Comm. Math. Sci., 1(1) (2003), 87-133.

[38] W. E and T. Y. Hou, *Homogenization and convergence of the vortex method for 2-D Euler equations with oscillatory vorticity fields*, Comm. Pure and Appl. Math., 43 (1990), 821-855.

[39] Y. R. Efendiev, Multiscale Finite Element Method (MsFEM) and its Applications, Ph. D. Thesis, Applied Mathematics, Caltech, 1999.

[40] Y. Efendiev, V. Ginting, T. Hou and R. Ewing, *Accurate multiscale finite element methods for two-phase flow simulations*, Comp. Physics. (to appear).

[41] Y. Efendiev, T. Hou and V. Ginting, *Multiscale finite element methods for nonlinear problems and their applications*, Comm. Math. Sci., 2 (2004), 553-589.

[42] Y. R. Efendiev, T. Y. Hou and X. H. Wu, *Convergence of a nonconforming multiscale finite element method*, SIAM J. Numer. Anal., 37 (2000), 888-910.

[43] Y. Efendiev and A. Pankov, *Homogenization of nonlinear random parabolic operators*, Advances in Differential Equations, 10(11) (2005), 1235-1260

[44] Y. Efendiev and A. Pankov, *Numerical homogenization of nonlinear random parabolic operators*, SIAM Multiscale Modeling and Simulation, 2(2) (2004), 237-268.

[45] Y. R. Efendiev and L. J. Durlofsky, *Numerical modeling of subgrid heterogeneity in two phase flow simulations*, Water Resour. Res.,

38(8) (2002), 1128.

[46] Y. R. Efendiev, L. J. Durlofsky and S. H. Lee, *Modeling of subgrid effects in coarse-scale simulations of transport in heterogeneous porous media,* Water Resour. Res., 36 (2000), 2031-2041.

[47] Y. R. Efendiev and B. Popov, *On homogenization of nonlinear hyperbolic equations*, Communications on Pure and Applied Analysis, 4(2) (2005), 295-309.

[48] B. Engquist, *Computation of oscillatory solutions to partial differential equations,* in: Proc. Conference on Hyperbolic Partial Differential Equations, C. Carasso, P. A. Raviart and D. Serre, eds., Lecture Notes in Mathematics, Vol. 1270, Springer-Verlag, 1987, 10-22.

[49] B. Engquist and T. Y. Hou, *Computation of oscillatory solutions to hyperbolic equations using particle methods,* in: Lecture Notes in Mathematics, Vol. 1360, C. R. Anderson and C. Greengard eds., Springer-Verlag, 1988, 68-82.

[50] B. Engquist and T. Y. Hou, *Particle method approximation of oscillatory solutions to hyperbolic differential equations,* SIAM J. Numer. Anal., 26 (1989), 289-319.

[51] B. Engquist and H. O. Kreiss, *Difference and finite element methods for hyperbolic differential equations,* Comput. Methods Appl. Mech. Engrg., 17/18 (1979), 581-596.

[52] B. Engquist and E.D. Luo, *Convergence of a multigrid method for elliptic equations with highly oscillatory coefficients,* SIAM J. Numer. Anal., 34 (1997), 2254-2273.

[53] R. Eymard, T. Gallouët and R. Herbin, *Finite volume methods*, in: Handbook of Numerical Analysis, Vol. VII, North-Holland, Amsterdam, 2000, 713-1020.

[54] R. Fetecau and T. Y. Hou, *A modified particle method for semilinear hyperbolic systems with oscillatory solutions,* Methods and Applications of Analysis, 11 (2004), 573-604.

[55] J. Fish and K. L. Shek, *Multiscale analysis for composite materials and structures*, Composites Science and Technology: An International Journal, 60 (2000), 2547-2556.

[56] J. Fish and Z. Yuan, *Multiscale enrichment based on the partition of unity*, International Journal for Numerical Methods in Engineering, 62 (2005), 1341-1359.

[57] D. Gilbarg and N. S. Trudinger, Elliptic Partial Differential Equations of Second Order, Springer, Berlin, New York, 2001.

[58] J. Glimm, H. Kim, D. Sharp and T. Wallstrom *A stochastic analysis of the scale up problem for flow in porous media,* Comput. Appl. Math., 17 (1998), 67-79.

[59] T. Y. Hou, *Homogenization for semilinear hyperbolic systems with oscillatory data,* Comm. Pure and Appl. Math., 41 (1988), 471-

495.

[60] T. Hou, X. Wu and Y. Zhang, *Removing the cell resonance error in the multiscale finite element method via a Petrov-Galerkin formulation*, Communications in Mathematical Sciences, 2(2) (2004), 185-205.

[61] T. Y. Hou, A. Westhead and D. P. Yang, *A framework for modeling subgrid effects for two-phase flows in porous media*, SIAM Multiscale Modeling and Simulation (to appear).

[62] T. Y. Hou and X. H. Wu, *A multiscale finite element method for elliptic problems in composite materials and porous media*, J. Comput. Phys., 134 (1997), 169-189.

[63] T. Y. Hou and X. H. Wu, *A multiscale finite element method for PDEs with oscillatory coefficients*, in: Proceedings of 13th GAMM-Seminar Kiel on Numerical Treatment of Multi-Scale Problems, Jan 24-26, 1997, Notes on Numerical Fluid Mechanics, Vol. 70, W. Hackbusch and G. Wittum, eds., Vieweg-Verlag, 1999, 58-69.

[64] T. Y. Hou, X. H. Wu and Z. Cai, *Convergence of a Multiscale Finite Element Method for Elliptic Problems With Rapidly Oscillating Coefficients*, Math. Comput., 68 (1999), 913-943.

[65] T. Y. Hou, D. -P. Yang and K. Wang, *Homogenization of incompressible Euler equation*, J. Comput. Math., 22 (2004), 220-229.

[66] T. Y. Hou, D. P. Yang and H. Ran, *Multiscale analysis in the Lagrangian formulation for the 2-D incompressible Euler equation*, Discrete and Continuous Dynamical Systems, 13 (2005), 1153-1186.

[67] T. Y. Hou, D. -P. Yang and H. Ran, *Multiscale computation of isotropic homogeneous turbulent flow*, in: Contemporary Mathematics, AMS Publ., 2006, Proceedings of the Conference on Inverse Problems, Multi-Scale Analysis, and Homogenization, H. Ammari and H. Kang, eds. (to appear).

[68] T. Y. Hou, D. -P. Yang and H. Ran, *Multiscale analysis and computation for the 3-D incompressible Navier-Stokes Equations*, Multiscale Modeling and Simulation (to appear).

[69] T. Y. Hou and X. Xin, *Homogenization of linear transport equations with oscillatory vector fields*, SIAM J. Appl. Math., 52 (1992), 34-45.

[70] Y. Huang and J. Xu, *A partition-of-unity finite element method for elliptic problems with highly oscillating coefficients*, 2000.

[71] T. J. R. Hughes, *Multiscale phenomena: Green's functions, the Dirichlet-to-Neumann formulation, subgrid scale models, bubbles and the origins of stabilized methods*, Comput. Methods Appl. Mech Engrg., 127 (1995), 387-401.

[72] T. J. R. Hughes, G. R. Feijóo, L. Mazzei, J. -B. Quincy, *The variational multiscale method—A paradigm for computational mechan-*

ics, Comput. Methods Appl. Mech Engrg., 166(1998), 3-24.

[73] M. Gerritsen and L. J. Durlofsky, *Modeling of fluid flow in oil reservoirs,* Annual Reviews in Fluid Mechanics, 37 (2005), 211-238.

[74] P. Jenny, S. H. Lee and H. Tchelepi, *Adaptive multi-scale finite volume method for multi-phase flow and transport in porous media,* Multiscale Modeling and Simulation, 3 (2005), 30-64.

[75] V. V. Jikov, S. M. Kozlov and O. A. Oleinik, Homogenization of Differential Operators and Integral Functionals, Springer-Verlag, 1994, Translated from Russian.

[76] I. G. Kevrekidis, C. W. Gear, J. M. Hyman, P. G. Kevrekidis, O. Runborg and C. Theodoropoulos, *Equation-free, coarse-grained multiscale computation: enabling microscopic simulators to perform system-level analysis,* Commun. Math. Sci., 1(4) (2003), 715-762.

[77] S. Knapek, *Matrix-dependent multigrid-homogenization for diffusion problems,* in: Proceedings of the Copper Mountain Conference on Iterative Methods, Vol. I, SIAM Special Interest Group on Linear Algebra, Cray Research, T. Manteuffal and S. McCormick, eds., 1996.

[78] P. Langlo and M. S. Espedal, *Macrodispersion for two-phase, immiscible flow in porous media,* Adv. Water Resources, 17 (1994), 297-316.

[79] A. M. Matache, I. Babuska and C. Schwab, *Generalized p-FEM in homogenization,* Numer. Math., 86 (2000), 319-375.

[80] A. M. Matache and C. Schwab, *Homogenization via p-FEM for problems with microstructure,* Appl. Numer. Math., 33 (2000), 43-59.

[81] J. F. McCarthy, *Comparison of fast algorithms for estimating large-scale permeabilities of heterogeneous media,* Transport in Porous Media, 19 (1995), 123-137.

[82] D. W. McLaughlin, G. C. Papanicolaou and L. Tartar, *Weak limits of semilinear hyperbolic systems with oscillating data,* in: Lecture Notes in Physics, Vol. 230, Springer-Verlag, Berlin, New York, 1985, 277-289.

[83] D. W. McLaughlin, G. C. Papanicolaou and O. Pironneau, *Convection of microstructure and related problems,* SIAM J. Applied Math, 45 (1985), 780-797.

[84] S. Moskow and M. Vogelius, *First order corrections to the homogenized eigenvalues of apPeriodic composite medium: A convergence proof,* Proc. Roy. Soc. Edinburgh, A, 127 (1997), 1263-1299.

[85] F. Murat, *Compacité par compensation,* Ann. Scuola Norm. Sup. Pisa, 5 (1978), 489-507.

[86] F. Murat, *Compacité par compensation II,* in: Proceedings of the

International Meeting on Recent Methods in Nonlinear Analysis, Rome, May 8-12, 1978, E. De Giorgi, E. Magenes and U. Mosco, eds., Pitagora Editrice, Bologna, 1979, 245-256.

[87] H. Neiderriter, *Quasi-Monte Carlo methods and pseudo-random numbers*, Bull. Amer. Math. Soc., 84 (1978), 957-1041.

[88] H. Owhadi and L. Zhang, *Homogenization of parabolic equations with a continuum of space and time scales*.

[89] H. Owhadi and L. Zhang, *Metric based up-scaling*, Comm. Pure and Applied Math. (to appear).

[90] A. Pankov, G-Convergence and Homogenization of Nonlinear Partial Differential Operators, Kluwer Academic Publishers, Dordrecht, 1997.

[91] W. V. Petryshyn, *On the approximation-solvability of equations involving A-proper and pseudo-A-proper mappings*, Bull. Amer. Math. Soc., 81 (1975), 293-312.

[92] O. Pironneau, *On the transport-diffusion algorithm and its application to the Navier-Stokes equations*, Numer. Math. 38 (1982), 309-332.

[93] R. E. Rudd and J. Q. Broughton, *Coarse-grained molecular dynamics and the atomic limit of finite elements*, Phys. Rev. B, 58 (1998), R5893.

[94] G. Sangalli, *Capturing small scales in elliptic problems using a residual-free bubbles finite element method*, Multiscale Modeling and Simulation, 1(3) (2003), 485-503.

[95] I. V. Skrypnik, Methods for Analysis of Nonlinear Elliptic Boundary Value Problems, Translations of Mathematical Monographs, Vol. 139, American Mathematical Society, Providence, RI, 1994. Translated from the 1990 Russian original by Dan D. Pascali.

[96] T. Strinopoulos, Upscaling of Immiscible Two-Phase Flows in an Adaptive Frame, Ph. D. Thesis, California Institute of Technology, Pasadena, 2005.

[97] T. Strouboulis, I. Babuška and K. Copps, *The design and analysis of the generalized finite element method*, Comput. Methods Appl. Mech. Engrg., 181 (2000), 43-69.

[98] L. Tartar, *Compensated compactness and applications to P.D.E.*, in: Nonlinear Analysis and Mechanics, Heriot-Watt Symposium, Vol. IV, R. J. Knops, ed., Research Notes in Mathematics, Vol. 39, Pitman, Boston, 1979, 136-212.

[99] L. Tartar, *Solutions oscillantes des équations de Carleman*, Seminaire Goulaouic-Meyer-Schwartz (1980-1981), exp. XII. Ecole Polytechnique (Palaiseau), 1981.

[100] L. Tartar, *Nonlocal effects induced by homogenization*, in: PDE and Calculus of Variations, F. Culumbini, et al. eds., Birkhäuser, Boston, 1989, 925-938.

[101] X. H. Wu, Y. Efendiev and T. Y. Hou, *Analysis of upscaling absolute permeability,* Discrete and Continuous Dynamical Systems, Series B, 2 (2002), 185-204.

[102] P. M. De Zeeuw, *Matrix-dependent prolongation and restrictions in a blackbox multigrid solver,* J. Comput. Applied Math., 33 (1990), 1-27.

[103] S. Verdiere and M. H. Vignal, *Numerical and theoretical study of a dual mesh method using finite volume schemes for two-phase flow problems in porous media,* Numer. Math., 80 (1998), 601-639.

[104] T. C. Wallstrom, M. A. Christie, L. J. Durlofsky and D. H. Sharp, *Effective flux boundary conditions for upscaling porous media equations,* Transport in Porous Media, 46 (2002), 139-153.

[105] T. C. Wallstrom, M. A. Christie, L. J. Durlofsky and D. H. Sharp, *Application of effective flux boundary conditions to two-phase upscaling in porous media,* Transport in Porous Media, 46 (2002), 155-178.

[106] T. C. Wallstrom, S. L. Hou, M. A. Christie, L. J. Durlofsky and D. H. Sharp, *Accurate scale up of two phase flow using renormalization and nonuniform coarsening,* Comput. Geosci, 3 (1999), 69-87.

[107] E. Zeidler, Nonlinear Functional Analysis and Its Applications, II/B, Nonlinear Monotone Operators, Springer-Verlag, New York, 1990. Translated from the German by the author and Leo F. Boron.

An Introduction of Elastic Complex Fluids: An Energetic Variational Approach

Chun Liu

Department of Mathematics, Pennsylvania State University
University Park, PA 16803, USA
E-mail: liu@math.psu.edu

Abstract

In this chapter, we consider an inverse problem of determining the corrosion occurring in an inaccessible interior part of a pipe from the measurements on the outer boundary. The problem is modeled by Laplace's equation with an unknown term γ in the boundary condition on the inner boundary. Based on the Maz'ya iterative algorithm, a regularized BEM method is proposed for obtaining approximate solutions for this inverse problem. The numerical results show that our method can be easily realized and is quite effective.

1 Introduction

Complex fluids such as polymeric solutions, liquid crystal solutions, pulmonary surfactant solutions, electro-rheological fluids, magneto-rheological fluids and blood suspensions exhibit many intricate rheological and hydrodynamic features that are very important to biological and industrial processes. Applications include the treatment of airway closure disease by surfactant injection; polymer additive to jets in inkjet printers, fuel injection, fire extinguishers; magneto-rheological damping of structural vibrations etc. The segregation, migration and aggregation of the particles and the stretching, coiling and entanglement of the molecules in the complex fluids that endows them with the unique rheological and hydrodynamic properties required for specific biological, physiological and industrial needs. One good example is the migration of blood cells in arteries towards the center axis (the Fahreus-Lynquist effect). This segregation leaves a low viscosity plasma marginal layer that helps reduces the overall resistance to blood flow. This complex physiological rheology has important implications in blood pressure, clotting, plaque formation and other cardiovascular diseases. An important goal of the large and multi-disciplinary field of fluid mechanics is to derive continuum partial differential equations (field equations) to describe the rheology of these various fluids and to solve these equations to explain and predict their macroscopic behavior.

The most common origin and manifestation of anomalous phenomena in complex fluids are different "elastic" effects. They can be the elasticity of deformable particles, elastic repulsion between charged liquid crystals, polarized colloids or multi-component phases, elasticity due to microstructures, or bulk elasticity endowed by polymer molecules in viscoelastic complex fluids. The physical properties are purely determined by the interplay of entropic and structural intermolecular elastic forces and interfacial interactions. These elastic effects can be represented in terms of certain internal variables, for example, the orientational order parameter in liquid crystals (related to their microstructures), the distribution density function in the dumb-bell model for polymeric materials, the magnetic field in magneto-hydrodynamic fluids, the volume fraction in mixture of different materials etc. The different rheological and hydrodynamic properties can be attributed to the special coupling between the transport of the internal variable and the induced elastic stress. In our energetic formulation, this represents a competition between the kinetic energy and the elastic energy. We look at the following system (a simplified Ericksen-Leslie system modeling the flow of nematic liquid crystals) as an example for such complex fluids:

$$u_t + (u \cdot \nabla)u + \nabla p - \nu \Delta u + \lambda \nabla \cdot (\nabla d \odot \nabla d) = 0, \qquad (1.1)$$

$$d_t + (u \cdot \nabla)d - \gamma(\Delta d - f(d)) = 0, \qquad (1.2)$$

with $\nabla \cdot u = 0$, where u represents the flow velocity, p the pressure, d represents the *normed* director, $f(d) = F'(d)$ where $F(d)$ is the bulk part of the elastic energy. It is the coupling between the transport of d (material derivative here) and the induced elastic stress $(\nabla d \odot \nabla d)_{ij} = \sum_{k=1}^{n}(\nabla_i d_k)(\nabla_j d_k)$ that yields the following energy law, which presents the dissipative nature of the system:

$$\frac{1}{2}\frac{d}{dt}\int_\Omega (|u|^2 + \lambda|\nabla d|^2 + 2\lambda F(d))dx = -\int_\Omega (\nu|\nabla u|^2 + \lambda\gamma|\Delta d - f(d)|^2)dx. \qquad (1.3)$$

On the other hand, the force balance (momentum) equation can be derived by the Least Action Principle, using the total energy functional and the way the internal variable d is transported. The competition between kinetic and elastic energy also produces the specific properties of the system, such as the stability and regularity of the hydrostatic configurations. When applied to micro-particles or molecules, the elastic energy determines the microstructures formation and how they interact with the fluid. The understanding of such underlying structures is also crucial in designing the accurate numerical algorithms in order to simulate the system, especially when the solutions involve singularities.

Most complex fluid behavior results from the multi-scale properties of the fluid material at the micro-structure scales. Hence, understanding complex fluid rheology and hydrodynamics must necessarily begin at the molecular and particulate level. The Fokker-Planck, Ginzburg-Landau or Liouville type statistical equations describing the nanoscale molecular dynamics or the microscale particulate dynamics are used to obtain rheological constitutive equations through least action principles, as have been done for viscoelastic polymeric fluids and liquid crystal solutions. The systems will satisfy the energy law (Second Law of Thermodynamics). The resulting partial differential equation system will involve multiple scales. In order to obtain the effective continuum equations at the macroscopic scale, mean field theories are often invoked to obtain closure in such field theoretic approaches. When these constitutive equations are inserted back into the Cauchy equation for force balance, the desired partial differential equation results.

The Navier-Stokes equation is the simplest of these, and fortunately, it does obey an energy law. On the other hand, the dumb-bell model equation for polymeric materials loses the energy law after closure, even for the simplest Hookean case. Recently, more and more studies show that this classical approach is inadequate due to several deficiencies. Pertinent physics at the particulate and molecular level remains elusive for many complex fluids. For example, blood cell segregation has been attributed to particle deformation, inertia, asymmetry and a host of other origins. Even when the physics are known, some microscale phenomena remain unexplored due to mathematical and/or numerical difficulties. For example, defects in the liquid crystal have been shown to produce bulk flow due to the elastic pressure gradient they generate. The resulting flow can also destroy the defects and hence change the bulk rheology.

In this lecture note, we intend to introduce some of the mathematical tools, modeling, analysis and numerics, that are useful in studying these important and complicated materials. The brief description of the contents will suffice to show that this note is in no sense a systematic study of the broad area of complex fluids. Many important topics are not touched at all here. Moreover, due to the time constraints, some of the materials were taken out from other papers. We hope this will just serve as an introduction and some reference for the students who will become interested in these fascinating subjects.

2   Calculus of variations

We begin the short course by reviewing some basic mathematical tools in the theory of calculus of variations. All the materials can be found in the following references [33, 34, 90].

- L. C. Evans. *Partial Differential Equations*. AMS, 1998.
- L. C. Evans. *Weak Convergence Methods for Nonlinear Partial Differential Equations*. AMS, 1990.
- M. Struwe. *Variational Methods*. Springer-Verlag, 1990.

2.1   Euler-Lagrange equations

For a given Banach space A and a functional

$$E : A \longrightarrow D, \tag{2.1}$$

the Euler-Lagrange equation is defined as

$$DE(u) = 0, \tag{2.2}$$

where $DE : A \to A^*$ is the Fréchet derivative defined by

$$\frac{d}{d\epsilon}|_{\epsilon=0} E(u + \epsilon v) = (DE(u), v) = 0. \tag{2.3}$$

Example.   For a functional $W(u, \nabla u)$, the corresponding Euler-Lagrange equation will be

$$-\nabla \cdot \frac{\partial W}{\partial \nabla u} + \frac{\partial W}{\partial u} = 0, \tag{2.4}$$

which in the weak form will be

$$\left(\frac{\partial W}{\partial \nabla u}, \nabla v \right) + \left(\frac{\partial W}{\partial u}, v \right) = 0 \tag{2.5}$$

for any test function v.

2.2   Direct methods

The following basic concepts are crucial for the direct method of calculus of variation:

- Lower semicontinuous: $\{u \in A | W(u) > a\}$ is open in A.
- Sequentially weak lower semicontinuous: If $u_n \to u$ weakly in A, then
$$W(u) \leqslant \lim_{n \to \infty} \inf W(u_n). \tag{2.6}$$

- Coercivity: If $|u_n|_A \to \infty$, then $W(u) \to \infty$.

With these concepts, we can state the following theorem.

Theorem 1. *If A is a reflexive Banach space, W is a nonnegative functional and is both coercive and lower semicontinuous, then W attains its infinimum in A.*

Proof. See Evan's or Struwe's book.

2.3   Convexity

Given a functional $W(x, u, \nabla u)$, it is convex if $W_{\nabla_i u \nabla_j u}(x, u, \nabla u)$ is nonnegative when u is a minimizer.

Theorem 2. *If W is bounded below, convex in ∇u. Then W is weakly lower semicontinuous.*

Proposition 1. *If W is coercive and convex, then there exists at least one minimizer.*

Theorem 3 (Uniqueness). *If W is strictly convex, then the minimizer is unique.*

Example. Given $u \in W^{1,q}(\Omega)$ and

$$|W(x, z, p)| \leqslant c(|p|^q + |z|^q + 1), \tag{2.7}$$
$$|\nabla_p W|, |\nabla_z W| \leqslant c(|p|^{q-1} + |z|^{q-1} + 1), \tag{2.8}$$

the weak solution of Euler-Lagrange equation is

- The minimizer satisfies the Euler-Lagrange equation.
- Coercivity: $W(x, z, p) \geqslant \alpha |p|^q - \beta$.
- Convexity in p gives the existence of minimizer.

2.4   Dynamics

The gradient flow (fastest decent): In the case when only the long time behavior of the solution are important, the gradient flow will determine the properties of the solution. Moreover, the gradient flow also gives a method to achieve the stationary solution of the Euler-Lagrange equations:

$$u_t = -\gamma \frac{\delta W}{\delta u}, \tag{2.9}$$

where γ represents the relaxation time. The solution of the above equation (with either Dirichlet or natural boundary conditions) satisfies the following dissipative law:

$$\frac{d}{dt}\int_{\Omega} W\,dx = -\frac{1}{\gamma}\int_{\Omega}|u_t|^2\,dx. \qquad (2.10)$$

Remark 1 (Long time behavior). For time $t \to \infty$, from Fubini's Theorem, there exists a subsequence t_i such that $u_t(\cdot, t_i) \to 0$ and

$$\frac{\delta W}{\delta u} \to 0, \qquad (2.11)$$

hence $u(\cdot, t_i)$ approaches to a stationary solution.

Damped wave equation:

$$\epsilon u_{tt} + u_t = -\gamma\frac{\delta W}{\delta u}. \qquad (2.12)$$

The energy law becomes

$$\frac{d}{dt}\int_{\Omega}\left(\gamma W + \frac{\epsilon}{2}|u_t|^2\right)dx = -\int_{\Omega}|u_t|^2\,dx. \qquad (2.13)$$

It is from this energy law that we can see the long time behavior of the solution are determined by the gradient flow.

2.5 Hamilton's principle

Hamilton's principle, which is also referred to as principle of virtual work or the least action principle, is the most fundamental principle in mechanics. In fact, it gives the momentum equations — the force balance equations.

The material presented here can be founded in the following classical references [1, 3, 74]:

- V. I. Arnold. *Mathematical Methods of Classical Mechanics.* Springer-Verlag, 1978.

- R. Abraham, J. E. Marsden. *Fundations of Mechanics.* Springer-Verlag, 1978.

- J. E. Marsden, T. S. Ratiu. *Introduction to Mechanics and Symmetry.* Springer-Verlag, 1999.

2.5.1 Flow map and deformation tensor

The evolution of all materials involves the following basic mechanical concepts:

- Lagrangian coordinate (original labelling): X.
  Eulerian coordinate (observer's coordinate): x.
- Flow map (trajectory): $x(X, t)$ such that

$$x_t = u(x(X, t), t), \quad x(X, 0) = X, \tag{2.14}$$

  where $u(x, t)$ is the velocity field.

Remark 2. If u is Lip in x then the flow map is uniquely determined.

- Deformation: $F_{ij}(X, t) = \frac{\partial x_i}{\partial X_j}$.
  Without ambiguity, we can define $F(x(X, t), t) = F(X, t)$. The simple application of the chain rule gives the following important transport equation of F:

$$F_t + u \cdot \nabla F = \nabla u F. \tag{2.15}$$

- Each of the following equivalent statements will represent the incompressibility of the material.
  1. $\det F = 1$;
  2. $\operatorname{div} u = 0$ (from the identity $\delta \det F = \det F \operatorname{tr}(F^{-1} \delta F)$).
  3. $\nabla \cdot F_t + (u \cdot \nabla)(\nabla \cdot F) = 0$.

2.5.2 Variation of the domain v.s. variation of the function

Given an energy functional $W(\phi, \nabla \phi)$, depending on some variable ϕ, in order to find the critical point, we can employ each of the following two methods:

- Euler-Lagrange equation (variation with respect to ϕ): $\frac{\delta W}{\delta \phi} = 0$, which is expressed in the weak form as

$$\left(\frac{\partial W}{\partial \nabla \phi}, \nabla \psi \right) + \left(\frac{\partial W}{\partial \phi}, \psi \right) = 0 \tag{2.16}$$

  for any test function ψ.
  We note that the usual energy estimates are derived by setting $\psi = \phi$.
- Variation with respect to domain: $\frac{\delta W}{\delta x} = 0$, and the result can be expressed in the weak form as

$$\left(\frac{\partial W}{\partial \nabla \phi} \otimes \nabla \phi - WI, \nabla y \right) = 0 \tag{2.17}$$

  for any test function y.

The formal equivalency of the two procedure is reflected in the following theorem.

Theorem 4. *Given an energy functional $W(\phi, \nabla\phi)$, all solutions of the Euler-Lagrangian equation*

$$-\nabla \cdot \frac{\partial W}{\partial \nabla\phi} + \frac{\partial W}{\nabla\phi} = 0$$

also satisfy the equation

$$\nabla \cdot \left(\frac{\partial W}{\partial \nabla\phi} \otimes \nabla\phi - WI \right) = 0.$$

The proof of the theorem is the consequences of direct computations. From this theorem, we can immediately make the following remarks:

- Pohozaev inequality: set $y = x$ (same as multiply Euler-Lagrange equation by $x \cdot \nabla\phi$). This extra inequality is very important in the study of semilinear elliptic equations [34, 90].

- The variation of the domain require more regularity than that of the normal weak solutions of the Euler-Lagrange equations. This is in connection of the stationary weak solution for harmonic maps [86].

2.5.3   Least action principle

The force balance equation (momentum conservation law) is the state such that the flow map minimizes the action functional:

$$\mathcal{A}(x) = \int_0^T \int_{\Omega_0} \left(\frac{1}{2}\rho|u|^2 - W(\phi(x)) \right) \det F \, dX \, dt. \qquad (2.18)$$

Here $W(\phi)$ is the elastic internal energy, $\frac{q}{2}\rho|u|^2$ is the kinetic energy.

2.6   Constraint problems

Most physical problems involve finding the minimizers (critical points) in a constraint class of functions. The method of Lagrange multiplier is the basic tool for such a problem. However, this brings some extra difficulties and we will illustrate this using the following examples.

2.6.1   Harmonic maps

This is a simple, but most classical example that can illustrate the role of constraint in the calculus of variations and the difficulties associated with it [86, 85, 5, 9, 44, 61].

For any function $u : \Omega \to B_1(0)$ with target space the unit sphere, we want to find the minimize the following Dirichlet energy:

$$W(u) = \int_\Omega \frac{1}{2}|\nabla u|^2 \, dx, \tag{2.19}$$

with Dirichlet boundary condition:

$$u|_{\partial\Omega} = u_0. \tag{2.20}$$

The Euler-Lagrange equation will be

$$-\Delta u = \lambda(x)u, \tag{2.21}$$

where the Lagrange multiplier $\lambda(x) = |\nabla u|^2$ (with the help of the identity $\Delta u u = \Delta \frac{|u|^2}{2} - |\nabla u|^2$). Notice the difficulty of high nonlinearity on the right-hand side of the equation. Moreover, the right-hand side does posses the property of being a form of total derivation (like a Jacobian). Using this, Helein obtained the regularity in 2-dimensional cases [90].

2.6.2   Liquid crystals

For the uniaxial nematic liquid crystal materials, the bulk energy density in Ericksen's model are assumed to depend on the orientation vector (optical director) n, with $|n| = 1$, and the orientational order $s \in [-\frac{1}{2}, 1]$. Again we do not consider the effects due to surface energies, applied fields, etc. The energy is given by

$$0 \leqslant w(s, n, \nabla s, \nabla n) \equiv w_0(s) + w_2(s, n, \nabla s, \nabla n). \tag{2.22}$$

Here the behavior of $w_0(s)$ is the bulk part of the energy.

The case $s = 1$ corresponding to the property that each molecules is perfectly aligned, and the case $s = -\frac{1}{2}$ means all molecules are lie in a plane perpendicular to the optical axis. Both situations are physically unrealistic and therefore we can have

$$w_0\left(-\frac{1}{2}\right) = w_0(1) = +\infty. \tag{2.23}$$

The term w_2 takes the form

$$2w_2 = k_1 |\operatorname{div} n|^2 + k_2 |n \cdot \operatorname{curl} n|^2 + k_3 |n \wedge \operatorname{curl} n|^2 \tag{2.24}$$
$$+ (k_2 + k_4)[\operatorname{tr} (\nabla n)^2 - (\operatorname{div} n)^2] + L_1 |\nabla s|^2 + L_2 (\nabla s \cdot n)^2 \tag{2.25}$$
$$+ L_3 (\nabla s \cdot n)\operatorname{div} n + L_4 \nabla s \cdot (\nabla n)n, \tag{2.26}$$

where k's and L's are functions of s as well as the temperature θ.

With the help of the following identity:

$$\text{div } (f[(\nabla n)n - (\text{div } n)n])$$
$$= f[\text{tr } (\nabla n)^2 - (\text{div } n)^2] + \nabla f \cdot [(\nabla n)n - (\text{div } n)n], \quad (2.27)$$

$$2W_2 = \overline{K}_1(\text{div } n)^2 + K_2|n \cdot \text{curl}n|^2 + \overline{K}_3|n \wedge \text{curl}n|^2$$
$$+(K_2 + K_4)[\text{tr } (\nabla n)^2 - (\text{div } n)^2]$$
$$+K_5|\nabla s - (\nabla s \cdot n)n - \nu(\nabla n)n|^2$$
$$+K_6(\nabla s \cdot n - \sigma \text{ div } n)^2, \quad (2.28)$$

where

$$\overline{K}_1 = K_1 - \sigma^2 K_6 = K_1 - \frac{L_3^2}{4(L_1+L_2)},$$
$$\overline{K}_3 = K_3 - \nu^2 K_5 = K_3 - \frac{L_4^2}{4L_1}, \quad (2.29)$$
$$K_5 = L_1, \quad K_6 = L_1 + L_2,$$
$$\nu = -\frac{L_4}{2L_1}, \quad \sigma = -\frac{L_3}{2(L_1+L_2)}.$$

The size of various constants are characterized below:

$$\overline{K}_1 > 0, \quad K_2 > |K_4|, \quad \overline{K}_3 > 0,$$
$$K_5 > 0, \quad K_6 > 0, \quad \text{for } s \neq 0,$$
$$\sigma, \nu \cong o(s), \quad \overline{K}_1, K_2, \overline{K}_3, K_4 \cong o(s^2), \quad (2.30)$$
$$K_6 \cong o(s).$$

The simplest form of the bulk energy density is

$$w_0(s) + k|\nabla s|^2 + s^2|\nabla n|^2. \quad (2.31)$$

Here one has $K_1 = K_3 = s^2$, $K_4 = L_2 = L_3 = L_4 = 0$, and $L_1 = k$. We shall see later this form of energy functional is closely related to the energy functional of maps from a domain in $\mathbb{R}^3$ to a circular cone in $\mathbb{R}^4$ or $\mathbb{R}^{3,1}$ the Minkowski space.

The classical Oseen-Frank model can be derived from the Ericksen's model by imposing the additional constraint on the orientational order, $s = s^*$.

We note that the simplest form of such energy densities is $2w = |\nabla n|^2$. This corresponds to the case $k_1 = k_2 = k_3 = 1$ and $k_4 = q = 0$. The corresponding mathematical problem is to study harmonic maps from a domain to S^2 or $\mathbb{R}P^2$.

Strong anchoring condition. When the surface of a container of liquid crystals is specially treated, the orientation of the liquid crystals molecules near the surface of container will aligned with the treatment and hence can be specified. This is usually referred to as the *strong anchoring condition.* Mathematically we can describe it as following Dirichlet boundary value problem.

2.6.3 Methods of penalty

We will just look at the harmonic problem. In order to avoid the non-linearity in the problem, we will introduce the following approximate problem [18,91]:

$$\min_{u \in H^1(R^3)} \int_\Omega \left(\frac{1}{2} |\nabla u|^2 + \frac{1}{4\epsilon^2} (|u|^2 - 1)^2 \right) dx. \tag{2.32}$$

The above functional is also called the Ginzburg-Landau functional, which arises from the theory of superconductivity [6,23,26].

We can see, as $\epsilon \to 0$, u will convergent to a unit vector. The Euler-Lagrange equation of the approximate problem is

$$-\Delta u + \frac{1}{\epsilon^2} (|u|^2 - 1)u = 0. \tag{2.33}$$

For each fixed ϵ, the solution is smooth. As ϵ approaches zero, the solution of the Ginzburg-Landau equation will convergence (weakly) to a solution of the harmonic map [91].

Finally, we will discuss a new type of relaxation that was discussed with M. Chipot and D. Kinderlehrer [19].

We will study the following minimization problem under relaxed constraint:

$$\min_{u \in A_\epsilon} E(u), \tag{2.34}$$

where $A_\epsilon = \{v \in H^1(\Omega), v|_{\partial\Omega} = g(x), \int_\Omega (|u|^2 - 1)^2 \, dx \leqslant \epsilon^2\}$. Notice, the relaxation is in the constraint, rather in the energy functional itself.

Lemma 1. *If u_ϵ is a minimizer of problem (2.34), then*

$$\int_\Omega (|u_\epsilon|^2 - 1)^2 \, dx = \epsilon^2. \tag{2.35}$$

Proof. We will prove this lemma by contradiction. If the statement is false, that is,

$$\int_\Omega (|u_\epsilon|^2 - 1)^2 \, dx < \epsilon^2, \tag{2.36}$$

then for variations $\delta\phi$ of small δ, we have $u_\epsilon + \delta\phi \in A_\epsilon$. Hence we have u_ϵ will satisfies the Euler-Lagrange equation

$$-\Delta u_\epsilon = 0, \tag{2.37}$$

with boundary condition $g(x)$. So we get $u_\epsilon = \hat{u}$ which is independent of ϵ. $\int_\Omega (|u_\epsilon|^2 - 1)^2 \, dx = \int_\Omega (|\hat{u}|^2 - 1)^2 \, dx > \epsilon$ for ϵ sufficiently small, we get a contradiction.

The following lemma is obvious from the definitions.

Lemma 2.

$$\min_{u \in A_\epsilon} E(u) \leqslant \min_{|u|=1} E(u) = M \qquad (2.38)$$

where M is a constant independent of ϵ.

Proof. Notice here we have $\min_{|u|=1} E(u) = \min_{u \in A_0} E(u)$ and the lemma follows immediately.

From the lemma we see that if u_ϵ is the minimizer of problem (2.34) for each ϵ, then they all satisfy the following equations:

$$-\Delta u_\epsilon = \lambda_\epsilon (|u_\epsilon|^2 - 1) u_\epsilon, \qquad (2.39)$$

with boundary condition

$$u_\epsilon|_{\partial\Omega} = g(x), \qquad (2.40)$$

and the uniform bound

$$\int_\Omega |\nabla u_\epsilon|^2 \, dx \leqslant M. \qquad (2.41)$$

Pass to the limit of $\epsilon \to 0$, we have (up to a subsequence) $u_\epsilon \to u^*$ weakly in $H^1(\Omega)$, strongly in $L^2(\Omega)$ and almost everywhere in Ω.

One takes the cross product of equation (2.39) by u_ϵ to get

$$\nabla \cdot (\nabla u_\epsilon \times u_\epsilon) = 0. \qquad (2.42)$$

Pass to the limit in the weak formulation, we can get that

$$\nabla \cdot (\nabla u^* \times u^*) = 0. \qquad (2.43)$$

On the other hand, since we have

$$\int_\Omega (|u_\epsilon|^2 - 1)^2 \, dx \to 0, \qquad (2.44)$$

by Fatou's lemma,

$$|u^*| = 1. \qquad (2.45)$$

We will have the following main theorem:

Theorem 5. *u^* satisfies the harmonic map equation:*

$$-\Delta u^* = |\nabla u^*|^2 u^*. \qquad (2.46)$$

Proof. To prove the theorem, we use (2.43) and get that

$$\Delta u^* \times u^* = \nabla \cdot (\nabla u^* \times u^*) = 0. \qquad (2.47)$$

This means that Δu^* is parallel to u^*. In a weak form, we see that, for any $v \in Hq_0(\Omega)$, $v - (v \cdot u^*)u^*$ is perpendicular to u^*. Then (2.47) is equivalent to

$$\int_\Omega \nabla u^* \nabla (v - (v \cdot u^*)u^*) \, dx = 0.$$

The left-hand side is equal to

$$\int_\Omega \left(\nabla u^* \nabla v - |\nabla u^*|^2 v \cdot u^* - \nabla (v \cdot u^*)u^* \cdot \nabla u^* \right) dx.$$

The last term is equal to 0 since u^* is unit length,

$$\int_\Omega \left(\nabla u^* \nabla v - |\nabla u^*|^2 v \cdot u^* \right) dx = 0,$$

which is exactly the weak form of (2.46).

The cross product method in proving the convergence of the sequence was well known in the studying of the harmonic maps with target space being a sphere [91].

Finally, the following lemma gives more detailed information of the Lagrange multiplier λ_ϵ in equation (2.39).

Lemma 3. *Suppose that Ω is strictly star-shaped with respect to 0 and the boundary $\partial \Omega$ is C^1. If λ_ϵ is the Lagrange multiplier as in (2.39), then*

$$M_1 \leqslant -\lambda_\epsilon \epsilon^2 \leqslant M_2, \tag{2.48}$$

where M_i are the constants independent of ϵ.

Proof. We use the Pohozaev type of argument. Let Ω_h, $0 \leqslant h \leqslant h_0$, be the star-shaped domain that are closed to the original domain. h is the distance between the boundaries. The existence of these neighbouring domains can be adjustified by the smoothness of the domain. We multiplier equation (2.39) by $(x \cdot \nabla)u_\epsilon$ and integrate over the domain Ω_h:

$$-\int_{\Omega_h} (\Delta u_\epsilon) \cdot (x \cdot \nabla)u_\epsilon \, dx = \int_{\Omega_h} \lambda_\epsilon (|u_\epsilon|^2 - 1)u_\epsilon \cdot (x \cdot \nabla)u_\epsilon \, dx. \tag{2.49}$$

The right-hand side is equal to

$$\int_{\Omega_h} \lambda_\epsilon (|u_\epsilon|^2 - 1)u_\epsilon \cdot (x \cdot \nabla)u_\epsilon \, dx = \frac{\lambda_\epsilon}{4} \int_{\Omega_h} (x \cdot \nabla)(|u_\epsilon|^2 - 1)^2 \, dx$$

$$= \frac{d\lambda_\epsilon}{4} \int_{\Omega_h} (|u_\epsilon|^2 - 1)^2 \, dx$$

$$= \frac{d\lambda_\epsilon}{4} \epsilon^2. \tag{2.50}$$

The last equality uses the integration by parts and constraint (2.35).

On the other hand, the left-hand side is equal to

$$-\int_{\Omega_h} (\Delta u_\epsilon) \cdot (x \cdot \nabla) u_\epsilon \, dx$$

$$= \frac{2-d}{2} \int_{\Omega_h} |\nabla u_\epsilon|^2 \, dx - \frac{1}{2} \int_{\partial\Omega_h} |\nabla u_{\epsilon\nu}|^2 (x \cdot \nu) \, dx$$

$$+ \int_{\partial\Omega_h} (x \cdot \tau) u_{\epsilon\tau} (x \cdot \nu) u_{\epsilon\nu} \, dx$$

$$> \frac{2-d}{2} M - \frac{1}{2} \int_{\partial\Omega_h} |\nabla u_{\epsilon\nu}|^2 (x \cdot \nu) \, dx$$

$$+ \int_{\partial\Omega_h} (x \cdot \tau) u_{\epsilon\tau} (x \cdot \nu) u_{\epsilon\nu} \, dx \qquad (2.51)$$

for $n \geqslant 2$.

Now, we can integrate in the normal direction. We get that

$$M_1(h_0) \leqslant -\lambda_\epsilon \epsilon^2 \leqslant M_2(h_0). \qquad (2.52)$$

Notice that we have used Cauchy's inequality and the property of strict star-shapeness of the domain.

The last theorem show that the Euler-Lagrange multiplier λ_ϵ is of order $O(\frac{1}{\epsilon^2})$. The constant h_0, hence the size of M_i, is determined by the smoothness of the boundary $\partial\Omega$.

3   Navier-Stokes equation

There are many references on the theory of Navier-Stokes equations [20, 53, 72, 94]. We will just list out some of them here:

- P. Constantin and C.Foias, *Navier-Stokes Equation.* University of Chicago Press, 1988.

- R. Temam, *Navier-Stokes Equation, Theory and Application.* AMS Chelsea Publishing, 1984.

- A. Majda and A. Bertozzi, *Vorticity and Incompressible Flow.* Cambridge University Press, 2002.

- L. D. Landau and E. M. Lifshitz, *Fluid Mechanics.* Pergamon Press, 1987.

- G. K. Bachelor *An Inroduction to Fluid Mechanics.* Cambridge University Press, 1967.

3.1   Newtonian fluids

The hydrodynamical systems for Newtonian fluids include the following equations:

Balance of mass:

$$\rho_t + \nabla \cdot (\rho u) = 0. \tag{3.1}$$

If $\rho = \rho_0$ is a constant, then $\nabla \cdot u = 0$. Notice, however that the reverse is not true.

Momentum equation (force balance equation):

$$\rho(u_t + u \cdot \nabla u) + \nabla p = \mu \Delta u. \tag{3.2}$$

This equation can be derived from the least action principle. Introduction of the viscosity through: postulating the dissipative term in energy law; or introduce random perturbation in the variation process (Peskin's work [79]).

Finally, the energy law:

$$\frac{d}{dt} \int_\Omega \frac{1}{2} \rho |u|^2 \, dx = -\int_\Omega \mu |\nabla u|^2 \, dx. \tag{3.3}$$

Notice in the incompressible fluids, with $\nabla \cdot u = 0$, the energy equation is not an independent equation. It can be derived from the conservation of mass and the conservation of momentum equations.

In the case that the initial term can be neglected, the system will be the Stokes equation, which is a linear equation in velocity. For the inviscid fluids, the system becomes the Euler equation.

All the system will be equipped with proper initial and boundary conditions.

3.1.1   Existence of global weak solution

The way to show the global existence of weak solutions (consistent with the energy laws), which is called the Larey-Hopf solution, is through the usual Galerkin scheme.

The goal of the Galerkin scheme is using the separation of variables method to approximate the original the problem (an infinite dimensional evolution problem) by finite dimensional ODE systems.

We first define the functional space:

$$\begin{cases} V = \{u \in H^1(\Omega) : \operatorname{div} u = 0\}, \\ H = \{u \in L^2(\Omega) : \operatorname{div} u = 0\}. \end{cases} \tag{3.4}$$

Also, V' is the dual space of V. We have V is a subset of H which is a subset of V'.

The Stokes operator A is defined as a map from H onto the space

$$D(A) = \{u \in H, \, \Delta u \in H\}, \tag{3.5}$$

such that for any given f, $u = A^{-1}(f)$ satisfies the following Stokes problem:

$$-\Delta u + \nabla p = 0, \, \nabla \cdot u = 0. \tag{3.6}$$

The operator A is positive, selfadjoint. Since the inverse of A is a linear continuous map from H to $D(A)$ and it is compact. A can be viewed as a selfadjoint operator in H and its eigenfunction ϕ_i form a basis of H.

Theorem 6. *For any $f \in L^2(0, T, V')$ and the initial condition $u_0 \in H$ given, there exists a weak solution u to the Navier-Stokes equation such that*

$$u \in L^2(0, T; V) \cap L^\infty(0, T; H). \tag{3.7}$$

Moreover, the solution is unique when the dimension is 2.

Sketch of the proof. We look at the approximation of the solution u in the finite dimensional subspace spanned by the eigenfunction of the Stokes operator A. For any given integer n,

$$u_n = \sum_{i=1}^{n} g_{in}(t)\phi_i. \tag{3.8}$$

The function u_m satisfies the system:

$$(u_n, v)_t + \mu((u_n, v)) - (u_m u_m, \nabla v) = (f, v), \tag{3.9}$$
$$u_n(0) = P_n u_0,$$

where P_n is the orthogonal projection in H onto span$\{\phi_i\}$ and the test function v is any function in this space.

The above system is equivalent to an ODE system of the coefficients g_{in}. The ODE system always has a local solution. Moreover, we still have the following a priori estimate:

$$\frac{d}{dt} \frac{1}{2} |u_n|^2 + \mu |\nabla u_n|^2 = (f, u_n). \tag{3.10}$$

Hence,

$$\sup |u_n|^2 + \int_0^T \mu |\nabla u_n|^2 \, dt \leqslant M, \tag{3.11}$$

where M is a constant depending on the initial condition and f. From this, first, we can see that the ODE solution exists all the time. Secondly we can extract a subsequence (still denote as u_n) which convergence weakly in the space $u \in L^2(0, T; V) \cap L^\infty(0, T; H)$. Finally, Aubin-Lions's compactness theorem shows that the weak limit is a solution of the original Navier-Stokes equation.

3.1.2 Existence of classical solution

Theorem 7. *For given* $f \in L^\infty(0, T; H)$ *and* $u_0 \in V$, *in 2 dimensional case, there exists a unique global solution*

$$u \in L^2(0, T; D(A) \cap L^\infty(0, T; V). \tag{3.12}$$

However, such a solution exists when μ *is large or* u_0 *is small.*

Sketch of the proof. We will prove the theorem using the higher order energy estimates. For this we multiply the equation by Au and integrate by parts, we have

$$\frac{1}{2}\frac{d}{dt}|\nabla u|^2 + \mu|Au|^2 + (uu, Au) = (f, Au). \tag{3.13}$$

The right-hand side can be bounded as

$$(f, AU) \leqslant \frac{\mu}{4}|Au|^2 + \frac{1}{\mu}|f|^2.$$

The trilinear term can be estimated by

$$(uu, Au) \leqslant |u|_4^2|Au|.$$

Using Ladyzhenskaya's inequalities to interpolate the L^4 norm by the L^2 and H^1 norms.

In 2-D, we have

$$|u|_4^2 \leqslant |u|_2|\nabla u|_2$$

and in 3-D

$$|u|_4^2 \leqslant |u|_2^{1/2}|\nabla u|_2^{3/2}.$$

Hence we can show that in 2-D, we have the global classical solution. In 3-D, we can use the situation that either μ is large or u_0 is small to get that $|\nabla u|_2$ is in fact monotone in time.

3.1.3 Regularity

Theorem 8. *If a solution* $u \in L^p(0, T; L^q(\Omega))$ *is a solution of Navier-Stokes equation and* $2/p + 3/q \leqslant 1$, *then the solution is a unique classical solution.*

3.1.4 Partial regularity

Theorem 9. *There exists a weak solution of the Navier-Stokes equation such that the 1-dimensional Hausdorff measure of the singularity set is zero.*

4   Viscoelastic materials

All complex fluids have distinguished viscoelastic properties.. The following references [7, 41, 54, 78, 82, 87] cover some of the most important area of the studies, both in mathematics and engineering/physics.

- R. B. Bird, R. C. Armstrong, and O. Hassager. *Dynamics of Polymeric Liquids,* volume 1 of *Fluid Mechanics.* Weiley Interscience, New York, 1987.

- M. E. Gurtin. *An Introduction to Continuum Mechanics,* volume 158 of *Mathematics in Science and Engineering.* Academic Press, 1981.

- R. G. Larson. *The Structure and Rheology of Complex Fluids.* Oxford, 1995.

- R. G. Owens and T. N. Phillips. *Computational Rheology.* Imperial College Press, London, 2002.

- M. Renardy, W. J. Hrusa, and J. A. Nohel. *Mathematical Problems in Viscoelasticity,* volume 35 of *Pitman Monographs and Surveys in Pure and Applied Mathematics.* Longman Scientific & Technical, Harlow, 1987.

- W. R. Schowalter. *Mechanics of Non-Newtonian Fluids.* Pergamon Press, 1978.

4.1   Flow map and deformation tensor

In the context of hydrodynamics, the basic variable are the flow map (particle trajectory) $x(X, t)$. X is the original labeling (Lagrangian coordinate) of the particle. It is also referred to as material coordinate. x is the current (Eulerian) coordinate and referred to as reference coordinate. For a given velocity field $v(x, t)$ the flow map is defined by the following ordinary differential equation:

$$x_t(X, t) = v(x(X, t), t), \quad x(X, 0) = X. \tag{4.1}$$

The deformation tensor $F(X, t)$ is defined as

$$F(X, t) = \frac{\partial x}{\partial X}. \tag{4.2}$$

When look in the Eulerian coordinate, we can define $\tilde{F}(x, t)$ such that $\tilde{F}(x(X, t), t) = F(X, t)$. With no ambiguity, we will not distinguish these two notations in this paper. Applying the chain rule, we see that $F(x, t)$ satisfies the following transport equation [41, 54, 70]:

$$F_t + v \cdot \nabla F = \nabla v F, \tag{4.3}$$

which stands for $F_{ij_t} + v_k \nabla_k F_{ij} = \nabla_k v_i F_{kj}$. This is a direct consequence of the chain rule. Here we point out that in this paper, we use the notation $F_{ij} = \frac{\partial x_i}{\partial X_j}$ and $(\nabla v)_{ij} = \frac{\partial v_i}{\partial x_j}$. This is different from notations in other papers by a transpose, for instance [54].

The incompressibility is represented as

$$\det F = 1. \tag{4.4}$$

By the identity of the variation of the determinant of a tensor,

$$\delta \det F = \det F \operatorname{tr}(F^{-1} \delta F), \tag{4.5}$$

we see that $\nabla \cdot v = 0$. Moreover, we assume that the density $\rho = \rho_0$ to be a constant. This will replace the conservation of mass equation:

$$\rho_t + \nabla \cdot (\rho v) = 0. \tag{4.6}$$

Finally, in this case, if we denote $(\nabla \cdot F)_j = (\nabla_i F_{ij})$, we have [41, 54, 70]

$$(\nabla_i F_{ij})_t + v_k \nabla_k (\nabla_i F_{ij}) + \nabla_i v_k (\nabla_k F_{ij}) = \nabla_k v_i (\nabla_i F_{kj}) + \nabla_i \nabla_k v_i F_{kj}.$$

Using the incompressibility and switch the indices i and k of the first term on the right-hand side, we have

$$(\nabla \cdot F)_t + v \cdot \nabla (\nabla \cdot F) = 0. \tag{4.7}$$

4.2 Force balance and Oldroyd-B systems

For general viscoelastic fluid, we start from the following conservation of momentum equation:

$$\rho(v_t + v \cdot \nabla v) = \nabla \cdot \tau, \tag{4.8}$$

where τ is total stress. In Newtonian flow, we have the constitutive equation $\tau = -pI + \mu D$, where p is the pressure, μ the viscosity and $D = \frac{\nabla v + \nabla^T v}{2}$ is the strain rate.

There have been many attempts to capture different non-Newtonian phenomena of the materials, such as those of Ericksen-Rivlin [87, 88] or high-grade fluid [48], Ladyzhenskaya where τ is nonlinear in the strain rate D [52] and by Necas's group where viscosity depending on both D and p [45, 73]. All these models only involve instantaneous constitutive relation between the stress and strain.

For the nonlocal (in time) constitutive equations, there are Maxwell model $\tau_t + \gamma \tau = \mu D$, the transport model $\tau_t + v \cdot \nabla \tau + \gamma \tau = \mu D$, and the Oldroyd (upper convective) models

$$\tau_t + v \cdot \nabla \tau - \nabla v \tau - \tau \nabla v^T + \gamma \tau = \mu D. \tag{4.9}$$

The constant γ in the above models represents the time scale for the elastic relaxation. It is associate to the Debra number $De = \frac{\mu}{\gamma}$, which indicates the relation between the characteristic flow time and the characteristic elastic time scales [7].

There are other types of Oldroyd models. Those are associated with the different ways the stress tensor is transported. For instance, the Johnson-Segaman model is just the linear combination of the upper convictive and the lower convective Oldroyd models.

We can also look at the following modified Oldroyd model:

$$\tau = -pI + \mu D + \tau_1, \tag{4.10}$$

and the elastic stress τ_1 satisfies the transport equation:

$$\tau_{1t} + v \cdot \nabla \tau_1 - \nabla v \tau_1 - \tau_1 \nabla v^T + \gamma \tau_1 = \delta I. \tag{4.11}$$

Equation (4.11) can be related to the modified Oldroyd model (4.9) by simply change of variable as $\tau_1 = \tau - \eta I$, where $\eta = \mu/2$ [78].

The tensor $C = FF^T$ is usually called the Cauchy-Green strain tensor and $B = C^{-1}$ is the finger tensor [41, 54, 78]. In particular, equation (4.11) is equivalent to

$$(F^{-1}\tau_1 F^{-T})_t + v \cdot \nabla(F^{-1}\tau_1 F^{-T}) = -\gamma(F^{-1}\tau_1 F^{-T}) + \delta F^{-1}F^{-T}.$$

Hence, we can implicitly write the solution in the form:

$$\begin{aligned}
\tau_1(x,t) &= \exp\{-\gamma t\}F(x,t)\tau_1(x,0)F^T(x,t) \\
&+ \delta \int_{-\infty}^t \exp\{-\gamma(t-s)\}F(x,t)F^{-1}(x,s) \\
&\quad F^{-T}(x,s)F^T(x,t)\, ds.
\end{aligned} \tag{4.12}$$

From here, it is obvious that τ_1 is positive definite. In fact, in this case, we can define the *induced deformation* tensor $F_1 = \sqrt{\tau_1}$.

Lemma 4. *If a tensor τ satisfies the equation*

$$\tau_t + v \cdot \nabla \tau - \nabla v \tau - \tau \nabla v^T = 0, \tag{4.13}$$

and the initial condition $\tau(x,0) = \tau_0(x)$ is positive definite, then

$$\tau(x,t) = F\tau_0 F^T. \tag{4.14}$$

Moreover, the induced deformation tensor $F_1 = \sqrt{\tau}$ satisfies the same equation as (4.3):

$$(F)_t + v \cdot \nabla F = \nabla v F.$$

We remark that the above result, together with the results in [70] will allow us to obtain a global weak (Larey) solution for a small (induced) strain viscoelasticity. We notice that this type of results are different from the existence results of [39, 40, 82] and the more recent ones in [16, 27, 66] which will be discussed in the later sections.

Finally, we see that, for the Oldroyd model, the system satisfies the following energy law:

$$\frac{d}{dt} \int_\Omega \left(\frac{1}{2} \rho |v|^2 + \frac{1}{2} \mathrm{tr}\, \tau_1 \right) dx = - \int_\Omega \mu |D|^2. \qquad (4.15)$$

4.3 Energetic variational formulation

In [70], in order to study the mixture of a fluid with a visco-elastic solid, we wrote the momentum equation for the viscoelastic materials in the Eulerian framework. Assuming that the elastic energy of the solid is $W(F)$, where $F = [\partial x/\partial X]$ is the deformation tensor (strain). The following system (in weak form) gives the force balance equations (linear momentum equations):

$$\int_\Omega [\rho(v_t + (v \cdot \nabla)v) \cdot u - p\nabla \cdot u + \tau \cdot \nabla u]\, dx = \int_\Omega \rho f \cdot u dx \qquad (4.16)$$

for any test function u, the elastic stress: $\tau = \mu D(v) + (1/J)S(F)F^T$, where $S(F) = [\partial W/\partial F]$ takes the Piola Kirchhoff form. Here we also adopt the constraint $J = \det(F) = 1$ for incompressibility. This momentum equation can be derived through the least action principle (Hamilton's principle). The action functional take the form

$$A(x) = \int_0^T \int_{\Omega_0} \left(\frac{1}{2} \rho |x_t(X, t)|^2 - W(F) \right) dX dt, \qquad (4.17)$$

where Ω_0 is the original domain occupied by the material. We use the fact that $J = \det F = 1$.

Now we take any one-parameter family of volume preserving flow map $x^\epsilon(X, t)$ with $\frac{dx^\epsilon}{d\epsilon}|_{\epsilon=0} = y$. From the fact that $J = \det F = 1$ and identity (4.5), we have $\nabla \cdot y = 0$. Now equation (4.16) (without the viscosity dissipation term) can be seen just following the variation of A with respect to x:

$$\frac{d}{d\epsilon} A(x^\epsilon)|_{\epsilon=0} = 0. \qquad (4.18)$$

We usually study the elasticity through the force balance equation, using the Lagrangian coordinate. Here we use the trajectory $x(X, t)$ as the unknown variable (or the displacement $x - X$). The equation reads as

$$\rho x_{tt} = -\frac{\delta W}{\delta x} = \nabla_X \cdot W_F + \nabla_X (F^{-T} p), \qquad (4.19)$$

where p is the Lagrangian multiplier to the incompressibility condition, and it satisfies the energy law:

$$\frac{d}{dt} \int_{\Omega_0} \left(\frac{1}{2} \rho |x_t|^2 + W(F) \right) dX = 0. \tag{4.20}$$

In the case of Hookean (linear) elasticity, $W(F) = |F|^2 = \operatorname{tr}(FF^T)$, it becomes the usual wave equation:

$$\rho x_{tt} = \nabla_X \cdot W_F = \nabla_X \cdot F + \nabla_X(F^{-T}p) = \Delta_X x + \nabla_X(F^{-T}p). \tag{4.21}$$

We point out that it will be difficult to input the frame indifferent viscosity term in the above equations.

Again, system (4.16) satisfies the energy estimate (second law of thermodynamics [41]):

$$\frac{d}{dt} \int_{\Omega} \left(\frac{1}{2} \rho |v|^2 + W(F) \right) dx = - \int_{\Omega} \mu |D|^2 \, dx. \tag{4.22}$$

We notice that even in the linear elasticity case, the elastic stress term $\tau_2 = W_F F^T = FF^T$ is still nonlinear. In fact, it is always the same order as the energy. This is the main difficulty of the current setting. On the other hand, we can make the following observation. Using the fact that F satisfies the transport equation (4.3), we have

$$\tau_{2t} + v \cdot \nabla \tau_2 - \nabla v \tau_2 - \tau_2 \nabla v^T = 0 \tag{4.23}$$

and we recover the Oldroyd system (without the damping). Notice in this case that $W(F) = \operatorname{tr} \tau_2$. Hence the two energy laws are also consistent with each other.

In fact, we can also start with the above energy law and derive the linear momentum equations (hence the constitutive equations). This is also the approach that was used by Ericksen in the study of liquid crystal materials [30] and Gurtin for phase transitions [35].

The linear transport equation $F_t + (v \cdot \nabla)F = \nabla v \, F$ in tensor case cannot be treated directly in the framework of [24] or [51]. We may apply the div-curl lemma [92] to obtain weak solutions [70]. However, this is not enough to achieve the convergence of the stress term. As an alternative, we used the polar decomposition (R be the rotation part and the symmetric U be small) and get the equations $R_t + u \cdot \nabla R = W(v)R$, $U_t + v \cdot \nabla U = R^T D(v)R$, where $F = R(I + U)$, $D(v)$, $W(v)$ are the symmetric skew components of ∇v. This was not the usual linear elastic formulation, rather, it was in the same sitting as the famous work by F. John [47] where he had applied the John-Nirenberg inequality [36] to study nonlinear elasticity for the static small strain cases. We linearized the elastic stress:

$$DW(F)F^T = R(DW(I) + DW(I)U + \mathcal{C}(U) + O(U^2))R^T, \tag{4.24}$$

where we used the notation $\mathcal{C}(U)_{j\beta} = D^2 \mathcal{W}(I)(U)_{j\beta} = \frac{\partial^2 \mathcal{W}}{\partial F_{i\alpha} \partial F_{j\beta}}(I)U_{i\alpha}$. The special form of the equation of R allowed us to get an approximate system for R and to generalize the tools for scalar transport equations [24] to this small strain case, and eventually leaded to the global existence of the approximate system [70].

Lin, Liu and Zhang studied the existence of the original system. According to different situations, we let Ω be a bounded domain in $\mathbb{R}^2$ (or $\mathbb{R}^3$) with smooth boundary, the whole domain or periodic boxes. The (linear) viscoelastic fluid system takes the following form:

$$
\begin{aligned}
F_t + v \cdot \nabla F &= \nabla v F, \\
v_t + v \cdot \nabla v + \nabla p &= \mu \Delta v + \nabla \cdot (F F^T), \\
\nabla \cdot v &= 0,
\end{aligned}
\tag{4.25}
$$

where the i-th component of $\nabla \cdot (F F^T)$ on the right-hand side of the momentum equation is $\nabla_j (F_{ik} F_{jk})$. The system has the initial condition:

$$
F(x,0) = F_0(x), \; v(x,0) = v_0(x).
\tag{4.26}
$$

In cases of bounded domain, we chose the boundary condition: for any x on the boundary $\partial\Omega$,

$$
F(x,t) = I, \; v(x,t) = 0.
\tag{4.27}
$$

The system satisfies the energy identity:

$$
\frac{d}{dt} \int_\Omega \left(\frac{1}{2}|v|^2 + \frac{1}{2}|F|^2 \, dx \right) = - \int_\Omega \mu |\nabla v|^2 \, dx.
\tag{4.28}
$$

From the identity $(\nabla \cdot F)_t + v \cdot \nabla(\nabla \cdot F) = 0$ for the incompressible materials, if we assume that $\nabla \cdot F_0 = 0$, we have $\nabla \cdot F = 0$ and $F = \nabla \times \phi$, where ϕ is a matrix. In 2-dimensional case, if we denote $\phi = (\phi_1, \phi_2)$, then the original system can be transformed (after adjusting the order and sign) into

$$
\begin{aligned}
\phi_t + v \cdot \nabla \phi &= 0, \\
v_t + v \cdot \nabla v + \nabla p &= \mu \Delta v - \sum_{i=1}^{2} \Delta \phi_i \nabla \phi_i, \\
\nabla \cdot v &= 0,
\end{aligned}
\tag{4.29}
$$

with initial condition:

$$
\phi(x,0) = \phi_0, \; v(x,0) = v_0(x),
\tag{4.30}
$$

and in case of bounded domain, the boundary conditions: for any x on the boundary $\partial\Omega$,

$$
\phi(x,t) = x, \; v(x,t) = 0.
\tag{4.31}
$$

And the energy law becomes:

$$\frac{d}{dt} \int_\Omega \left(\frac{1}{2}|v|^2 + \frac{1}{2}|\nabla\phi|^2 \, dx \right) = - \int_\Omega \mu |\nabla v|^2 \right) dx. \qquad (4.32)$$

The following theorems are proved in [65].

Theorem 10. *Let $k \geqslant 2$ be a positive integer, $\nabla\phi_0 \in H^k(\Omega)$, $v_0 \in H^k(\Omega)$. Then there exists a positive time T, which depends only on $|\nabla\phi_0|_{H^2}$ and $|v_0|_{H^2}$, such that the system possesses a unique solution in the time interval $[0, T]$ with*

$$\partial_t^j \nabla_x^\alpha v \in L^\infty([0,T]; H^{k-2j-|\alpha|}(\Omega)) \cap L^2([0,T]; H^{k-2j-|\alpha|+1}(\Omega)),$$
$$\partial_t^j \nabla_x^\alpha \nabla\phi \in L^\infty([0,T]; H^{k-2j-|\alpha|}(\Omega)) \qquad (4.33)$$

for all j, α satisfying $2j + |\alpha| \leqslant k$. Moreover, if T^* is the maximal time of existence, then

$$\int_0^{T^*} |\nabla v|_{H^2}^2 \, ds = +\infty. \qquad (4.34)$$

Theorem 11. *Let Ω be a periodic box or the whole space $\mathbb{R}^2$, $k \geqslant 2$ be a positive integer, $\nabla\phi_0 \in H^k(\Omega)$ and $v_0 \in H^k(\Omega)$. Furthermore, for some large enough constant C, we assume that*

$$|\nabla v_0|_{H^2} + |\nabla\psi_0|_{H^2} \leqslant \frac{\mu}{C(1 + \frac{1}{\mu})^3(1 + \mu + \frac{1}{\mu})}. \qquad (4.35)$$

Then system (4.29) will have a unique global classical solution, such that

$$|v|_{H^2}^2 + |\nabla\psi|_{H^2}^2 + \int_0^\infty (\mu |\nabla v|_{H^2}^2 + \frac{1}{\mu}|\nabla\Delta\psi|_{L^2}^2) \, ds \leqslant \frac{\mu}{C(1 + \mu + \frac{1}{\mu})}, \qquad (4.36)$$

and (4.33) holds for $T = \infty$.

The results has been generalized to the general system in [56] and the small strain viscoelastic materials [55].

5   Liquid crystal flows

5.1   Ericksen-Leslie theory

The hydrodynamical theory must describe not only orientation, as represented by the director field $n(x, t)$, but macroscopic motion, represented by the velocity field $u(x, t)$.

We shall present the Ericksen-Leslie's set up (the corresponding static theory will be that of Oseen-Frank).

As usual, for liquids idealized as incompressible, we have the equation of continuity

$$\text{div } u = 0\,,\ u = (u_1, u_2, u_3), \tag{5.1}$$

representing conservation of mass. In general terms, equations of motion for u are of conventional form, i.e.,

$$\rho\left(\frac{\partial u_i}{\partial t} + u_{i,j}u_j\right) = t_{ij,j} + f_i, \tag{5.2}$$

where ρ is the (constant) mass density, f the body force, t the stress tensor. Then the stress tensor t can be written as

$$t = t^s + t^D, \tag{5.3}$$

the superscript s indicating a part covered by static theory, D indicating a dissipative part, vanishing when there is no motion.

Under various physical considerations, Leslie and Ericksen derived that

$$\begin{aligned} t_{ij}^s &= -p\delta_{ij} + W\delta_{ij} - \tau_{kj}n_{k,i}, \\ \tau_{ij} &= \frac{\partial W}{\partial n_{i,j}}, \end{aligned} \tag{5.4}$$

where W is the Oseen-Frank energy density (with $q = 0$ in nematics) and p is the pressure. Regarding the motion of n, as suggested by static theory,

$$n \wedge h = 0, \tag{5.5}$$

h being the total molecular field. There is an equivalent formulation, rephrasing this in terms of a balancing of moments. (Again one ignores the effect of the electromagnetic field.)

Similarly,

$$\begin{aligned} h &= h^s + h^D, \\ h_i^s &= \frac{\partial W}{\partial n_i} - \tau_{ij,j}. \end{aligned} \tag{5.6}$$

The terms t^D, h^D in dynamics have been treated from various viewpoints. In the purely dissipative model (parabolic system), the constitutive assumption presumes that t^D, h^D are linear functions of ∇u and

$$\dot{n} = \frac{\partial n}{\partial t} + u \cdot \nabla n, \tag{5.7}$$

with coefficients depending on n. Under further symmetry assumptions, as well as thermodynamics and mechanical arguments, the constitutive equations reduce to the form

$$t^D = \frac{\partial \Delta}{\partial \nabla u}, \tag{5.8}$$

$$h^D = \frac{\partial \Delta}{\partial \dot{n}}, \tag{5.9}$$

Δ being a dissipation function. In terms of the variables A and N given by

$$2A = \nabla u + \nabla u^t, \tag{5.10}$$

$$N = \dot{n} - \frac{1}{2}(\nabla u - \nabla u^t)n, \tag{5.11}$$

this function has the form

$$2\Delta = \alpha_1(n \cdot An)^2 + \alpha_4(tr A^2) + (\alpha_5 + \alpha_6)\|n \otimes An\|^2$$
$$+\gamma_1\|N\|^2 + 2\gamma_2 N \cdot An \geqslant 0. \tag{5.12}$$

Here the scalar α's and γ's, the measure of viscosity, depend on the material and the temperature.

Based on a somewhat different argument, Leslie obtained more general parabolic-hyperbolic systems. This system can be written in the following more concise form:

$$\rho\dot{u} = \text{div}\left(-pI + \nabla n^T \cdot W_q + \frac{\partial \Delta}{\partial \nabla u}\right) + F, \tag{5.13}$$

$$\sigma\ddot{n} = -W_n + \text{div } W_q + \frac{\partial \Delta}{\partial \dot{n}} + \gamma n + G. \tag{5.14}$$

Here p is (as before) the pressure, γ is a Lagrange multiplier due to the constraint $|n| = 1$, and F, G are external forces.

System (5.14) is derived from the conservation law of the form (proposed by J. Ericksen)

$$\frac{d}{dt}\int_\Omega (\rho u^2 + W + \sigma|\dot{n}|^2)dx$$

$$= -\int_\Omega \Delta dx + \text{boundary terms and harmless terms.} \tag{5.15}$$

The first system (with $\sigma = 0$) is parabolic, and can be thought of as nonlinear coupling between harmonic maps heat flow and Navier-Stokes equations.

The second system is a parabolic-hyperbolic coupled system. Here one has a nonlinear coupling between wave maps with dumping effect and Navier Stokes equations.

Remark 3. In Ericksen's equations, if we choose $k_1 = k_2 = k_3 = 1, k_4 = q = 0, \alpha_1 = \alpha_4 = \alpha_5 + \alpha_6 = \gamma_2 = 0, \gamma_1 = 1$, then the coupled systems can be written as

$$\begin{cases} \frac{\partial}{\partial t}u^i + u \cdot \nabla u^i = \Delta u^i + \nabla_i p - (n_{x_i} \cdot n_{x_j})_{x_j}, & i = 1,2,3, \\ \frac{\partial}{\partial t}n^i + u \cdot \nabla n^i - \Omega_j^i n^j = \Delta n^i + |\nabla n|^2 n^i, & i = 1,2,3. \end{cases} \tag{5.16}$$

In addition, we have two constraint $\operatorname{div} u = 0$ and $|n| = 1$. Where $\Omega^i_j = \frac{1}{2}[u^i_{x_j} - u^j_{x_i}]$. There is also a similar version for Leslie's equations.

If $u \equiv 0$, then we have

$$\frac{\partial n}{\partial t} = \Delta n + |\nabla n|^2 n, \ |n| = 1, \tag{5.17}$$

which is the equation of heat flow of harmonic maps from $\Omega \to S^2$.

Remark 4. The first equation concerning balance of linear moments becomes

$$(n_{x_i} \cdot n_{x_j})_{x_j} = \frac{\partial}{\partial x_i} P. \tag{5.18}$$

Not all weak solutions of (5.17) satisfy (5.18), those weak solutions of (5.17) and, in addition (5.18), have to satisfy so called *energy-monotonicity inequality*. In the static case, those solutions are exactly those called [SU] stationary solutions. They satisfy the energy monotonicity inequality.

5.2   Existence and regularity

In order to understand the Ericksen-Leslie theory, we will look at the following system. Although the system is simplified, it retained most mathematical and physical difficulties of the original system. Moreover, it emphasizes the special coupling between the director and the flow field.

$$u_t + (u \cdot \nabla)u + \nabla p - \nu \operatorname{div} D(u) + \lambda \operatorname{div}(\nabla d \odot \nabla d) = 0, \tag{5.19}$$

$$\nabla \cdot u = 0, \tag{5.20}$$

$$d_t + (u \cdot \nabla)d - \gamma(\Delta d - f(d)) = 0, \tag{5.21}$$

with initial conditions

$$u|_{t=0} = u_0, \quad d|_{t=0} = d_0, \tag{5.22}$$

satisfying either the Dirichlet boundary condition [62]

$$u = 0, \quad d = d_0, \tag{5.23}$$

or the *free surface* boundary conditions

$$u \cdot n = 0, \quad ((\nabla \times u) \times n) \times n = 0, \quad \frac{\partial d}{\partial n} = 0, \tag{5.24}$$

on the boundary $\partial\Omega$ of the domain with n being the outward normal.

In the above system, u represents the velocity of the liquid crystal fluid, p the pressure, d represents the *normed* director of the molecule.

The vectors $u, d : \Omega \times \mathbb{R}^+ \to \mathbb{R}^n$, and the function $p : \Omega \times \mathbb{R}^+ \to \mathbb{R}$, where $\Omega \subset \mathbb{R}^n$ is a bounded smooth domain (or a polygonal domain) with boundary $\partial\Omega$. $D(u) = (1/2)(\nabla u + (\nabla u)^T)$ is the stretching tensor, $\sigma^v = pI + \nu D(u)$ is the fluid viscosity part of the stress tensor, $(\nabla d \odot \nabla d)_{ij} = \sum_{k=1}^n (\nabla_i d_k)(\nabla_j d_k)$, and finally, $f(d)$ is a polynomial of d which satisfies $f(d) = F'(d)$ where $F(d)$ is the bulk part of the elastic energy. The choice of $F(d)$ is such that the maximal principle for $|d|$ holds in equation (5.21), that is, if $|d| \leqslant 1$ on the boundary and in the initial data, the $|d| \leqslant 1$ is true everywhere at any time. Usually, we choose $F(d)$ to be the Ginzburg-Landau penalization

$$F(d) = \frac{1}{4\epsilon^2}(|d|^2 - 1)^2. \tag{5.25}$$

Again, we see that (5.20) represents the fact that the fluid is incompressible. (5.21) is the evolution of the director, the left-hand side shows that d is transported by the flow. We want to point out that the momentum equation (5.19) can be derived through the following least action principle.

The above highly simplified system, in fact, captures all the mathematical difficulties (as shown in the later discussions) and the physical characteristics of the original model. To demonstrate the later one, we will derive the linear momentum equation (force balance) equation using the least action (Hamiltonian) principle.

Let us begin by computing the variations of the following "elastic" part of the "action" functional among all the volume preserving flow maps:

$$A(x) = \int_0^T \int_\Omega \left(\frac{\lambda}{2}|\nabla d|^2 - h(d) \right) dx dt. \tag{5.26}$$

We look at the volume preserving flow maps $x(X, t)$ such that

$$x_t(X, t) = v(x(X, t), t), \quad x(X, 0) = X. \tag{5.27}$$

Here we can view X as the Lagrangian (initial) material coordinate and $x(X, t)$ the Eulerian (reference) coordinate.

In order to perform the variation, we look at the one parameter family of such maps x^ϵ such that

$$x^0 = x, \quad \frac{dx^\epsilon}{d\epsilon} = y \tag{5.28}$$

for any y such that $\nabla_x \cdot y = 0$.

Now we computer the variation of $A(x^\epsilon) = A(d(x^\epsilon, t))$ with respect to ϵ:

$$
\begin{aligned}
0 &= \frac{d}{d\epsilon}\big|_{\epsilon=0} A(x^\epsilon) \\
&= \int_0^T \int_{\Omega_0} \left(\lambda \nabla_x^i d \frac{d}{d\epsilon}\big|_{\epsilon=0} (\nabla_{x^\epsilon}^i d(x^\epsilon, t)) + f'(d) \nabla_x^j dy^j \right) dX\, dt \\
&= \int_0^T \int_{\Omega_0} \left(\lambda \nabla_x^i d \frac{d}{d\epsilon}\big|_{\epsilon=0} (\nabla_x^j d(x^\epsilon, t) \nabla_{x^\epsilon}^i x^j) + f'(d) \nabla_x^j dy^j \right) dX\, dt \\
&= \int_0^T \int_{\Omega_0} \left(\lambda \nabla_x^i d \nabla_x^j \nabla_x^i d(x, t) + \lambda \nabla_x^i d(x, t) \nabla_x^j d(x, t) \nabla_x^i y^j \right. \\
&\qquad\qquad \left. + f'(d) \nabla_x^j dy^j \right) dX\, dt.
\end{aligned}
$$

Here we have used the fact that $\nabla_{x^\epsilon} x$ is the inverse matrix of $\nabla_x x^\epsilon$.

Since y is an arbitrary divergence free vector field, integration by parts gives the following equation:

$$u_t + (u \cdot \nabla)u + \nabla p + \lambda \operatorname{div}(\nabla d \odot \nabla d) = 0. \tag{5.29}$$

We point out that the viscosity and other types of dissipation are due to other term.

From this derivation, it is easy to see the energy law of the system. Moreover, the hydrodynamic equilibrium will be a special kind of stationary solution of d, as we remarked in the last section. This can be viewed in the following spatial Noether theorem.

In [62], we study the wellposedness of the system by establishing the following basic energy law:

$$\frac{1}{2}\frac{d}{dt}\int_\Omega (\|u\|^2 + \lambda\|\nabla d\|^2 + 2\lambda F(d))dx = -\int_\Omega (\nu\|\nabla u\|^2 + \lambda\gamma\|\Delta d - f(d)\|^2)dx \tag{5.30}$$

for all $t \in (0, T]$.

This energy law together with a modified Galerkin method enables us to prove the existence of a weak solution of the above system:

Theorem 12. *Under the assumptions that $u_0(x) \in L^2$, and that $d_0(x) \in H^1(\Omega)$ with $d_0|_{\partial\Omega} \in H^{3/2}(\partial\Omega)$, the system has a global weak solution (v, d) satisfying*

$$
\begin{aligned}
u &\in L^2(0, T; H^1(\Omega)) \cap L^\infty(0, T; L^2(\Omega)), \\
d &\in L^2(0, T; H^2(\Omega)) \cap L^\infty(0, T; H^1(\Omega))
\end{aligned}
\tag{5.31}
$$

for all $T \in (0, \infty)$.

We also proved the existence of the classical solution.

Theorem 13. *The system has a unique global classical solution (u,d) provided that $u_0(x) \in H^1(\Omega)$, $d_0(x) \in H^2(\Omega)$ and, that either $\dim \Omega = 2$ or $\dim \Omega = 3$ and $\nu \geqslant \nu(\gamma, \lambda, u_0, d_0)$.*

The stability result is established in the following theorem.

Theorem 14. *Suppose that $d^* \in H^2(\Omega)$ is an absolute minimum of the functional*

$$E(d) \equiv \frac{1}{2} \int_\Omega (\|\nabla d\|^2 + F(d))dx \qquad (5.32)$$

in the sense that $E(d^*) \leqslant E(d)$ whenever $d = d^*$ on $\partial\Omega$. Then there is an $\epsilon > 0$, possibly depending on the system such that: whenever $\|d_0 - d^*\|_{H^2(\Omega)} + \|u_0\|_{H^1(\Omega)} < \epsilon$, then the original system has a unique global solution (u,d) with $u(x,t) \to 0$ in $H^1(\Omega)$, as $t \to +\infty$. Moreover, for any sequence $t_i \to +\infty$, $d(x, t_i') \to \tilde{d}$ in $H^2(\Omega)$ for a subsequence $\{t_i'\}$. Here $\tilde{d}$ is a critical point of E.

Theorem 15. *If the domain and the initial-boundary conditions in system are smooth enough, then there exists a suitable weak solution whose singular set has one-dimensional Hausdorff measure zero in space-time.*

6   Free interface motion in mixtures

The interfacial dynamics in the mixture of different fluids, solids or gas have attracted attentions for more than two centuries. Many surface properties, such as capillarity, are associated with the surface tension through special boundary conditions on the interfaces [14, 29, 50, 80].

In classical approaches, the interface is usually considered to be a free surface that evolves in time with the fluid (the kinematic boundary condition). The dynamics of the interface at each time is determined by the following stress (force) balance condition:

$$[T] \cdot n = mHn, \qquad (6.1)$$

where $[T] = [\nu D(u) - pI]$ is the jump of the stress across the interface Γ_t, n is its normal, $D(u) = \frac{\nabla u + (\nabla u)^T}{2}$ is the stretching tensor, H is the mean curvature of the surface and m is the surface tension constant. This is the usual Young-Laplace junction condition (see, for instance, [4, 29, 50, 80]). The hydrodynamic system describing the mixture of two Newtonian fluids with a free interface will be the usual Navier-Stokes equations in each of the fluid domains (possibly with different density and viscosity) together with the kinematic and force balance (traction free) boundary conditions on the interface. The weak form of such a system when the density ρ and viscosity ν may vary in the mixture can be represented exactly in the following form [68]:

$$\int_0^T \int_\Omega [-\rho u v_t - \rho u u \cdot \nabla v + \nu \nabla u \nabla v - p \nabla \cdot v] \, dx dt$$
$$= \int_0^T \int_{\Gamma_t} m H n \cdot v \, ds dt \qquad (6.2)$$

for any test function v.

One classical method to study the moving interfaces is to employ a mesh that has grid points on the interfaces, and deforms according to the motion of the boundary, such as the boundary integral and boundary element methods (cf. [21,49,95] and their references). Keeping track of the moving mesh may entail computational difficulties and large displacement in internal domains may cause mesh entanglement. Typically, sophisticated remeshing schemes have to be used in these cases.

As an alternative, fixed-grid methods that *regularize* the interface have been highly successful in treating deforming interfaces. These include the volume-of-fluid (VOF) method [58, 59], the front-tracking method [37, 38] and the level-set method [15, 77]. Instead of formulating the flow of two domains separated by an interface, these methods represent the interfacial tension as a body-force or bulk-stress spreading over a narrow region covering the interface. Then a single set of governing equations can be written over the entire domain, and solved on a fixed grid in a purely Eulerian framework.

The energetic phase field model can be viewed as a physically motivated level-set method. Instead of choosing an artificial smoothing function for the interface, the diffuse-interface model describes the interface by a mixing energy. This idea can be traced to van der Waals [96], and is the foundation for the phase-field theory for phase transition and critical phenomena (see [12,13,25,75,76,93] and the references therein). The phase field models allow topological changes of the interface [71] and over the years, they have attracted a lot of interests in the field of nonlinear analysis (cf. [2,11,17,84,89]). Similar to the popular level set formulations (see [77] for an extensive discussion), they have many advantages in numerical simulations of the interfacial motion (cf. [15]). When the transition width approaches zero, the phase field model with diffuse-interface becomes identical to a sharp-interface level-set formulation and it can also be reduced properly to the classical sharp-interface model.

In this paper, we will illustrate some basic features and general approaches of this method.

6.1 An energetic variational approach with phase field method

Here we will present the simplest case of the phase field method to study the mixture of two incompressible Newtonian fluids.

Introduce a "phase" function $\phi(x, t)$ to identify the two fluids ($\{x : \phi(x, t) = 1\}$ is occupied by fluid 1 and $\{x : \phi(x, t) = -1\}$ by fluid 2). Looking at the following Ginzburg-Landau type of mixing energy:

$$\tilde{W}(\phi, \nabla\phi) = \int_\Omega \left[\frac{\eta}{2} |\nabla\phi|^2 + \frac{1}{4\eta} (\phi^2 - 1)^2 \right] dx.$$

We can view ϕ as volume fraction. The mixing density and viscosity will be functions of ϕ. The part of bulk energy represents the interaction of different volume fractions of individual species (like Flory-Huggins free energy [28, 54]). The gradient part plays the role of regularization (relaxation). The combination represents the competition between the (hydro)phobic and (hydro)philic effects between different species. The interface is represented by $\{x : \phi(x, t) = 0\}$, with the fixed transition layer of thickness η. The dynamics of ϕ can be driven by either Allen-Cahn or Cahn-Hillard types of gradient flow, depending on the choice of different dissipative mechanism. The later one preserves the overall volume fraction of two fluids. For Chan-Hillard case (where the volume is preserved):

$$\phi_t + u \cdot \nabla\phi = -\gamma \frac{\delta W}{\delta \phi} = -\gamma \Delta(\Delta\phi - f(\phi)), \tag{6.3}$$

where $f(\phi) = F'(\phi) = \frac{1}{\eta}(\phi^2 - 1)\phi$. u is the velocity field. The right-hand side can be viewed as the variation with respect to ϕ is the regular L^2 space. The left-hand side indicates that the variable ϕ is transported by the flow, on top of the energy decent dynamics. As $\gamma \to 0$, where γ represents the elastic relaxation time, the limiting ϕ satisfies the transport equation, which is equivalent to the mass transport equation (for incompressible fluids). Hence this formulation can also be viewed as the link (relaxation) between the mass average (in the kinetic energy) and the volume average (in the elastic energy).

In case the variational space is taken to be H^{-1}, then ϕ will take the dynamics as the Allen-Cahn equation:

$$\phi_t + u \cdot \nabla\phi = \gamma(\Delta\phi - f(\phi)). \tag{6.4}$$

Combining this elastic energy with the kinetic energy, we have the total energy $E = \int_\Omega [\frac{\rho}{2}|u|^2 + \frac{\lambda}{2}|\nabla\phi|^2 + \lambda F(\phi)] \, dx$. Using the least action principle (the principle of virtual work), we can derive the following linear momentum equation (balance of force equation) [42,67,68,81]:

$$\rho(u_t + (u \cdot \nabla)u) + \nabla p - \nu \Delta u + \lambda \nabla \cdot (\nabla \phi \otimes \nabla \phi) = g(x). \qquad (6.5)$$

The density will satisfy the following transport equation:

$$\rho_t + u \cdot \rho = 0, \qquad (6.6)$$

under the incompressibility condition for the velocity field:

$$\nabla \cdot u = 0. \qquad (6.7)$$

The final system (6.3), (6.5)–(6.7)(together with the suitable boundary and initial conditions) will then possess the following energy law:

$$\frac{d}{dt} \int_\Omega \left[\frac{\rho}{2}|u|^2 + \frac{\lambda}{2}|\nabla \phi|^2 + \lambda F(\phi) \right] dx$$
$$= - \int_\Omega [\nu|\nabla u|^2 + \gamma \lambda |\nabla(\Delta \phi - f(\phi))|^2] \, dx. \qquad (6.8)$$

We can see that as $\eta \to 0$, the elastic force $\lambda \nabla \cdot (\nabla \phi \otimes \nabla \phi)$ converges to a measure supported only on the interface, with magnitude proportional to the mean curvature [70]. Hence we recover the traction-free boundary condition with surface tension. Moreover, we can also derive the relation of our parameters into the sharp interface ones as the following:

Consider a one-dimensional interface. We require that the diffuse mixing energy in the region be equal to the traditional surface energy:

$$\sigma = \lambda \int_{-\infty}^{+\infty} \left[\frac{1}{2} \left(\frac{d\phi}{dx} \right)^2 + f_0(\phi) \right] dx. \qquad (6.9)$$

Let us further assume that the diffuse interface is at equilibrium, and thus has zero chemical potential,

$$\frac{\delta F_{\text{mix}}}{\delta \phi} = \lambda \left[-\frac{d^2 \phi}{dx^2} + f_0'(\phi) \right] = 0. \qquad (6.10)$$

Since $f_0(\pm\infty) = 0$ and $\left. \frac{d\phi}{dx} \right|_{x=\pm\infty} = 0$, this equation can be integrated once to give

$$\frac{1}{2} \left(\frac{d\phi}{dx} \right)^2 = f_0(\phi), \qquad (6.11)$$

which implies equal partition of the free energy between the two terms at equilibrium.

Equation (6.11) can be solved together with the boundary condition $\phi(0) = 0$, and we obtain the equilibrium profile for $\phi(x)$:

$$\phi(x) = \tanh \frac{x}{\sqrt{2}\epsilon}. \tag{6.12}$$

Thus, the capillary width ϵ is a measure of the thickness of the diffuse interface. More specifically, 90% of variation in ϕ occurs over a thickness of 4.1641ϵ, while 99% of the variation corresponds to a thickness of 7.4859ϵ.

Substituting Eq. (6.12) into Eq. (6.9), we arrive at the following matching condition for the interfacial tension σ:

$$\sigma = \frac{2\sqrt{2}}{3} \frac{\lambda}{\epsilon}. \tag{6.13}$$

As the interfacial thickness ϵ shrinks toward zero, so should the energy density parameter λ; their ratio gives the interfacial tension in the sharp interface limit.

Obviously, the correspondence between the diffuse-and sharp-interface models is meaningful only when the former is at equilibrium. During the relaxation of the diffuse interface (cf. Eq. (6.3)), one cannot speak of a constant interfacial tension. Although one may view this as a deficiency of the diffuse-interface model, it in fact reflects the reality that the interface has its own dynamics which cannot be summarized by a constant σ except under limiting conditions. To anticipate the results in Section 3.3, we note that f_{anch} may also contribute to the surface energy, thus giving rise to an anisotropic "interfacial tension" that is not encompassed by the traditional version of the concept.

These models allow for topological changes of the interface ([8, 43, 46, 71]) and have many other advantages in numerical simulations of the interfacial motion, and have seen many applications in the physics and engineering literature [81, 97].

The conservative dynamics of the above diffuse-interface model can be formulated in the classical procedures of Lagrangian mechanics [1, 3]. The starting point is the Lagrangian $L = T - F$, where T and F are the kinetic and potential energies of the system. The least action principle requires that the action integral $I = \int L \, dt$ be stationary under variations of "paths". This will lead to a momentum equation, with elastic stresses arising from the microstructural changes embodied in the free energy F, and evolution equations for the field variables whose momenta are included in T.

The least action principle (variation on the flow maps), which gives the momentum equation, and the fastest decent dynamics or other types of gradient flows (variation on the phase variables) are due to different physical principles. However, they are related in the static case: the first one is equivalent to the variation with respect to the domain and the second one is the variation of the same functional with respect to the function. It is clear that if the solutions are smooth (or regular enough), they are equivalent. The discrepancy between these two equations requires the presence of the singularities and defects.

The existence of the hydrodynamic equilibrium states for the coupled systems (the static solution with the velocity $u = 0$) can be viewed as a direct consequence of the special relation between the solution of the Euler-Lagrange equation of the elastic energy and the solution of the equation from variation of the domain to such an energy. Formally, it can be summarized into the following simple theorem (see, for example, [64]):

Theorem 16. *Given an energy functional $W(\phi, \nabla\phi)$, all solutions of the Euler-Lagrangian equation*

$$-\nabla \cdot \frac{\partial W}{\partial \nabla \phi} + \frac{\partial W}{\nabla \phi} = 0 \tag{6.14}$$

also satisfy the equation

$$\nabla \cdot \left(\frac{\partial W}{\partial \nabla \phi} \otimes \nabla \phi - W I \right) = 0. \tag{6.15}$$

This theorem guarantees the existence of the hydrodynamic equilibrium states for most systems. It also gives the stability results [62] and shows that all solutions of system (6.24)–(6.28) will approach to an equilibrium state as $t \to +\infty$. One can also derive from Theorem 16 the usual Pohozaev identity [90] by writing equations (6.14) and (6.15) in weak forms.

It is the generality of this energetic variational procedure, especially in accommodating microstructured fluids via the free energy F, that has made the diffuse-interface (phase field) method our choice for tackling interfacial problems of complex fluids. Conceivably, any complex fluid with a properly defined free energy can be included in this formulation. In this paper, we will be dealing with two kinds: The thermo-induced Marangoni-Benard convection, and the mixture involving nematic liquid crystals, which are described by a regularized Leslie-Ericksen model [22, 30,31,57]. The latter also introduces the issue of surface anchoring.

- Level set formulation: work of Hamilton, Evans-Spruck, Evans-Soner-Souganidis, Xinfu Chen.

- Giga and Solonnikov: Classical results.

- General solutions.

6.2 Marangoni-Benard convection

The conventional Marangoni-Benard convection is described by the following two phase fluids with a sharp interface, involving the Boussinesq approximation:

$$\nabla \cdot u = 0, \tag{6.16}$$

$$\rho_0(u_t + (u \cdot \nabla)u) + \nabla p - \nu \operatorname{div} D(u) = -\rho g j, \tag{6.17}$$

$$\theta_t + u \cdot \nabla \theta = k \Delta \theta. \tag{6.18}$$

Here ρ is the temperature dependent density, g is the gravitational acceleration, j is the upward direction. u and p stands for the fluid velocity and the pressure, θ is the temperature. k is the thermal diffusion. Moreover, we assume that

$$\rho = \rho_0[1 - \alpha(\theta - \theta_0)]. \tag{6.19}$$

With the usual initial and boundary conditions, the interface conditions take the form:

$$\eta_t + u \cdot \eta = 0, \tag{6.20}$$

$$[T] \cdot n = -\sigma K n + (t \cdot \nabla \sigma)t. \tag{6.21}$$

Equation (6.20) is the kinematic condition, representing the surface ($\eta = 0$) evolve with the fluid. (6.21) is the traction (T) free boundary (balance of forces) condition. The surface tension depends on the temperature $\sigma = \sigma_0 - \sigma_1 \theta$.

In order to incorporate this effect in the phase field model and still maintain the energy law, we consider the action function:

$$
A(x) = \int_0^T \int_{\Omega_0} \left[\frac{1}{2}\rho_0 |x_t(X,t)|^2 - \frac{\lambda(x(X,t))}{2} \right.
$$
$$
\left. \times (|\nabla_x \phi(x(X,t),t)|^2 + F(\phi(x(X,t),t))) \right] dX dt. \tag{6.22}
$$

Here we can view X as the Lagrangian (initial) material coordinate and $x(X,t)$ the Eulerian (reference) coordinate. Ω_0 is the initial domain occupied by the fluid. The notion that $\phi(x(X,t),t)$ indicated that ϕ is transported by the flow field. The special feature in this case is the spatial dependence of λ. In fact, it can be a function of temperature that is transported by the flow.

For incompressible materials, we look at the volume preserving flow maps $x(X,t)$ such that

$$x_t(X,t) = v(x(X,t),t), \quad x(X,0) = X. \tag{6.23}$$

We arrive at the following system:

$$\rho_0(u_t + (u \cdot \nabla)u) + \nabla p - \nu \operatorname{div} D(u)$$

$$= -\nabla \cdot \left(\lambda \nabla \phi \otimes \nabla \phi - \frac{\lambda}{2}|\nabla \phi|^2 - \frac{\lambda}{4\epsilon^2}(\phi^2 - 1)^2 \right)$$

$$-(1 + \phi)g(\rho_1 - \rho_0)j - (1 - \phi)g(\rho_2 - \rho_0)j, \qquad (6.24)$$

$$\phi_t + (u \cdot \nabla)\phi + \gamma \Delta(\Delta \phi - f(\phi)) = 0, \qquad (6.25)$$

$$\nabla \cdot u = 0, \qquad (6.26)$$

$$\theta_t + u \cdot \nabla \theta = k \Delta \theta, \qquad (6.27)$$

with initial conditions

$$u|_{t=0} = u_0, \quad d|_{t=0} = d_0, \qquad (6.28)$$

and appropriate boundary conditions. In our simulations, we choose the period boundary conditions. The parameter λ is a linear function of the temperature θ.

Here we used the classical Boussinesq approximation, which is the linear version of all different types of average approaches. The "background" density can be treated as a constant ρ_0 and the difference between the actual density and ρ_0 will contribute only to the buoyancy force [60].

Moreover, we see that

$$-\nabla \cdot \left(\lambda \nabla \phi \otimes \nabla \phi - \frac{\lambda}{2}|\nabla \phi|^2 - \frac{\lambda}{4\epsilon^2}(\phi^2 - 1)^2 \right)$$

$$= -\lambda \Delta \phi \nabla \phi - \frac{\lambda}{2}\nabla|\nabla \phi|^2 - (\nabla \lambda \cdot \nabla \phi)\nabla \phi$$

$$+ \frac{\nabla \lambda}{2}|\nabla \phi|^2 + \frac{\lambda}{2}\nabla|\nabla \phi|^2 + \frac{1}{4\epsilon^2}\nabla \lambda(\phi^2 - 1)^2 + \frac{1}{4\epsilon^2}\lambda \nabla(\phi^2 - 1)^2.$$

And the right-hand side convergence to $-\sigma H n + \nabla \sigma - (\nabla \sigma \cdot n)n = -\sigma H n + (\nabla \sigma \cdot t)t$ where t is the tangential direction of the interface. This recovers the traction free boundary condition (6.21).

In order to avoid the using of the Boussinesq approximation, which is only valid when the density of the mixture does not vary much, we can solve the transport equation (6.6) instead. The momentum equation will becomes

$$\rho(u_t + (u \cdot \nabla)u) + \nabla p - \nu \operatorname{div} D(u)$$

$$= -\nabla \cdot \left(\lambda \nabla \phi \otimes \nabla \phi - \frac{\lambda}{2}|\nabla \phi|^2 - \frac{\lambda}{4\epsilon^2}(\phi^2 - 1)^2 \right) - \rho g j. \quad (6.29)$$

More generally, the viscosity in different components can also be different.

Since we can view ϕ as the approximation of the volume fraction, an alternative is to use the "average" density and viscosity as follows:

$$\begin{aligned}
\frac{1}{\rho(\phi)} &= \frac{1+\phi}{2\rho_1} + \frac{1-\phi}{2\rho_2}, \\
\frac{1}{\nu(\phi)} &= \frac{1+\phi}{2\nu_1} + \frac{1-\phi}{2\nu_2},
\end{aligned} \tag{6.30}$$

where ρ_1, ρ_2 are the corresponding density and ν_1, ν_2 are the viscosity constants. The reason to choose the harmonic average as in (6.30) is that the solution of the Cahn-Hilliard equation (6.3) does not satisfy the maximal principle. Hence, the linear average cannot be guaranteed to be bounded away from zero. However, due to the L^∞-bound of the solution [10], the harmonic averages lead to desired properties. This approach can be replaced using the normal linear averages in the case when (6.3) is replaced by the Allen-Cahn equation (6.4) for which the solution satisfies the maximal principle.

The modified momentum equation with variable density and viscosity takes the form

$$\begin{aligned}
(\rho(\phi)u)_t + (u \cdot \nabla)(\rho(\phi)u) + \nabla p \\
- \operatorname{div}(\nu(\phi)D(u)) + \lambda \nabla \cdot (\nabla\phi \otimes \nabla\phi) = -\rho(\phi)gj,
\end{aligned} \tag{6.31}$$

where $g(x)$ is the external body force. As equation (6.25) converges to the pure transport equation, together with the incompressibility condition (6.25), the density ρ will satisfy the continuity equation:

$$\rho_t + \nabla \cdot (\rho u) = 0. \tag{6.32}$$

6.3   Mixtures involving liquid crystals

In an immiscible blend of a nematic liquid crystal and a Newtonian fluid, there are three types of elastic energies: mixing energy of the interface, bulk distortion energy of the nematic, and the anchoring energy of the liquid crystal molecules on the interface.

We again use the previously discussed Ginzburg-Landau energy for the mixing energy

$$f_{\text{mix}}(\phi, \nabla\phi) = \frac{1}{2}\lambda|\nabla\phi|^2 + f_0(\phi), \tag{6.33}$$

with a double-well potential for F,

$$f_0 = \frac{\lambda}{4\epsilon^2}(\phi^2 - 1)^2. \tag{6.34}$$

The nematic has rod-like molecules whose orientation can be represented by a unit vector $n(x)$ known as the director. When the director field is not uniform, the nematic has an Oseen-Frank distortion energy [22]:

$$f_{\text{bulk}} = \frac{1}{2}K_1(\nabla \cdot n)^2 + \frac{1}{2}K_2(n \cdot \nabla \times n)^2 + \frac{1}{2}K_3(n \times \nabla \times n)^2, \quad (6.35)$$

where K_1, K_2, K_3 are elastic constants for the three canonical types of orientational distortion: splay, twist and bend. We will adopt the customary one-constant approximation: $K = K_1 = K_2 = K_3$, so that the Frank energy simplifies to $f_{\text{bulk}} = \frac{K}{2}\nabla n : (\nabla n)^{\text{T}}$. Liu and Walkington [69] used a modified model by allowing a non-unity director whose length indicates the order parameter. Thus, the regularized Frank elastic energy becomes

$$f_{\text{bulk}} = K \left[\frac{1}{2}\nabla n : (\nabla n)^{\text{T}} + \frac{(|n|^2 - 1)^2}{4\delta^2} \right]. \quad (6.36)$$

The second term on the right-hand side serves as a penalty whose minimization is simply the Ginzburg-Landau approximation of the constraint $|n| = 1$ for small δ. The advantage of this regularized formulation is that the energy is now bounded for orientational defects, which are non-singular points where $|n| = 0$. This makes the numerical treatment much easier. Note that the regularization is based on the same idea as in Cahn-Hilliard's mixing energy. It is also related to Ericksen's theory of uniaxial nematics with a variable order parameter [32].

Depending on the chemistry of the two components, the rod-like molecules of the nematic phase prefer to orient on the interface in a certain direction known as the easy direction. The two most common types of anchoring are planar anchoring, where all directions in the plane of the interface are easy directions, and homeotropic anchoring, where the easy direction is the normal to the interface.

In the classical sharp interface picture, the anchoring energy is a surface energy. In our diffuse-interface model, however, we write it as a volumetric energy density in the same vein as the mixing energy:

$$f_{\text{anch}} = \frac{A}{2}(n \cdot \nabla\phi)^2 \quad (6.37)$$

for planar anchoring, and

$$f_{\text{anch}} = \frac{A}{2}[|n|^2|\nabla\phi|^2 - (n \cdot \nabla\phi)^2] \quad (6.38)$$

for homeotropic anchoring. In these two equations, the positive parameter A indicates the strength of the anchoring.

Finally, the total free energy density for the two-phase material is written as

$$f(\phi, \boldsymbol{n}, \nabla\phi, \nabla\boldsymbol{n}) = f_{\text{mix}} + \frac{1+\phi}{2} f_{\text{bulk}} + f_{\text{anch}}, \qquad (6.39)$$

where $\frac{1+\phi}{2}$ is the volume fraction of the nematic component, and $\phi = 1$ in the purely nematic phase. This energy is equivalent to that of Rey [83], and contains all the physics discussed there.

The induced elastic energy will be

$$\boldsymbol{\sigma}^e = -\lambda(\nabla\phi \otimes \nabla\phi) - K\frac{1+\phi}{2}(\nabla\boldsymbol{n}) \cdot (\nabla\boldsymbol{n})^{\text{T}} - \boldsymbol{G}, \qquad (6.40)$$

where $\boldsymbol{G} = A(\boldsymbol{n}{\cdot}\nabla\phi)\boldsymbol{n}{\otimes}\nabla\phi$ for planar anchoring and $\boldsymbol{G} = A[(\boldsymbol{n}{\cdot}\boldsymbol{n})\nabla\phi - (\boldsymbol{n} \cdot \nabla\phi)\boldsymbol{n}] \otimes \nabla\phi$ for homeotropic anchoring. Note that the asymmetry of $\boldsymbol{G}$ reflects the fact that surface anchoring exerts a net torque on the fluid. Bulk distortion will give rise to an asymmetric stress as well if the elastic constants are unequal [22]. Moreover, from the derivation of the previous section, we see that the anchoring energy f_{anch}, hence the term $\boldsymbol{G}$, induces a Marangoni force along isotropic-nematic interfaces.

For our model system of a blend of a nematic and a Newtonian fluid, the field variables are velocity $\boldsymbol{v}$, pressure p, phase function ϕ and director $\boldsymbol{n}$. We write the continuity and momentum equations in the usual form:

$$\nabla \cdot \boldsymbol{v} = 0, \qquad (6.41)$$

$$\rho\left(\frac{\partial \boldsymbol{v}}{\partial t} + \boldsymbol{v} \cdot \nabla\boldsymbol{v}\right) = -\nabla p + \nabla \cdot \boldsymbol{\sigma}, \qquad (6.42)$$

where $\boldsymbol{\sigma}$ is the deviatoric stress tensor.

Based on the free energy in equation (6.39), a generalized chemical potential can be defined as $\delta F/\delta\phi$. If one assumes a generalized Fick's law that the mass flux be proportional to the gradient of the chemical potential, the Cahn-Hilliard equation is obtained as an evolution equation for ϕ [13]:

$$\frac{\partial \phi}{\partial t} + \boldsymbol{v} \cdot \nabla\phi = \nabla \cdot \left[\gamma_1 \nabla\left(\frac{\delta F}{\delta\phi}\right)\right], \qquad (6.43)$$

where γ_1 is the mobility, taken to be a constant in this paper. The diffusion term on the right-hand side has contributions from all three forms of free energy.

The rotation of $\boldsymbol{n}$ is determined by the balance between a viscous torque and an elastic torque. The latter, also known as the molecular field [22], arises from the free energies of the system:

$$\boldsymbol{h} = -\frac{\delta F}{\delta \boldsymbol{n}} = K\left[-\nabla \cdot \left(\frac{1+\phi}{2}\nabla\boldsymbol{n}\right) + \frac{1+\phi}{2}\frac{(\boldsymbol{n}^2 - 1)\boldsymbol{n}}{\delta^2}\right] + \boldsymbol{g}, \qquad (6.44)$$

where $\boldsymbol{g} = A(\boldsymbol{n} \cdot \nabla\phi)\nabla\phi$ for planar anchoring, and $\boldsymbol{g} = A[(\nabla\phi \cdot \nabla\phi)\boldsymbol{n} - (\boldsymbol{n} \cdot \nabla\phi)\nabla\phi]$ for homeotropic anchoring. Now the evolution equation of $\boldsymbol{n}$ is written as

$$\frac{\partial \boldsymbol{n}}{\partial t} + \boldsymbol{v} \cdot \nabla\boldsymbol{n} = \gamma_2 \boldsymbol{h}, \qquad (6.45)$$

where the constant γ_2 determines the relaxation time of the director field. Equation (6.45) is a simplified version of the Leslie-Ericksen equation [22].

Equations (6.41)–(6.43) and (6.45) form the complete set of equations governing the evolution of the nematic-Newtonian two-phase system.

In this paper, we assume that the two phases have the same constant density, with negligible volume change upon mixing. Thus, the mixture is incompressible with a solenoidal velocity. In general, however, the diffuse-interface method is not restricted to equal-density components. When the two phases have differing densities, one approach is to view the mixture as a compressible fluid with $\nabla \cdot \boldsymbol{v} \neq 0$ in the mixing layer, where $\boldsymbol{v}$ is a mass-averaged velocity [71]. As an alternative, [67] have proposed a picture in which the components mix by advection only without diffusion. Thus, the velocity at a spatial point is defined as that of the component occupying that point; it is spatially continuous and remains solenoidal. An inhomogeneous average density is established from the initial condition, which is later transported by the velocity field. Finally, if the density difference is small, the Boussinesq approximation can be employed [67].

A solution to the above governing equations obeys an energy law:

$$\frac{d}{dt} \int_\Omega \left(\frac{\rho}{2} |\boldsymbol{v}|^2 + f \right) d\Omega$$
$$= -\int_\Omega \left(\mu \nabla\boldsymbol{v} : \nabla\boldsymbol{v}^{\mathrm{T}} + \gamma_1 \left| \nabla \frac{\delta F}{\delta\phi} \right|^2 + \gamma_2 \left| \frac{\delta F}{\delta \boldsymbol{n}} \right|^2 \right) d\Omega, \quad (6.46)$$

where f is the system's potential energy density. Physically, the law states that the total energy of the system (excluding thermal energy) will decrease from internal dissipation. The work of Lin and Liu [62,63] can be used to rigorously prove the well-posedness of such a system. Under such an energy law, a finite-dimensional approximation to the governing equations, such as a finite-element or spectral scheme, can be shown to be guaranteed to converge [69]. This constitutes one of the advantages of our method over previous methods that do not maintain the system's total energy budget.

7   Magneto hydrodynamics (MHD)

7.1   Introduction

We are interested in the following unsteady incompressible magneto hydrodynamics (MHD) system:

$$\rho(\partial_t \boldsymbol{u} + \boldsymbol{u} \cdot \nabla \boldsymbol{u}) + \nabla p = \mu \Delta \boldsymbol{u} + \frac{1}{c} \boldsymbol{j} \times \boldsymbol{b}, \tag{7.1}$$

$$\nabla \cdot \boldsymbol{u} = 0, \tag{7.2}$$

$$\frac{1}{c} \partial_t \boldsymbol{b} + \nabla \times \boldsymbol{e} = \boldsymbol{0}, \tag{7.3}$$

$$\nabla \times \boldsymbol{b} = \frac{4\pi}{c} \boldsymbol{j}, \tag{7.4}$$

$$\sigma \left(\boldsymbol{e} + \frac{1}{c} \boldsymbol{u} \times \boldsymbol{b} \right) = \boldsymbol{j}, \tag{7.5}$$

in a smooth bounded domain in either $\mathbb{R}^2$ or $\mathbb{R}^3$. We also equip the system with no-slip and perfectly conducting wall conditions

$$\boldsymbol{u} = \boldsymbol{0}, \quad \boldsymbol{e} \times \boldsymbol{n} = \boldsymbol{0} \quad \text{on } \Gamma = \partial\Omega, \tag{7.6}$$

where ρ is the fluid density, $\boldsymbol{u}$ the fluid velocity, $\boldsymbol{b}$, $\boldsymbol{j}$ and $\boldsymbol{e}$ are the magnetic field, the electric current density and the electric field respectively. μ, σ and c are physical constants representing the viscosity coefficient, the electric conductivity and the speed of light.

We see that (7.1) consists of the Navier-Stokes equation governing the motion of the solenoidal fluid motion, coupled with the Faraday equation describing the evolution of the magnetic field, through the Lorentz force $\boldsymbol{j} \times \boldsymbol{b}$ and the electro transport $\boldsymbol{u} \times \boldsymbol{b}$.

7.2   The evolution of the magnetic field

First, let us look at the transport equation of $\boldsymbol{b}$, and take divergence of the equation with respect to the spatial variable. We have

$$(\nabla \cdot \boldsymbol{b})_t = 0. \tag{7.7}$$

Hence we see that $\boldsymbol{b}$ is a divergence free vector field (assume this is true for the initial datum). This, combined with the fact that $\boldsymbol{u}$ is also divergence free, we have

$$\nabla \times (\boldsymbol{u} \times \boldsymbol{b}) = (\boldsymbol{b} \cdot \nabla)\boldsymbol{u} - (\boldsymbol{u} \cdot \nabla)\boldsymbol{b} + \boldsymbol{u}(\nabla \cdot \boldsymbol{b}) - \boldsymbol{b}(\nabla \cdot \boldsymbol{u})$$
$$= (\boldsymbol{b} \cdot \nabla)\boldsymbol{u} - (\boldsymbol{u} \cdot \nabla)\boldsymbol{b}.$$

We can now rewrite the evolution equation (7.4) of the magnetic field $\boldsymbol{b}$ as

$$\boldsymbol{b}_t + (\boldsymbol{u} \cdot \nabla)\boldsymbol{b} - (\boldsymbol{b} \cdot \nabla)\boldsymbol{u} = -\frac{c}{\sigma} \nabla \times \boldsymbol{j}. \tag{7.8}$$

The right-hand side of (7.8) is the dissipation term, while the left-hand side can be written as $b_t + \mathcal{L}_u b$, where $\mathcal{L}_u b$ is the Lie derivative of b with respect to the flow field u.

Again we look at F, the deformation tensor $\frac{\partial x}{\partial X}$, where x is the Eulerian (reference) coordinate and X is the Lagrangian (material) coordinate. The flow trajectory of a partial will be $x(X, t)$, where

$$\frac{\partial}{\partial t} x(X, t) = u(x(X, t), t), \quad x(X, 0) = X. \tag{7.9}$$

In case of incompressible materials, x is a volume preserving diffeomorphism and $\det F = 1$. In this case, we have the following identity [41,70]:

$$\frac{d}{dt} F(x(X, t), t) = \frac{\partial}{\partial t} F(x, t) + u(x, t) \cdot \nabla F(x, t) = \nabla u(x, t) F(x, t). \tag{7.10}$$

The transport $b_t + (u \cdot \nabla)b - (b \cdot \nabla)u$ represents exactly the relation $b(x(X, t), t) = F^{-1} b_0(X)$, where b_0 is the initial magnetic fields. From here, we see that F carries all the transport information of b. This makes the MHD system very much related to the viscoelastic system discussed in the earlier sections.

7.3   The energy law

System (7.1) admits the following energy law:

$$\frac{d}{dt} \int_\Omega \left(\frac{1}{2}\rho|u|^2 + \frac{1}{8\pi}|b|^2 \right) dx = -\int_\Omega \left(\mu|\nabla u|^2 + \frac{4\pi}{c\sigma}|j|^2 \right) dx. \tag{7.11}$$

This energy law can be derived by multiply (7.2) by u and (7.4) by b, add the results together, and integration by parts. The special cancellation of the term from the Lorentz force and the term from the transport of the magnetic field shows the special coupling that will be discussed in the later sections.

7.4   The linear momentum equation

The linear momentum equation (without the dissipation term μu) in (7.1) can be derived through the variations of the following "action" functional in the space of volume preserving flow maps:

$$A(x) = \int_0^T \int_\Omega \left[\frac{1}{2}\rho|u(x, t)|^2 + \frac{1}{8\pi}|b(x, t)|^2 \right] dx\, dt$$

$$= \int_0^T \int_{\Omega_0} \left[\frac{1}{2}\rho|x_t(X, t)|^2 + \frac{1}{8\pi}|F(X, t)b(x(X, t))|^2 \right] dX\, dt$$

$$= A_1(x) + \frac{1}{4\pi} A_2(x). \tag{7.12}$$

Here we use the fact that the Jacobian of x with respect to X is 1, due to the incompressibility.

In order to perform the variation, we look at the one parameter family of such maps x^ϵ such that

$$x^0 = x, \qquad \frac{dx^\epsilon}{d\epsilon} = y \tag{7.13}$$

for any y such that $\nabla_x \cdot y = 0$.

We computer the variation of $A(x^\epsilon) = A(\phi(x^\epsilon, t))$ with respect to ϵ. The contribution of the kinetic part $\frac{1}{2}\rho|u(x,t)|^2$ will give the Euler part of equation (7.2). Suppose that $F^\epsilon = \frac{x^\epsilon}{X}$, the contribution of the magnetic field will be

$$\frac{d}{d\epsilon}|_{\epsilon=0} A_2(x^\epsilon)$$

$$= \int_0^T \int_{\Omega_0} \left(F(X,t)b(x(X,t)), \frac{d}{d\epsilon}|_{\epsilon=0}(F^\epsilon(X,t)b(x^\epsilon(X,t))) \right) dX dt$$

$$= \int_0^T \int_{\Omega_0} \left(F(X,t)b(x(X,t)), F(X,t)(y \cdot \nabla_x)b \right)$$
$$+ \left(F(X,t)b(x(X,t)), \nabla_X y b(x(X,t)) \right) dX dt.$$

Since y is an arbitrary divergence free vector field, integration by parts gives the following equation:

$$\rho(\partial_t u + u \cdot \nabla u) + \nabla p = \frac{1}{4\pi}\nabla \cdot (b \otimes b) + \frac{1}{2\pi}\nabla|b|^2. \tag{7.14}$$

Since we have the identity

$$(\nabla \times b) \times b = (b \cdot \nabla)b - \frac{1}{2}\nabla|b|^2 = \nabla \cdot (b \otimes b) - \frac{1}{2}\nabla|b|^2, \tag{7.15}$$

we have

$$\rho(\partial_t u + u \cdot \nabla u) + \nabla p_1 = \frac{1}{4\pi}(\nabla \times b) \times b = \frac{1}{c}j \times b. \tag{7.16}$$

From the above derivation, we see that the Lorentz force comes from the fact that the magnetic field is transport by the flow map as a co-variant 1-form and the inclusion of the magnetic energy of b in the total energy.

7.5   The dynamics of magnetic field lines

In the case that $\sigma = \infty$, then we are in the case that

$$e = -\frac{1}{c}u \times b. \tag{7.17}$$

Taking the $\nabla \times$ on the both sides of the equation yields that

$$(\nabla \times \boldsymbol{b})_t + (\boldsymbol{u} \cdot \nabla)(\nabla \times \boldsymbol{b}) - ((\nabla \times \boldsymbol{b}) \cdot \nabla)\boldsymbol{u}$$
$$- (\boldsymbol{b} \cdot \nabla)(\nabla \times \boldsymbol{u}) + ((\nabla \times \boldsymbol{u}) \cdot \nabla)\boldsymbol{b} = 0. \qquad (7.18)$$

In the 2-dimensional case, that is, both $\boldsymbol{u}$ and $\boldsymbol{b}$ depend only on the first two coordinate x_1, x_2 and the third component $\boldsymbol{u}^3 = \boldsymbol{b}^3 = 0$, then we have

$$(\nabla \times \boldsymbol{b})_t + (\boldsymbol{u} \cdot \nabla)(\nabla \times \boldsymbol{b}) - (\boldsymbol{b} \cdot \nabla)(\nabla \times \boldsymbol{u}) = 0. \qquad (7.19)$$

If we are in a irrotational flow field, we have

$$(\nabla \times \boldsymbol{b})_t + (\boldsymbol{u} \cdot \nabla)(\nabla \times \boldsymbol{b}) = 0. \qquad (7.20)$$

Hence we can see that if $(\nabla \times \boldsymbol{b})$ concentrate on a curve (interface) at the initial time, it will also be concentrated a curve.

Finally, since $\nabla \cdot \boldsymbol{b} = 0$, we have

$$\boldsymbol{b} = \nabla \times \mathbf{A}, \qquad (7.21)$$

where $\mathbf{A}$ is the electric potential with the Coulomb gauge $\nabla \cdot \mathbf{A} = 0$. The electric current $\boldsymbol{j} = -\Delta \mathbf{A}$. Moreover, it satisfies the transport equation:

$$\mathbf{A}_t + (\boldsymbol{u} \cdot \nabla)\mathbf{A} = 0. \qquad (7.22)$$

In the 2-dimensional case, we have the simpler form

$$\boldsymbol{b} = \nabla^{\perp}\phi = (\phi_y, -\phi_x) \qquad (7.23)$$

for a scalar function $\phi(x, y)$. The current $\boldsymbol{j} = -\Delta \phi \mathbf{e}_3$ and it satisfies the transport equation:

$$\phi_t + (\boldsymbol{u} \cdot \nabla)\phi = 0. \qquad (7.24)$$

Moreover, the Lorentz force

$$\boldsymbol{j} \times \boldsymbol{b} = \Delta \phi \nabla \phi. \qquad (7.25)$$

The level set of ϕ, $\{\phi = c\}$ is the magnetic field lines. The dynamics of such curves is very important in understanding the MHD equations.

References

[1] R. Abraham and J. E. Marsden. *Foundations of Mechanics*. Benjamin/Cummings Publishing Co. Inc. Advanced Book Program, Reading, Mass., 1978. Second edition, revised and enlarged, With the assistance of Tudor Ratiu and Richard Cushman.

[2] N. D. Alikakos, P. W. Bates, and X. F. Chen. Convergence of the Cahn-Hilliard equation to the Hele-Shaw model. *Arch. Rational Mech. Anal.*, 128(2): 165–205, 1994.

[3] V. I. Arnold. *Mathematical Methods of Classical Mechanics.* Springer-Verlag, New York, 1989. Translated from the 1974 Russian original by K. Vogtmann and A. Weinstein, Corrected reprint of the second (1989) edition.

[4] G. K. Batchelor. *An Introduction to Fluid Dynamics.* Cambridge University Press, Cambridge, paperback edition, 1999.

[5] F. Bethuel and H. Brezis. Regularity of minimizers of relaxed problems for harmonic maps. *J. Funct. Anal.*, 101(1): 145–161, 1991.

[6] F. Bethuel, H. Brezis, and F. Helein. *Ginzburg–Landau Vorticies.* Klumer, 1995.

[7] R. B. Bird, R. C. Armstrong, and O. Hassager. *Dynamics of Polymeric Liquids, Volume 1: Fluid Mechanics.* Weiley Interscience, New York, 1987.

[8] F. Boyer. A theoretical and numerical model for the study of incompressible mixture flows. *Computers and Fluids*, 31: 41–68, 2002.

[9] H. Brezis, J. Coron, and E. Lieb. Harmonic maps with defects. *Comm. Math. Phys.*, 107(4): 649–705, 1986.

[10] L. A. Caffarelli and N. E. Muler. An L^∞ bound for solutions of the Cahn-Hilliard equation. *Arch. Rational Mech. Anal.*, 133(2): 129–144, 1995.

[11] G. Caginalp and X. F. Chen. Phase field equations in the singular limit of sharp interface problems. In *On the Evolution of Phase Boundaries (Minneapolis, MN, 1990–91)*, 1–27. Springer, New York, 1992.

[12] J. W. Cahn and S. M. Allen. A microscopic theory for domain wall motion and its experimental varification in Fe-Al alloy domain growth kinetics. *J. Phys. Colloque*, C7–C51, 1977.

[13] J. W. Cahn and J. E. Hillard. Free energy of a nonuniform system. I. Interfacial free energy. *J. Chem. Phys.*, 28: 258–267, 1958.

[14] P. M. Chaikin and T. C. Lubensky. *Principles of Condensed Matter Physics.* Cambridge, 1995.

[15] Y. C. Chang, T. Y. Hou, B. Merriman, and S. Osher. A level set formulation of eulerian interface capturing methods for incompressible fluid flows. *J. Comput. Phys.*, 124(2): 449–464, 1996.

[16] J. Y. Chemin and N. Masmoudi. About lifespan of regular solutions of equations related to viscoelastic fluids. *SIAM J. MAth. Anal.*, 33(1): 84–112, 2001.

[17] X. F. Chen. Generation and propagation of interfaces in reaction-diffusion systems. *Trans. Amer. Math. Soc.*, 334, 1992.

[18] Y. M. Chen and M. Struwe. Regularity for heat flow for harmonic maps. *Math. Z.*, 201: 83–103, 1989.

[19] M. Chipot, D. Kinderlehrer, and C. Liu. Some variational problems with relaxed constraints. *unpublished*, 1997.

[20] P. Constantin and C. Foias. *Navier-Stokes Equations*. Chicago Lectures in Mathematics. University of Chicago Press, Chicago, IL, 1988.

[21] V. Cristini, J. Blawzdziewicz, and M. Loewenberg. Drop breakup in three-dimensional viscous flows. *Phys. Fluids*, 10: 1781–1783, 1998.

[22] P. G. de Gennes and J. Prost. *The Physics of Liquid Crystals*. Oxford University Press, 1993.

[23] J. Deang, Q. Du, M. Gunzburger, and J. Peterson. Vortices in superconductors: modelling and computer simulations. *Philos. Trans. Roy. Soc. London*, 355(1731): 1957–1968, 1997.

[24] R. J. DiPerna and P. L. Lions. Ordinary differential equations, transport theory and Sobolev spaces. *Invent. Math.*, 98(3): 511–547, 1989.

[25] J. E. Dunn and J. Serrin. On the thermomechanics of interstitial working. *Arch. Rational Mech. Anal.*, 88(2): 95–133, 1985.

[26] W. E. Dynamics of vortices in Ginzburg-Landau theories with applications to superconductivity. *Phys. D*, 77(4): 383–404, 1994.

[27] W. E, T. Li, and P. Zhang. Well-posedness for the dumbbell model of polymeric fluids. *Comm. in Math. Phy.*, 248 (2): 409–427, 2004.

[28] W. E and P. Palffy-Muhoray. Phase separation in incompressible systems. *Phys. Rev. E (3)*, 55(4): R3844–R3846, 1997.

[29] D. A. Edwards, H. Brenner, and D. T. Wasan. *Interfacial Transport Process and Rheology*. Butterworth-Heinemann, 1991.

[30] J. Ericksen. Conservation laws for liquid crystals. *Trans. Soc. Rheol.*, 5: 22–34, 1961.

[31] J. Ericksen. Continuum theory of nematic liquid crystals. *Res. Mechanica*, 21: 381–392, 1987.

[32] J. Ericksen. Liquid crystals with variable degree of orientation. *Arch Rath. Mech. Anal.*, 113: 97–120, 1991.

[33] L. C. Evans. *Weak Convergence Methods for Nonlinear Partial Differential Equations*, volume 74. A.M.S. Regional Conference Series in Mathematics, 1989.

[34] L. C. Evans. *Partial Differential Equations*. A.M.S., 1998.

[35] E. Freid and M. E. Gurtin. Continuum theory of thermally induced phase transitions based on an order parameter. *Physica D*, 68: 326–343, 1993.

[36] D. Gilbarg and N. S. Trudinger. *Elliptic Partial Differential Equations of Second Order*. Springer-Verlag, Berlin, New York, 1983.

[37] J. Glimm, J. W. Grove, X. L. Li, and D. C. Tan. Robust computational algorithms for dynamic interface tracking in three dimensions. *SIAM J. Sci. Comput.*, 21(6): 2240–2256 (electronic), 2000.

[38] J. Glimm, X. L. Li, Y. Liu, and N. Zhao. Conservative front tracking and level set algorithms. *Proc. Natl. Acad. Sci. USA*, 98(25): 14198–14201 (electronic), 2001.

[39] C. Guillopé and J. C. Saut. Existence results for the flow of viscoelastic fluids with a differential constitutive law. *Nonlinear Anal.*, 15(9): 849–869, 1990.

[40] C. Guillopé and J. C. Saut. Global existence and one-dimensional nonlinear stability of shearing motions of viscoelastic fluids of Oldroyd type. *RAIRO Modél. Math. Anal. Numér.*, 24(3): 369–401, 1990.

[41] M. E. Gurtin. *An Introduction to Continuum Mechanics*, volume 158 of *Mathematics in Science and Engineering*. Academic Press, 1981.

[42] M. E. Gurtin. Multiphase thermodynamics with interfacial structure, 1. Heat conduction and the capillary balance law. *Archive for Rational Mechanics and Analysis*, 104: 195–221, 1988.

[43] M. E. Gurtin, D. Polignone, and J. Viñals. Two-phase binary fluids and immiscible fluids described by an order parameter. *Math. Models Methods Appl. Sci.*, 6(6): 815–831, 1996.

[44] R. Hardt and F. H. Lin. Stability of singularities of minimizing harmonic maps. *J. Differential Geom.*, 29(1): 113–123, 1989.

[45] J. Hron, J. Málek, J. Nevcas, and K. R. Rajagopal. Numerical simulations and global existence of solutions of two-dimensional flows of fluids with pressure- and shear-dependent viscosities. *Math. Comput. Simulation*, 61(3-6): 297–315, 2003. MODELLING 2001 (Pilsen).

[46] D. Jacqmin. Calculation of two-phase Navier-Stokes flows using phase-field modeling. *J. Comput. Phys.*, 155(1): 96–127, 1999.

[47] F. John. Rotation and strain. *Comm. Pure Appl. Math.*, 14: 391–413, 1961.

[48] D. D. Joseph. Instability of the rest state of fluids of arbitrary grade greater than one. *Arch. Rational Mech. Anal.*, 75(3): 251–256, 1980/81.

[49] R. Khayat. Three-dimensional boundary-element analysis of drop deformation for newtonian and viscoelastic systems. *Int. J. Num. Meth. Fluids*, 34: 241–275, 2000.

[50] V. V. Krotov and A. I. Rusanov. *Physicochemical Hydrodynamics of Capillary Systems*. Inperial College Press, London, 1999.

[51] S. N. Kruzkov. First order quasilinear equations in several independent variables. *Math. USSR Sbornik*, 10(2): 217–243, 1970.

[52] O. A. Ladyzhenskaya and G. A. Seregin. On the regularity of solutions of two-dimensional equations of the dynamics of fluids with nonlinear viscosity. *Zap. Nauchn. Sem. S.-Peterburg. Otdel. Mat. Inst. Steklov. (POMI)*, 259(Kraev. Zadachi Mat. Fiz. i Smezh. Vopr. Teor. Funkts. 30): 145–166, 298, 1999.

[53] L. D. Landau and E. M. Lifshitz. *Fluid Mechanics*. Pergamon Press, 2 edition, 1987.

[54] R. G. Larson. *The Structure and Rheology of Complex Fluids*. Oxford, 1995.

[55] Z. Lei, C. Liu, and Y. Zhou. Existence of small strain incompressible viscoelasticity. *preprint*, 2005.

[56] Z. Lei, C. Liu, and Y. Zhou. Incompressible viscoelasticity fluids: Existence of classical solutions with near equilibrium initial datum. *Submitted to Comm. Math. Phys.*, 2005.

[57] F. Leslie. Some constitutive equations for liquid crystals. *Archive for Rational Mechanics and Analysis*, 28: 265–283, 1968.

[58] J. Li and Y. Renardy. Numerical study of flows of two immiscible liquids at low reynolds number. *SIAM Review*, 42: 417–439, 2000.

[59] J. Li and Y. Renardy. Shear-induced rupturing of a viscous drop in a bingham liquid. *J. Non-Newtonian Fluid Mech.*, 95: 235–251, 2000.

[60] J. Lighthill. *Waves in Fluids*. Cambridge, 1978.

[61] F. H. Lin. Mapping problems, fundamental groups and defect measures. *Acta Math. Sin. (Engl. Ser.)*, 15(1): 25–52, 1999.

[62] F. H. Lin and C. Liu. Nonparabolic dissipative systems, modeling the flow of liquid crystals. *Comm. Pure Appl. Math.*, XLVIII(5): 501–537, 1995.

[63] F. H. Lin and C. Liu. Global existence of solutions for the Ericksen Leslie–system. *Arch. Rat. Mech. Ana.*, 154(2): 135–156, 2001.

[64] F. H. Lin and C. Liu. Static and dynamic theories of liquid crystals. *Journal of Partial Differential Equations*, 14(4): 289–330, 2001.

[65] F. H. Lin, C. Liu, and P. Zhang. On a micro-macro model for polymeric fluids near equilibrium. *Comm. Pure Appl. Math.*, LIX: 1–29, 2005.

[66] P. L. Lions and N. Masmoudi. Global solutions for some Oldroyd models of non-Newtonian flows. *Chinese Ann. Math. Ser. B*, 21(2): 131–146, 2000.

[67] C. Liu and J. Shen. A phase field model for the mixture of two incompressible fluids and its approximation by a fourier-spectral method. *Physica D*, 179: 211–228, 2003.

[68] C. Liu and S. Shkoller. Variational phase field model for the mixture of two fluids. *preprint*, 2001.

[69] C. Liu and N. J. Walkington. Approximation of liquid crystal flows. *SIAM Journal on Numerical Analysis*, 37(3): 725–741, 2000.

[70] C. Liu and N. J. Walkington. An Eulerian description of fluids containing visco-hyperelastic particles. *Arch. Rat. Mech. Ana.*, 159: 229–252, 2001.

[71] J. Lowengrub and L. Truskinovsky. Quasi-incompressible Cahn-Hilliard fluids and topological transitions. *R. Soc. Lond. Proc. Ser. A Math. Phys. Eng. Sci.*, 454(1978): 2617–2654, 1998.

[72] A. J. Majda and A. L. Bertozzi. *Vorticity and Incompressible Flow*, volume 27 of *Cambridge Texts in Applied Mathematics*. Cambridge University Press, Cambridge, 2002.

[73] J. Málek, J. Nevcas, and K. R. Rajagopal. Global analysis of the flows of fluids with pressure-dependent viscosities. *Arch. Ration. Mech. Anal.*, 165(3): 243–269, 2002.

[74] J. E. Marsden and T. S. Ratiu. *Introduction to Mechanics and Symmetry*, volume 17 of *Texts in Applied Mathematics*. Springer-Verlag, New York, second edition, 1999. A basic exposition of classical mechanical systems.

[75] G. B. McFadden, A. A. Wheeler, R. J. Braun, S. R. Coriell, and R. F. Sekerka. Phase-field models for anisotropic interfaces. *Phys. Rev. E (3)*, 48(3): 2016–2024, 1993.

[76] W. W. Mullins and R. F. Sekerka. On the thermodynamics of crystalline solids. *J. Chem. Phys.*, 82, 1985.

[77] S. Osher and J. Sethian. Fronts propagating with curvature dependent speed: Algorithms based on Hamilton Jacobi formulations. *Journal of Computational Physics*, 79: 12–49, 1988.

[78] R. G. Owens and T. N. Phillips. *Computational Rheology*. Imperial College Press, London, 2002.

[79] C. S. Peskin. A random-walk interpretation of the incompressible Navier-Stokes equations. *Comm. Pure Appl. Math.*, 38(6): 845–852, 1985.

[80] R. F. Probstein. *Physicochemical Hydrodynamics, An Introduction*. John Wiley and Sons, INC., 1994.

[81] T. Qian, X. P. Wang, and P. Sheng. Molecular scale contact line hydrodynamics of immiscible flows. *preprint*, 2002.

[82] M. Renardy, W. J. Hrusa, and J. A. Nohel. *Mathematical Problems in Viscoelasticity*, volume 35 of *Pitman Monographs and Surveys in Pure and Applied Mathematics*. Longman Scientific & Technical, Harlow, 1987.

[83] A. D. Rey. Vescoelastic theory for nematic interfaces. *Physical Review E*, 61(2): 1540–1549, 2000.

[84] J. Rubinstein, P. Sternberg, and J. B. Keller. Fast reaction, slow diffusion, and curve shortening. *SIAM J. Appl. Math.*, 49(1): 116–133, 1989.

[85] R. Schoen and K. Uhlenbeck. Correction to: "A regularity theory for harmonic maps". *J. Differential Geom.*, 18(2): 329, 1983.

[86] R. Schoen and K. Uhlenbeck. Regularity of minimizing harmonic maps into the sphere. *Invent. Math.*, 78(1): 89–100, 1984.

[87] W. R. Schowalter. *Mechanics of Non-Newtonian Fluids*. Pergamon Press, 1978.

[88] M. Slemrod. Constitutive relations for Rivlin-Ericksen fluids based on generalized rational approximation. *Arch. Ration. Mech. Anal.*, 146(1): 73–93, 1999.

[89] H. M. Soner. Convergence of the phase-field equations to the Mullins-Sekerka problem with kinetic undercooling [97d: 80007]. In *Fundamental Contributions to the Continuum Theory of Evolving Phase Interfaces in Solids*, 413–471. Springer, Berlin, 1999.

[90] M. Struwe. *Variational Methods, Applications to Nonlinear Partial Differential Equations and Hamiltonian Systems*. Springer-Verlag, 1990.

[91] M. Struwe. *Geometric Evolution Problems*. AMS, 1992.

[92] L. Tartar. Compensated compactness and applications to partial differential equations. In R. J. Knops, editor, *Research Notes in Mathematics, Nonlinear Analysis, and Mechanics, Heriot Watt Symposium*, volume 4. Pitman Press, 1979.

[93] J. E. Taylor and J. W. Cahn. Linking anisotropic sharp and diffuse surface motion laws via gradient flows. *J. Statist. Phys.*, 77(1-2): 183–197, 1994.

[94] R. Temam. *Navier-Stokes Equations*. North Holland, 1977.

[95] E. M. Toose, B. J. Geurts, and J. Kuerten. A boundary integral method for two-dimensional (non)-newtonian drops in slow viscous flow. *J. Non-Newtonian Fluid Mech.*, 60: 129–154, 1995.

[96] J. van der Waals. The thermodynamic theory of capillarity under the hypothesis of a continuous density variation. *J. Stat. Phys.*, 20: 197–244, 1893.

[97] P. Yue, J. Feng, C. Liu, and J. Shen. A diffuse-interface method for simulating two-phase flows of complex fluids. *Journal of Fluid Mechanics*, 515: 293–317, 2004.

Introduction to Kinetic Theory for Complex Fluids*

Qi Wang

Department of Mathematics, Florida State University
Tallahassee, FL 32306-4510, USA
Email: wang@math.fsu.edu

Abstract

Complex fluids/soft matter are materials whose material properties are dominated by their meso-scale structures. In this chapter, we introduce some basic theories useful in modeling complex fluids/soft matter materials. We selectively focus on the kinetic theory approach to modeling complex fluids. The basic topics included are equilibrium thermodynamics, statistical mechanics, polymer physics, and the configurational space kinetic theory for transport phenomena in complex fluids.

1 Introduction

Complex fluids are fluids that are homogeneous at macroscopic scales and disordered at microscopic scales, but possess structure on a mesoscopic length scale. They are also known in the physics community as the soft matter, the matter between fluids and ideal solids. Complex fluids consist of polymer solutions, polymer melts, particle suspension fluids, gels, surfactant, emulsions, and many manmade and biological fluids etc. Polymer solutions are made of polymers dissolved in solutions or solvent; polymer melts are molten polymers; particle suspension fluids consist of solid particles suspended in a matrix fluid which may be a viscous or viscoelastic fluid; biological fluids such as blood and mucus consisting of live macromolecules and solvent. Given the large molecular weight and size in the complex fluids, they are capable of forming a variety of meso-phases, in which partial positional as well as orientational order can be present. These meso-phases of partial order are often identified

---

*The author is partly supported by the Air Force Office for Scientific Research (AFOSR) through grant F49550-05-1-0025 and the National Science Foundation (NSF) through grants DMS-0605029, DMS-0626180 and DMS-0724273.

as liquid crystalline phases. The polymeric liquids in the meso-phases are called liquid crystalline polymers. When the meso-phases are created above some critical concentration in solutions, the polymeric liquids are called lyotropic liquid crystal polymers. When the phases are attained at certain low temperature in melts, the materials are called thermotropic liquid crystal polymers. Not only miscible polymeric solutions and melts are capable of forming the liquid crystalline phase, immiscible polymer blends, emulsions, polymer-particle nano-composites, which are liquid mixtures of polymer solutions or melts and solid nano-sized particles, are all candidates for forming liquid crystalline phases.

Complex fluids exhibit a host of distinctive features from the isotropic liquids made up of small molecules like water, cooking oils, etc. in flows due to their large molecular weight and the variety of conformations. Their material properties are exclusively determined by their meso-structures. There have been several hallmark phenomena in the polymeric fluids well-documented such as rod-climbing, extruded swell, and tubeless siphon etc. [2]; when a gas bubble is trapped within a polymeric liquid, its geometry is distinct from that of a gas bubble trapped within a Newtonian fluid. These fascinating phenomena, distinctive of the polymeric liquids, have spurted a significant amount of research activities over the years. Theories and models developed for complex fluids are now applied to biofluids and biomaterials, nanocomposite materials, making it a fast growing interdisciplinary research area.

In the lecture, we will give a crash course on the basics needed to develop models for complex fluids in fundamental thermodynamics, statistic mechanics, polymer physics and eventually kinetic theories, and explore a systematic approach for flexible polymers and liquid crystal polymers within the framework of the kinetic theory. We hope the notes give you a brief introduction to the vast and vibrant subject. More detailed exhibition of the subject can be found in the references cited at the end.

2 A primer for equilibrium thermodynamics

We introduce the basic thermodynamic variables that will be used through out the lecture, especially, the free energy potentials and the basic thermodynamic laws which will guide our development of dynamic equations for complex fluid systems. In a simple notion, a thermodynamical system is any macroscopic matter/substance system; any measurable macroscopic quantities associated with the system can be called thermodynamical variables. For instance, the temperature T, pressure P, and volume V of the matter system are thermodynamic variables. The space of the thermodynamic variables is called the phase space, i.e., the space of $(T, P, V, \cdots)$ in the previous example. The equation of state, which characterizes the material properties of the thermodynamical system, is a functional relationship among the thermodynamical variables in the phase space. It is normally given as a level surface in the phase space. For example, in a system where the thermodynamical variables are T, P, V, the equation of state is expressed as an algebraic equation

$$f(P, V, T) = 0. \tag{2.1}$$

The equation of state dictates that the thermodynamic process defined therein can only take place in a confined subspace or manifold in the phase space.

A scalar function or functional can be associated to the thermodynamical system as the internal energy. Heat Q is a form of energy, which usually varies with the temperature. The rate of change of the heat with respect to the temperature is an important thermodynamic quantity, which is termed the heat capacity or specific heat

$$C = \frac{\Delta Q}{\Delta T}, \tag{2.2}$$

where $\Delta(\bullet)$ represents an infinitesimal amount of change in the quantity $(\bullet)$. In another word, the heat capacity is the amount of heat the system absorbs (or releases) when temperature is raised (or lowered) by one degree. Some thermodynamical variables are called extensive if they are proportional to the amount of the substance in the system, e.g., energy and volume, etc.; others, independent of the amount of the substance, are called intensive like temperature, pressure, etc. The extensive variables are additive while the intensive ones are not.

In equilibrium thermodynamics, the thermodynamic transformation describes a change of state in the phase space. If the initial state in the phase space is an equilibrium state, the transformation can be brought about only by changes in external conditions of the system. The transformation is said to be quasi-static if the external condition changes so slowly that at any moment, the system is approximately in equilibrium. In the phase space, a thermodynamical transformation from point (or state) A to point B is said to be reversible if when the external condition is reversed quasi-statically, the state is transformed back from B to A.

The equilibrium thermodynamics is built upon a set of basic thermodynamical laws which are developed based on the accumulative empirical evidence. The first law of thermodynamics reveals the essential relationship among the heat, the internal energy and the work done to the system.

The first law of thermodynamics. Let U be the internal energy of the substance system (an extensive thermodynamical variable), dQ the heat absorbed by the system, dW the work done by the system (to its surroundings). Then, the change in the internal energy is given by

$$dU = dQ - dW \qquad (2.3)$$

and the change in the internal energy only depends on the initial and the final state of the system. Thus, $dQ = dU + dW$.

The first law implies that dU is a complete differential, i.e., the integral

$$\int_L dU \qquad (2.4)$$

is independent of the path L in the phase space of the thermodynamic variables. We remark that the thermodynamical variables are not necessarily defined in the entire phase space. Therefore, the physical laws we discuss here apply to the common domain of the thermodynamical variables. This should be understood throughout the notes since we will not make any special efforts to articulate this again. Clearly dW is not a complete differential since the work done does depend on the path. Therefore, dQ is not a complete differential either. The first law also indicates that a part of the heat absorbed by the matter system is used to increase the internal energy while the other part is used to do work to its surroundings, which can be associated with the energy loss due to frictions, etc. The first law defines the internal energy as a potential function by

$$U(A) = \int_O^A (dQ - dW), \qquad (2.5)$$

where O and A are two points in the phase space.

The second law of thermodynamics is much more subtle. It addresses the issue of dissipation in a thermodynamical system. There is not a single version that is easy to use in practice or even to phrase. One version of the second law is phrased as follows [7].

The second law of thermodynamics. There exists a unique thermodynamic transformation whose sole effect is to extract a quantity of heat from a cooler reservoir and deliver it to a hotter reservoir.

Clearly, this is not easy to apply directly to modeling a thermodynamic process. The Clausius' theorem, which is based on the second law, gives an important consequence of it that one often uses. The proof of the theorem can be found for example in [7].

Clausius' theorem.   In any cyclic transformation, where a cyclic transformation means a thermodynamic transformation in the phase space whose starting and ending point coincide,

$$\int_C \frac{dQ}{T} \leqslant 0, \tag{2.6}$$

where C is a closed path in the phase space. For reversible processes however, the equality holds:

$$\int_C \frac{dQ}{T} = 0. \tag{2.7}$$

It follows from the Clausius' theorem that the ratio $\frac{dQ}{T}$ is a complete differential despite that dQ is not. We introduce another thermodynamic variable called entropy S whose differential is defined as

$$dS = \frac{dQ}{T}. \tag{2.8}$$

Then, $dQ = TdS$. The second law guarantees that dS is a complete differential so that the entropy is well defined

$$S(A) = \int_O^A \frac{dQ}{T}, \tag{2.9}$$

where O is a fixed point in the phase space and OA is any reversible path in the phase space joining O and A.

Corollary.   (i) For an arbitrary thermodynamic transformation from A to B in the phase space,

$$\int_A^B \frac{dQ}{T} \leqslant S(B) - S(A) \tag{2.10}$$

and the equality is held if the path is reversible.
(ii) For a thermally isolated system ($dQ = 0$),

$$dS \geqslant 0 \text{ or } \int_A^B \frac{dQ}{T} = 0 \leqslant S(B) - S(A). \tag{2.11}$$

This implies that entropy is always increasing in an isolated thermodynamic system. It reaches a local maximum at the equilibrium state.
In fact, it follows from the Clausius theorem that

$$\left(\int_O^A + \int_A^B + \int_B^O \right) \frac{dQ}{T} \leqslant 0. \tag{2.12}$$

This implies that

$$\int_A^B \frac{dQ}{T} \leqslant - \left(\int_O^A + \int_B^O \right) \frac{dQ}{T} = S(B) - S(A). \qquad (2.13)$$

It proves (i). (ii) follows from (i).

Another direct consequence of the second law is the Clausius-Duhem inequality for an open system. In a compact material domain $\mathcal{P}$ embedded in a much larger domain or universe, we denote S the entropy of the material volume, r the rate of heat production per unit volume, $\mathbf{h}$ the heat flux through the boundary $\partial\mathcal{P}$. The Clausius-Duhem inequality states that the time rate of change of the entropy in the volume is greater than or equal to the heat production per unit temperature minus the heat flux flown out of the domain per unit temperature:

$$\frac{d}{dt} S \geqslant \int_{\mathcal{P}} \frac{r}{T} d\mathbf{x} - \int_{\partial\mathcal{P}} \frac{\mathbf{h}}{T} \cdot d\mathbf{S}. \qquad (2.14)$$

This inequality is used exclusively for the second law in continuum mechanics theories.

Now, we have introduced two thermodynamic laws which in turn define two thermodynamical variables: internal energy and the entropy. Both are potential functions and intensive. In addition to the internal energy and entropy, there are other scalar extensive thermodynamic variables that may be more convenient to use at certain situations. It turns out the Legendre transformation plays an essential role in relating the extensive thermodynamic variables to internal energy and entropy.

Definition. Let $y = f(x)$ be a function with $f''(x) \neq 0$. Then, $f'(x)$ is a monotonic function. The equation $p = f'(x)$ has a unique solution $x = f'^{(-1)}(p)$ for any given values of p. We define $g(p) = x(p)p - f(x(p))$ as the Legendre transformation of the function $f(x)$.

A good property of the Legendre transformation is that it transforms a convex function into another convex function. In addition, the Legendre transformation of $g(p)$ is $f(x)$.

When the internal energy is given as a function of the entropy and other thermodynamic variables, its Legendre transformation gives the Helmholtz free energy A :

$$A(T, V, \cdots) = U(S, V, \cdots) - TS. \qquad (2.15)$$

The Legendre transformation of the internal energy yields a function of the absolute temperature along with other thermodynamic variables. The Gibbs free energy is defined as

$$G(T, P, \cdots) = A(T, V, \cdots) + PV, \qquad (2.16)$$

which can be thought of a Legendre transformation of the Helmholtz. When the internal energy is a function of $(T, V, \cdots)$, the enthalpy, as a function of $(T, P, \cdots)$, is the Legendre transformation of the internat energy:

$$H(T, P, \cdots) = U(T, V, \cdots) + PV. \tag{2.17}$$

For a mechanically isolated system kept at a constant temperature, the second law yields the following two useful results.

Theorem. The Helmholtz free energy never increases during any thermodynamical processes in a mechanically isolated system kept at constant temperature, i.e., $\Delta A \leqslant 0$. Therefore, the state of the thermodynamical equilibrium, if it exists, corresponds to a state of the minimum Helmholtz free energy.
Proof.

$$\Delta A = \Delta U - T\Delta S = \Delta Q - \Delta W - T\Delta S = \Delta Q - T\Delta S \leqslant 0. \tag{2.18}$$

Theorem. For the mechanically isolated thermodynamical system kept at a constant temperature and pressure, the Gibbs free energy never increases during any thermodynamical processes. Thus, the state of equilibrium is at the state of minimum Gibbs free energy.
Proof.

$$\Delta G = \Delta A + P\Delta V = \Delta U - T\Delta S + p\Delta V = \Delta Q - T\Delta S \leqslant 0. \tag{2.19}$$

The above theorems give the criteria for thermal equilibrium. When the pressure is not constrained, the thermal equilibrium is given by the minimum of the Helmholtz free energy. When the pressure is a constant, the equilibrium can be obtained by minimizing the Gibbs free energy. This is why you see one always calculates the system's free energy and minimize it to study the equilibrium and thermodynamical properties of a mechanically isolated, isothermal material system.

Using the potentials, we can represent the thermodynamic variables as derivatives of potentials. From the first law, we have

$$dU = dQ - dW = TdS - pdV, \tag{2.20}$$

where the work is assumed given by $dW = pdV$. It follows that

$$T = \frac{\partial U}{\partial S}, \quad p = -\frac{\partial U}{\partial V}. \tag{2.21}$$

The temperature is the derivative of the internal energy with respect to the entropy while the negative of the pressure is the derivative of the internal energy with respect to the matter system's volume.

From the definition of the Helmholtz free energy, it follows that

$$dA = dU - dTS - TdS = dU - SdT - dQ = -SdT - pdV. \quad (2.22)$$

Then,

$$p = -\frac{\partial A}{\partial V}, \quad S = -\frac{\partial A}{\partial T}. \quad (2.23)$$

So, the entropy is the negative of the derivative of the Helmholtz free energy with respect to the temperature while the pressure is given by the derivative of the free energy with respect to the volume. Similarly, it follows from the definition of the Gibbs free energy that

$$dG = dA + dpV + pdV = dU - TdS - SdT + pdV + Vdp$$
$$= -dW - SdT + pdV + Vdp = -SdT + Vdp. \quad (2.24)$$

So,

$$S = -\frac{\partial G}{\partial T}, \quad V = \frac{\partial G}{\partial p}. \quad (2.25)$$

It follows from the definition of enthalpy,

$$dH = dU + pdV + Vdp = dQ + Vdp = TdS + Vdp. \quad (2.26)$$

So,

$$T = \frac{\partial H}{\partial S}, \quad V = \frac{\partial H}{\partial p}. \quad (2.27)$$

When the thermodynamic variable is given by the density instead of the volume, the partial derivative

$$\frac{\partial(\bullet)}{\partial V} = -\rho^2 \frac{\partial(\bullet)}{\partial \rho}, \quad (2.28)$$

where V is understood as the specific volume, i.e., the volume occupied by a unit mass of the matter. Then, $\rho = \frac{1}{V}$.

The third law of thermodynamics sets the reference value for the entropy.

The third law of thermodynamics. The entropy of a system at absolute zero temperature is a universal constant, which may be taken to be zero.

In the above discussion, the matter system is assumed closed, i.e., the number of the particles or molecules in the system is a constant. When the macroscopic matter system is not closed, the number of the particles in the system is not conserved. The number density of the particles becomes a thermodynamic variable, an intensive variable. Corresponding to the variation in number density, a new potential called chemical potential must be introduced. This is the work done to system by adding one additional particle to the system. The first law is then modified as

$$dU = dQ - dW + \mu dN, \tag{2.29}$$

where N is the number of particles and μ is the chemical potential. Using the various potentials, the chemical potential can be calculated by

$$\mu = \left(\frac{\partial U}{\partial N}\right)_{S,V} = \left(\frac{\partial A}{\partial N}\right)_{T,V} = \left(\frac{\partial G}{\partial N}\right)_{p,T}. \tag{2.30}$$

Incorporating the chemical potential, the grand potential, the fourth potential that we have seen so far, is defined by

$$\mathcal{A} = U - TS - \mu N. \tag{2.31}$$

It yields

$$d\mathcal{A} = -SdT - pdV - Nd\mu. \tag{2.32}$$

The chemical potential can also be calculated from the derivative of the grand potential,

$$\mu = -\left(\frac{\partial \mathcal{A}}{\partial N}\right)_{T,V}. \tag{2.33}$$

The ideal gas is the most studied substance system in thermodynamics and statistical mechanics. The equation of state for the ideal gas is given by [6]

$$pV = RT, \tag{2.34}$$

where R is the gas constant. For non-ideal gases, the relation is no longer valid. A volume extension of the pressure can be written as

$$pV/RT = 1 + B(T)/V + C(T)/V^2 + D(T)/V^3 + \cdots, \tag{2.35}$$

where $B, C, D, \cdots$ are referred to as the second, third, fourth, $\cdots$ virial coefficients. The virial coefficients demonstrate the departure of the matter system away from the ideal gas. In particular, the second virial coefficient $B(T)$ is the most important since it is the leading order perturbation from the ideal gas law.

Thermodynamics are empirical. A more fundamental theory which eventually justifies the thermodynamical laws is the Statistical Mechanics, a subject in which statistical methods are used to study the behavior of an ensemble of particles in the matter system.

3   Basics of statistical mechanics

In statistical mechanics, any matter system is regarded as comprising of discrete particles. In the particle system, we assume there are N particles with their center of mass positions at $\mathbf{r}_i, i = 1, \cdots, N$, and momenta $\mathbf{p}_i, i = 1, \cdots, N$, in a domain Ω in a canonical coordinate $\mathbb{R}^n$. This is a closed system when the number of the particles is fixed. In contrast, the number of particles varies in an open system. The Hamiltonian of the system is denoted by $H(\{\mathbf{r}_i\}, \{\mathbf{p}_i\})$, which yields the total energy of the system. Here $\{\mathbf{r}_i\}$ and $\{\mathbf{p}_i\}$ are the shorthand notations for the sets $\{\mathbf{r}_i, i = 1, \cdots, N\}$ and $\{\mathbf{p}_i, i = 1, \cdots, N\}$, respectively. The equation of motion for each particle is given by the Hamilton equation

$$\begin{cases} \dot{\mathbf{r}}_i = \dfrac{\partial H}{\partial \mathbf{p}_i}, \\[2mm] \dot{\mathbf{p}}_i = -\dfrac{\partial H}{\partial \mathbf{r}_i}, \end{cases} \quad i = 1, \cdots, N. \tag{3.1}$$

In a Cartesian coordinate system, the Hamiltonian for a conservative system with potential Φ is given by $H(\{\mathbf{r}_i\}, \{\mathbf{p}_i\}) = \sum_{i=1}^{N} \frac{\|\mathbf{p}_i\|^2}{m_i} + \Phi(\mathbf{r}_i)$. The Hamilton equation yields the Newton's second law in classical mechanics:

$$m_i \frac{d^2 \mathbf{r}_i}{dt^2} = -\nabla_{\mathbf{r}_i} \Phi, \quad i = 1, \cdots, N. \tag{3.2}$$

We define a term called the virial of the force for the ith particle by

$$-\frac{1}{2} \mathbf{r}_i \cdot \mathbf{F}_i, \tag{3.3}$$

where $\mathbf{F}_i$ is the external force acting on the ith particle located at $\mathbf{r}_i$. The total virial of the force for the system is given by

$$\Theta = -\frac{1}{2} \sum_i \mathbf{r}_i \cdot \mathbf{F}_i. \tag{3.4}$$

In classical mechanics,

$$\frac{d}{dt} \mathbf{p}_i = m_i \frac{d^2 \mathbf{r}_i}{dt^2} = \mathbf{F}_i. \tag{3.5}$$

Multiplying both sides by $\mathbf{r}_i$, it follows that

$$\mathbf{r}_i \cdot \mathbf{F}_i = m_i \mathbf{r}_i \cdot \frac{d^2 \mathbf{r}_i}{dt^2} = \frac{m_i}{2} \left[\frac{d}{dt} \left(\mathbf{r}_i \cdot \frac{d\mathbf{r}_i}{dt} \right) - \left(\frac{d\mathbf{r}_i}{dt} \cdot \frac{d\mathbf{r}_i}{dt} \right) \right]. \tag{3.6}$$

Taking a time-average of (3.6) in $[0, T]$ and assuming

$$\lim_{T \to \infty} \frac{1}{T} \left[\mathbf{r}_i \cdot \frac{d\mathbf{r}_i}{dt} |_T - \mathbf{r}_i \cdot \frac{d\mathbf{r}_i}{dt} |_0 \right] = 0, \tag{3.7}$$

we arrive at

$$-\frac{1}{2} \sum_i \overline{\mathbf{r}_i \cdot \mathbf{F}_i} = -\frac{1}{2} \frac{1}{T} \int_0^T \sum_i (\mathbf{r}_i \cdot \mathbf{F}_i) dt$$
$$= \sum_i \frac{m_i}{2} \overline{\left(\frac{d\mathbf{r}_i}{dt} \cdot \frac{d\mathbf{r}_i}{dt} \right)} = \overline{K}. \tag{3.8}$$

Then, the virial theorem states that

$$\overline{\Theta} = \overline{K} = \sum_i \frac{m_i}{2} \overline{\frac{d\mathbf{r}_i}{dt} \cdot \frac{d\mathbf{r}_i}{dt}}. \tag{3.9}$$

It indicates that the time-averaged kinetic energy equals to the time averaged total virial of the force. In a conservative system,

$$\mathbf{F}_i = -\nabla_{\mathbf{r}_i} \Phi = -\frac{\partial \Phi}{\partial \mathbf{r}_i}, \tag{3.10}$$

where $\Phi(\mathbf{r}_i)$ is the external potential. If the potential is a homogeneous function of the coordinates of degree n,

$$\overline{\Theta} = \frac{1}{2} \sum_i \overline{\mathbf{r}_i \cdot \frac{\partial \Phi}{\partial \mathbf{r}_i}} = \frac{n}{2} \overline{\Phi} = \overline{K}. \tag{3.11}$$

This is the virial theorem represented in another form.

For the N-particle matter system, we define the Γ-space also known as the phase space as

$$\Gamma = \{ (\mathbf{r}_1, \mathbf{p}_1), \cdots, (\mathbf{r}_N, \mathbf{p}_N) \}. \tag{3.12}$$

A collection of points in the Γ-space in a volume V is called an ensemble of particles. We denote short-handedly $\mathbf{r} = \{\mathbf{r}_i\}, \mathbf{p} = \{\mathbf{p}_i\}$. We note that if the particles are all distinguishable, there exists a one-to-one correspondence between the particles in physical position-momentum space $\mathbb{R}^3 \times \mathbb{R}^3$ and a point in the phase space; if, on the other hand, the particles are not all distinguishable, the correspondence is not one-to-one. In the latter case, the particles in the physical space correspond multiple points in the phase space. The multiple correspondence will be addressed whenever it arises.

Let $\rho(\mathbf{r}, \mathbf{p}, t)$ denote the number density function in the Γ-space for the phase point $(\mathbf{r}, \mathbf{p})$, i.e., $\int_{\Delta V} \rho d\mathbf{r} d\mathbf{p}$ gives the number of particles in the volume $\Delta V \in V$, where V is the entire domain the system occupies. The following theorem shows that the number density obeys a conservation law.

Liouville's theorem.

$$\frac{\partial \rho}{\partial t} + \sum_{i=1}^{3N} \left(\frac{\partial \rho}{\partial p_i} \dot{p}_i + \frac{\partial \rho}{\partial r_i} \dot{r}_i \right) = \frac{\partial \rho}{\partial t} - [H, \rho] = 0, \qquad (3.13)$$

where $[H, \rho] = \frac{\partial H}{\partial \mathbf{r}} \cdot \frac{\partial \rho}{\partial \mathbf{p}} - \frac{\partial \rho}{\partial \mathbf{r}} \cdot \frac{\partial H}{\partial \mathbf{p}}$. It follows from the equation of motion in Hamiltonian mechanics. The bracket $[,]$ is known as the Poisson bracket [1].

As a matter of fact, the density function satisfies the conservation law

$$\frac{\partial \rho}{\partial t} + \sum_{i=1}^{N} \left[\frac{\partial}{\partial \mathbf{r}_i}(\dot{\mathbf{r}}_i \rho) + \frac{\partial}{\partial \mathbf{p}_i}(\dot{\mathbf{p}}_i \rho) \right] = 0. \qquad (3.14)$$

The Liouville's theorem is obtained after applying the Hamilton's equation of motion.

For a function of $g = g(\mathbf{r}, \mathbf{p})$ defined in the phase space, we define the ensemble average as

$$\langle g \rangle = \frac{\int_V g \rho \, d\mathbf{r} \, d\mathbf{p}}{\int_V \rho \, d\mathbf{r} \, d\mathbf{p}}. \qquad (3.15)$$

The normalized number density function

$$p(\mathbf{r}, \mathbf{p}) = \frac{\rho(\mathbf{r}, \mathbf{p})}{\int_V \rho(\mathbf{r}, \mathbf{p}) d\mathbf{r} d\mathbf{p}} \qquad (3.16)$$

is a probability density or distribution function defined in the Γ-space.

In the $6N$-dimensional Γ-space, we will introduce three ensembles applicable in statistical sampling. These ensembles correspond to a collection of particle realizations in the physical space dictated by their total energy levels.

Microcanonical ensemble. The microcanonical ensemble is one with the uniform probability density in a shell in the Γ-space given by

$$p(\mathbf{r}, \mathbf{p}) = \begin{cases} \text{const,} & \text{if } E < H(\mathbf{r}, \mathbf{p}) < E + \Delta, \\ 0, & \text{otherwise,} \end{cases} \qquad (3.17)$$

where E is the energy level and $\Delta \ll 1$. The distribution of the particles is concentrated in a energy shell in the phase space. As $\Delta \to 0$, the density function approaches the δ-function centered around the level surface defined by the energy level E.

If we introduce

$$\Gamma(E) = \int_{E<H<E+\Delta} d\mathbf{r} d\mathbf{p} \tag{3.18}$$

defining the volume of the energy shell in the phase space, then

$$S = k_B \ln \Gamma(E) \tag{3.19}$$

is identifiable with the entropy defined in the equilibrium thermodynamics following the second law, where k_B is the Boltzmann constant [7]. This is because S can be shown to be an intensive variable and also $\frac{\partial S}{\partial E}$ is well-defined which can be identified with the temperature. Define the temperature by

$$\frac{1}{T} = \frac{\partial S}{\partial E}, \tag{3.20}$$

and the pressure of the system by

$$p = T \frac{\partial S}{\partial V}. \tag{3.21}$$

Hence,

$$dS = \frac{\partial S}{\partial E} dE + \frac{\partial S}{\partial V} dV = \frac{1}{T}[dE + pdV] \tag{3.22}$$

or

$$dE = TdS - pdV. \tag{3.23}$$

This is the first law of thermodynamics. Therefore, the first and second law of thermodynamics are recovered from the quantities defined in statistical mechanics with the entropy defined in (3.19).

We denote $\mathbf{x}_i$ either $\mathbf{r}_i$ or $\mathbf{p}_i$ and evaluate the ensemble average

$$\left\langle \mathbf{x}_i \frac{\partial H}{\partial \mathbf{x}_j} \right\rangle = \frac{1}{\Gamma(E)} \int_{E<H<E+\Delta} \mathbf{x}_i \frac{\partial H}{\partial \mathbf{x}_j} d\mathbf{r} d\mathbf{p}$$

$$= \frac{\Delta}{\Gamma(E)} \frac{\partial}{\partial E} \int_{H<E} \mathbf{x}_i \frac{\partial H}{\partial \mathbf{x}_j} d\mathbf{r} d\mathbf{p} + O(\Delta^2). \tag{3.24}$$

Notice that

$$\int_{H<E} \mathbf{x}_i \frac{\partial H}{\partial \mathbf{x}_j} d\mathbf{r} d\mathbf{p}$$

$$= \int_{H<E} \frac{\partial(\mathbf{x}_i(H-E))}{\partial \mathbf{x}_j} d\mathbf{r} d\mathbf{p} - \delta_{ij} \int_{H<E} (H-E) d\mathbf{r} d\mathbf{p}. \tag{3.25}$$

Then, assuming the surface term vanishes,

$$\left\langle x_i \frac{\partial H}{\partial x_j} \right\rangle = \delta_{ij} \frac{\Delta}{\Gamma(E)} \frac{\partial}{\partial E} \int_{H<E} dr d\mathbf{p}(E-H) = \frac{\delta_{ij}\Delta}{\Gamma(E)} \int_{H<E} dr d\mathbf{p}$$

$$= \delta_{ij} \frac{\int_{H<E} dr d\mathbf{p}}{\frac{\partial}{\partial E} \int_{H<E} dr d\mathbf{p} + O(\Delta)}. \tag{3.26}$$

We ignore the higher order terms of $O(\Delta)$ since $\Delta \ll 1$. It can be shown that [7]

$$\ln \int_{H<E} dr d\mathbf{p} - \ln \int_{E<H<E+\Delta} dr d\mathbf{p} \tag{3.27}$$

is a constant. Thus, the denominator in the above equation can be recognized as

$$\frac{\int_{H<E} dr d\mathbf{p}}{\frac{\partial}{\partial E} \int_{H<E} dr d\mathbf{p}} = k_B \left(\frac{\partial S}{\partial E} \right)^{-1} = k_B T. \tag{3.28}$$

Finally, we arrive at

$$\left\langle x_i \frac{\partial H}{\partial x_j} \right\rangle = \delta_{ij} k_B T. \tag{3.29}$$

This is called the generalized equipartition theorem. The virial theorem follows

$$-\left\langle \sum_i r_i \frac{\partial H}{\partial r_i} \right\rangle = \left\langle \sum_i r_i \dot{p}_i \right\rangle = 3N k_B T. \tag{3.30}$$

$k_B T$ equals the kinetic energy for the ith particle. Many matter systems have Hamiltonian of the form

$$H = \sum_i (\alpha_i \|r_i\|^2 + \beta_i \|\mathbf{p}_i\|^2), \tag{3.31}$$

where α_i, β_i are the coefficients. For this type of Hamiltonian,

$$\sum_i r_i \frac{\partial H}{\partial r_i} + \mathbf{p}_i \frac{\partial H}{\partial \mathbf{p}_i} = 2H. \tag{3.32}$$

This equality along with the generalized equipartition theorem implies

$$\langle H \rangle = \frac{n}{2} k_B T, \tag{3.33}$$

where n is the number of nonzero coefficients α_i, β_i. This is known as the theorem of equipartition of energy [7].

Canonical ensemble. A canonical ensemble is given by the probability density function defined by

$$p(\mathbf{r}, \mathbf{p}) = \frac{1}{Z_N N! h^{3N}} e^{-\frac{H(\mathbf{r}, \mathbf{p})}{k_B T}}, \tag{3.34}$$

where h is the Planck constant, k_B is the Boltzmann constant, N is the number of particle in the system, and Z_N is the partition function given by

$$Z_N = \int \frac{1}{N! h^{3N}} e^{-\frac{H(\mathbf{r}, \mathbf{p})}{k_B T}} d\mathbf{r} d\mathbf{p}. \tag{3.35}$$

This is motivated by the Maxwell-Boltzmann most probable distribution theory [7].

The Helmholtz free energy is defined as

$$A = -k_B T \ln Z_N. \tag{3.36}$$

Differentiate (3.36) with respect to $\frac{1}{k_B T}$, yielding

$$A - \int H p d\mathbf{r} d\mathbf{p} - T \frac{\partial A}{\partial T} = 0. \tag{3.37}$$

Define the entropy by

$$S = -\frac{\partial A}{\partial T} \tag{3.38}$$

and identify the internal energy by

$$U = \langle H \rangle = \int p H d\mathbf{r} d\mathbf{p}. \tag{3.39}$$

This gives the first law of thermodynamics

$$A = E - T d S. \tag{3.40}$$

The other thermodynamic variables are calculated from the Maxwell relations [7] or see the previous section on equilibrium thermodynamics, for instance,

$$p = -\frac{\partial A}{\partial V}. \tag{3.41}$$

Grand canonical ensemble. When the system is an open system (such as a system where chemical reactions take place), the grand canonical ensemble is more appropriate, in which the probability density function for the subsystem with N particles is given by

$$p(\mathbf{r}, \mathbf{p}, N) = \frac{1}{N! h^{3N}} e^{-\frac{H(\mathbf{r}, \mathbf{p})}{k_B T}} e^{\frac{N\mu}{k_B T}} e^{-\frac{pV}{k_B T}}, \tag{3.42}$$

where

$$\mu = \frac{\partial A}{\partial N}, \quad p = -\frac{\partial A}{\partial V} \tag{3.43}$$

define the chemical potential and pressure, respectively. We define Z as the grand partition function,

$$Z = \sum_{N=0}^{\infty} e^{\frac{\mu N}{k_B T}} Z_N. \tag{3.44}$$

It follows from (3.42) and $\sum_{N=0}^{\infty} \int p(\mathbf{r}, \mathbf{p}, N) d\mathbf{r} d\mathbf{p} = 1$,

$$\frac{pV}{k_B T} = \ln Z \quad \text{or} \quad Z = e^{\frac{pV}{k_B T}}. \tag{3.45}$$

The probability density function for the subsystem of N particles in grand canonical ensemble can be rewritten into

$$p(\mathbf{r}, \mathbf{p}, N) = \frac{1}{Z N! h^{3N}} e^{-\frac{H(\mathbf{r}, \mathbf{p})}{k_B T}} e^{\frac{N\mu}{k_B T}}. \tag{3.46}$$

The internal energy is defined by

$$U = \langle H \rangle = \sum_{N}^{\infty} \int H p(\mathbf{r}, \mathbf{p}, N) d\mathbf{r} d\mathbf{p}. \tag{3.47}$$

All others are obtained from the Maxwell relations, for example,

$$S = \int_0^T \frac{1}{T} \frac{\partial U}{\partial T} dT \tag{3.48}$$

defines the entropy.

These ensembles are frequently used in statistical and stochastic simulations such as Monte Carlo simulations. The statistical mechanics reformulates the equilibrium thermodynamics on a solid foundation. We next look into the transport properties of particles when there are collisions involved.

Boltzmann Transport Equation.   In the position-momentum space, known as the μ-space, $G = \{(\mathbf{r}, \mathbf{p}) | \mathbf{r}, \mathbf{p} \in \mathbb{R}^3\}$, where $\mathbf{r}$ denotes the position and $\mathbf{p}$ the momentum vector. We denote the number density function in the space G by $f(\mathbf{r}, \mathbf{p}, t)$. Thus,

$$\int_V f \, d\mathbf{r} d\mathbf{p} = N \tag{3.49}$$

is the number of molecules/particles in the domain $V \in G$. We consider an external force depending on position only acting on each molecule in V. The velocity at $\mathbf{r}$ for the molecule of momentum $\mathbf{p}$ is given by $\mathbf{v} = \frac{\mathbf{p}}{m}$, where m is the mass of the molecule. Consider an infinitesimal time Δt. If there is no collision, a molecule originally at $(\mathbf{r}, \mathbf{p})$ shows up at $(\mathbf{r} + \mathbf{v}\Delta t, \mathbf{p} + \mathbf{F}\Delta t)$ after Δt. The Jacobian for the nonlinear transformation is given by

$$\begin{pmatrix} I & \dfrac{\Delta t}{m} I \\ \dfrac{\partial \mathbf{F}}{\partial \mathbf{r}} \Delta t & I \end{pmatrix}. \tag{3.50}$$

The determinant is $1 - O(\Delta t^2)$. Hence,

$$f(\mathbf{r} + \mathbf{v}\Delta t, \mathbf{p} + \mathbf{f}\Delta t) = f(\mathbf{r}, \mathbf{p}) + O(\Delta t^2). \tag{3.51}$$

If there are collisions, this has to be modified by

$$f(\mathbf{r} + \mathbf{v}\Delta t, \mathbf{p} + \mathbf{f}\Delta t) = f(\mathbf{r}, \mathbf{p}) + \left(\frac{\partial f}{\partial t}\right)_{\text{coll}} \Delta t + O(\Delta t^2), \tag{3.52}$$

where $\left(\frac{\partial f}{\partial t}\right)_{\text{coll}}$ is the rate of change for the number of molecules during Δt due to collision. Expanding the left-hand side up to order $O(\Delta t)$, we have

$$\left(\frac{\partial}{\partial t} + \mathbf{v} \cdot \nabla_r + \mathbf{F} \cdot \nabla_p\right) f(\mathbf{r}, \mathbf{p}, t) = \left(\frac{\partial f}{\partial t}\right)_{\text{coll}}, \tag{3.53}$$

where ∇_r and ∇_p are the gradient operators with respect to $\mathbf{r}$ and $\mathbf{p}$, respectively. This is called the Boltzmann transport equation.

The Boltzmann transport equation can be derived from the Liouville equation systematically [5]. We assume the particles are indistinguishable and define the single particle distribution function in the μ-space by

$$f(\mathbf{r}, \mathbf{p}, t) = \left\langle \sum_{i=1}^{N} \delta(\mathbf{r} - \mathbf{r}_i)\delta(\mathbf{p} - \mathbf{p}_i) \right\rangle$$

$$= \frac{\int \sum_{i=1}^{N} \delta(\mathbf{r}-\mathbf{r}_i)\delta(\mathbf{p}-\mathbf{p}_i)\rho(\mathbf{r}_1,\mathbf{r}_2,\cdots,\mathbf{r}_n,\mathbf{p}_1,\mathbf{p}_2,\cdots,\mathbf{p}_n,t)d\mathbf{r}_1\cdots d\mathbf{r}_n d\mathbf{p}_1\cdots d\mathbf{p}_n}{\int \rho(\mathbf{r}_1,\mathbf{r}_2,\cdots,\mathbf{r}_n,\mathbf{p}_1,\mathbf{p}_2,\cdots,\mathbf{p}_n,t)d\mathbf{r}_1\cdots d\mathbf{r}_n d\mathbf{p}_1\cdots d\mathbf{p}_n} \quad (3.54)$$

$$= N \frac{\int \rho(\mathbf{r},\mathbf{r}_2,\cdots,\mathbf{r}_n,\mathbf{p},\mathbf{p}_2,\cdots,\mathbf{p}_n,t)d\mathbf{r}_2\cdots d\mathbf{r}_n d\mathbf{p}_2\cdots d\mathbf{p}_n}{\int \rho(\mathbf{r}_1,\mathbf{r}_2,\cdots,\mathbf{r}_n,\mathbf{p}_1,\mathbf{p}_2,\cdots,\mathbf{p}_n,t)d\mathbf{r}_1\cdots d\mathbf{r}_n d\mathbf{p}_1\cdots d\mathbf{p}_n}.$$

Analogously, the m-particle distribution function or correlation function is defined by

$$f(\mathbf{r}^1,\cdots,\mathbf{r}^m,\mathbf{p}^1,\cdots,\mathbf{p}^m,t) = \left\langle \sum_{i=1}^{N} \Pi_{j=1}^{m}\delta(\mathbf{r}^j - \mathbf{r}_i)\delta(\mathbf{p}^j - \mathbf{p}_i) \right\rangle$$

$$= \frac{\int \sum_{i=1}^{N} \Pi_{j=1}^{m}\delta(\mathbf{r}^j-\mathbf{r}_i)\delta(\mathbf{p}^j-\mathbf{p}_i)\rho(\mathbf{r}_1,\mathbf{r}_2,\cdots,\mathbf{r}_n,\mathbf{p}_1,\mathbf{p}_2,\cdots,\mathbf{p}_n,t)d\mathbf{r}_1\cdots d\mathbf{r}_n d\mathbf{p}_1\cdots d\mathbf{p}_n}{\int \rho(\mathbf{r}_1,\mathbf{r}_2,\cdots,\mathbf{r}_n,\mathbf{p}_1,\mathbf{p}_2,\cdots,\mathbf{p}_n,t)d\mathbf{r}_1\cdots d\mathbf{r}_n d\mathbf{p}_1\cdots d\mathbf{p}_n}$$

$$\quad (3.55)$$

$$= \frac{N!}{(N-m)!} \frac{\int \rho(\mathbf{r}^1,\cdots,\mathbf{r}^m,\mathbf{r}_{m+1},\cdots,\mathbf{r}_n,\mathbf{p}^1,\cdots,\mathbf{p}^m,\mathbf{p}_{m+1},\cdots,\mathbf{p}_n,t)d\mathbf{r}_{m+1}\cdots d\mathbf{r}_n d\mathbf{p}_{m+1}\cdots d\mathbf{p}_n}{\int \rho(\mathbf{r}_1,\mathbf{r}_2,\cdots,\mathbf{r}_n,\mathbf{p}_1,\mathbf{p}_2,\cdots,\mathbf{p}_n,t)d\mathbf{r}_1\cdots d\mathbf{r}_n d\mathbf{p}_1\cdots d\mathbf{p}_n}.$$

We assume the Hamiltonian is given by

$$\mathcal{H} = \sum \frac{\|\mathbf{p}_i\|}{2m} + \sum_{i=1}^{N} U_i(\mathbf{r}_i) + \sum_{i<j} V(\|\mathbf{r}_i - \mathbf{r}_j\|), \quad (3.56)$$

where $U(r)$ is the external force potential and $V(r)$ denotes the mutual interaction force potential. Then, the external force exerted on the ith particle is given by

$$\mathbf{F}_i^e = -\frac{\partial U}{\partial \mathbf{r}_i}, \quad i = 1, \cdots, N, \quad (3.57)$$

while the mutual interaction force exerted by the jth particle on the ith one is given by

$$\mathbf{K}_{ij} = -\frac{\partial V(\|\mathbf{r}_i - \mathbf{r}_j\|)}{\partial \mathbf{r}_i}. \quad (3.58)$$

The total force exerted on the ith particle is

$$\mathbf{F}_i = \mathbf{F}_i^e + \sum_{j=1,i\neq j}^{N} \mathbf{K}_{ij}. \quad (3.59)$$

We note that

$$\dot{\mathbf{r}}_i = \frac{\mathbf{p}_i}{m}, \quad \dot{\mathbf{p}}_i = \mathbf{F}_i = \mathbf{F}_i^e + \sum_{j=1,j\neq i}^{N} \mathbf{K}_{ij},$$

$$N = \int \rho(\mathbf{r}_1,\cdots,\mathbf{r}_n,\mathbf{p}_1,\cdots,\mathbf{p}_n,t)d\mathbf{r}_1\cdots d\mathbf{r}_n d\mathbf{p}_1\cdots d\mathbf{p}_n,$$

$$\quad (3.60)$$

where m is the mass of a single particle. From the Liouville equation in the conservative form

$$\frac{\partial \rho}{\partial t} + \sum_{i=1}^{N} \left[\frac{\partial}{\partial \mathbf{r}_i} (\dot{\mathbf{r}}_i \rho) + \frac{\partial}{\partial \mathbf{p}_i} (\dot{\mathbf{p}}_i \rho) \right] = 0 \qquad (3.61)$$

and the definition of the m-particle distribution function, we integrate over $(\mathbf{r}_{m+1}, \cdots, \mathbf{r}_n, \mathbf{p}_{m+1}, \cdots, \mathbf{p}_n)$. Assuming ρ vanishes at the limits of the integration, we arrive at

$$\frac{\partial}{\partial t} f(\mathbf{r}_1, \cdots, \mathbf{r}_m, \mathbf{p}_1, \cdots, \mathbf{p}_m, t) + \frac{(N-1)!}{(N-m)!} \sum_{i=1}^{m} \frac{\partial}{\partial \mathbf{r}_i} \Big(\dot{\mathbf{r}}_i \int \rho(\mathbf{r}_1, \cdots,$$

$$\mathbf{r}_m, \cdots, \mathbf{r}_n, \mathbf{p}_1, \cdots, \mathbf{p}_m, \cdots, \mathbf{p}_n) \Big) d\mathbf{r}_{m+1} \cdots d\mathbf{r}_n d\mathbf{p}_{m+1} \cdots d\mathbf{p}_n$$

$$+ \frac{\partial}{\partial \mathbf{p}_i} \Big(\int \dot{\mathbf{p}}_i \rho(\mathbf{r}_1, \cdots, \mathbf{r}_m, \cdots, \mathbf{r}_n, \mathbf{p}_1, \cdots, \mathbf{p}_m, \cdots, \mathbf{p}_n) \Big)$$

$$d\mathbf{r}_{m+1} \cdots d\mathbf{r}_n d\mathbf{p}_{m+1} \cdots d\mathbf{p}_n$$

$$= \frac{\partial}{\partial t} f(\mathbf{r}_1, \cdots, \mathbf{r}_m, \mathbf{p}_1, \cdots, \mathbf{p}_m, t) + \sum_{i=1}^{m} \Big(\frac{\mathbf{p}_i}{m} \cdot \frac{\partial}{\partial \mathbf{r}_i} f(\mathbf{r}_1, \cdots, \mathbf{r}_m,$$

$$\mathbf{p}_1, \cdots, \mathbf{p}_m, t) + \hat{\mathbf{F}}_i \cdot \frac{\partial}{\partial \mathbf{p}_i} f(\mathbf{r}_1, \cdots, \mathbf{r}_m, \mathbf{p}_1, \cdots, \mathbf{p}_m, t) \Big)$$

$$+ \frac{(N-1)!}{(N-m)!} \sum_{i=1}^{m} \sum_{j=m+1}^{N} \int \mathbf{K}_{ij} \cdot \frac{\partial}{\partial \mathbf{p}_i} \rho(\mathbf{r}_1, \cdots, \mathbf{r}_m, \cdots, \mathbf{r}_n,$$

$$\mathbf{p}_1, \cdots, \mathbf{p}_m, \cdots, \mathbf{p}_n) d\mathbf{r}_{m+1} \cdots d\mathbf{r}_n d\mathbf{p}_{m+1} \cdots d\mathbf{p}_n = 0, \qquad (3.62)$$

where

$$\hat{\mathbf{F}}_i = \mathbf{F}_i^e + \sum_{j=1, i\neq j}^{m} \mathbf{K}_{ij}. \qquad (3.63)$$

Then

$$\frac{\partial}{\partial t} f(\mathbf{r}_1, \cdots, \mathbf{r}_m, \mathbf{p}_1, \cdots, \mathbf{p}_m, t) + \sum_{i=1}^{m} \Big(\frac{\mathbf{p}_i}{m} \cdot \frac{\partial}{\partial \mathbf{r}_i} \cdot f + \hat{\mathbf{F}}_i \cdot \frac{\partial}{\partial \mathbf{p}_i} f \Big)$$

$$= -\frac{(N-1)!}{(N-m+1)!} \sum_{i=1}^{m} \int \mathbf{K}_{im+1} d\mathbf{r}_{m+1} d\mathbf{p}_{m+1} \cdot \frac{\partial}{\partial \mathbf{p}_i} \rho$$

$$(\mathbf{r}_1, \cdots, \mathbf{p}_1, \cdots) d\mathbf{r}_{m+2} \cdots d\mathbf{r}_n d\mathbf{p}_{m+2} \cdots d\mathbf{p}_n$$

$$= -\sum_{i=1}^{m} \int \mathbf{K}_{im+1} \cdot \frac{\partial}{\partial \mathbf{p}_i} f$$

$$(\mathbf{r}_1, \cdots, \mathbf{r}_{m+1}, \mathbf{p}_1, \cdots, \mathbf{p}_{m+1}, t) d\mathbf{r}_{m+1} d\mathbf{p}_{m+1}. \qquad (3.64)$$

Equation (3.64) is the well-known BBGKY hierarchy [5, 7]. This indicates that in order to solve the m-th particle distribution function, the $(m+1)$-th particle distribution function must be known. Therefore, the hierarchy imposes a system of equations for the distribution function that is not closed before m reaches N.

The first two equations in the hierarchy are given by

$$\frac{\partial}{\partial t} f(\mathbf{r}_1, \mathbf{p}_1, t) + \frac{\mathbf{p}_1}{m} \cdot \frac{\partial}{\partial \mathbf{r}_1} f + \mathbf{F}_1^e \cdot \frac{\partial}{\partial \mathbf{p}_1} f$$

$$= -\int \mathbf{K}_{12} \cdot \frac{\partial}{\partial \mathbf{p}_1} f(\mathbf{r}_1, \mathbf{r}_2, \mathbf{p}_1, \mathbf{p}_2, t) d\mathbf{r}_2 d\mathbf{p}_2,$$

$$\frac{\partial}{\partial t} f(\mathbf{r}_1, \mathbf{r}_2, \mathbf{p}_1, \mathbf{p}_2, t) + \sum_{i=1}^{2} \left(\frac{\mathbf{p}_i}{m} \cdot \frac{\partial}{\partial \mathbf{r}_i} \cdot f + \hat{\mathbf{F}}_i \cdot \frac{\partial}{\partial \mathbf{p}_i} f \right)$$

$$= -\int (\mathbf{K}_{13} \frac{\partial}{\partial \mathbf{p}_1} + \mathbf{K}_{23} \frac{\partial}{\partial \mathbf{p}_2}) \cdot f(\mathbf{r}_1, \cdots, \mathbf{r}_3, \mathbf{p}_1, \cdots, \mathbf{p}_3, t) d\mathbf{r}_3 d\mathbf{p}_3. \quad (3.65)$$

The term on the right-hand of the first equation is the collision term

$$\left(\frac{\partial}{\partial t} f \right)_{coll} = -\int \mathbf{K}_{12} \cdot \frac{\partial}{\partial \mathbf{p}_1} f(\mathbf{r}_1, \mathbf{r}_2, \mathbf{p}_1, \mathbf{p}_2, t) d\mathbf{r}_2 d\mathbf{p}_2$$

$$= -\int \left(\mathbf{K}_{12} \cdot \frac{\partial}{\partial \mathbf{p}_1} - \mathbf{K}_{12} \cdot \frac{\partial}{\partial \mathbf{p}_2} \right) f(\mathbf{r}_1, \mathbf{r}_2, \mathbf{p}_1, \mathbf{p}_2, t) d\mathbf{r}_2 d\mathbf{p}_2. \quad (3.66)$$

We assume that

$$\mathbf{K}_{12} = 0, \quad \text{if } \|\mathbf{r}_1 - \mathbf{r}_2\| > r_0 \quad (3.67)$$

and the time scale on which the 2-particle distribution function varies is much faster than that for the single particle distribution. Thus, on the time scale the single particle distribution varies, the 2-particle distribution is already reached steady state, i.e.,

$$\sum_{i=1}^{2} \left[\frac{\mathbf{p}_i}{m} \cdot \frac{\partial}{\partial \mathbf{r}_i} f(\mathbf{r}_1, \mathbf{r}_2, \mathbf{p}_1, \mathbf{p}_2) + \hat{\mathbf{F}}_i \cdot \frac{\partial}{\partial \mathbf{p}_i} f(\mathbf{r}_1, \mathbf{r}_2, \mathbf{p}_1, \mathbf{p}_2) \right] = 0. \quad (3.68)$$

Moreover, we assume that the external forces vanish, i.e.,

$$\hat{\mathbf{F}}_1 = \mathbf{K}_{12}, \quad \hat{\mathbf{F}}_2 = \mathbf{K}_{21} = -\mathbf{K}_{12}. \quad (3.69)$$

It follows from (3.68) that

$$\left(\frac{\partial}{\partial t} f \right)_{coll} \quad (3.70)$$

$$= -\iint_{\|\mathbf{r}_2 - \mathbf{r}_1\| \leqslant r_0} \left(\mathbf{F}_1 \cdot \frac{\partial}{\partial \mathbf{p}_1} + \mathbf{F}_2 \cdot \frac{\partial}{\partial \mathbf{p}_2} \right) f(\mathbf{r}_1, \mathbf{r}_2, \mathbf{p}_1, \mathbf{p}_2) d\mathbf{r}_2 d\mathbf{p}_2$$

$$= \frac{1}{m} \iint_{\|\mathbf{r}_2 - \mathbf{r}_1\| \leqslant r_0} \left(\mathbf{p}_1 \cdot \frac{\partial}{\partial \mathbf{r}_1} + \mathbf{p}_2 \cdot \frac{\partial}{\partial \mathbf{r}_2} \right) f(\mathbf{r}_1, \mathbf{r}_2, \mathbf{p}_1, \mathbf{p}_2) d\mathbf{r}_2 d\mathbf{p}_2.$$

We introduce a local coordinate

$$\mathbf{R} = \frac{\mathbf{r}_1 + \mathbf{r}_2}{2}, \quad \mathbf{r} = \frac{\mathbf{r}_2 - \mathbf{r}_1}{2}, \quad \mathbf{P} = \mathbf{p}_1 + \mathbf{p}_2, \quad \mathbf{p} = \frac{\mathbf{p}_2 - \mathbf{p}_1}{2}. \quad (3.71)$$

Then

$$\mathbf{p}_1 \cdot \frac{\partial}{\partial \mathbf{r}_1} + \mathbf{p}_2 \cdot \frac{\partial}{\partial \mathbf{r}_2} = \frac{\mathbf{P}}{2} \cdot \frac{\partial}{\partial \mathbf{R}} + \mathbf{p} \cdot \frac{\partial}{\partial \mathbf{r}}. \quad (3.72)$$

We assume that $f(\mathbf{r}_1, \mathbf{r}_2, \mathbf{p}_1, \mathbf{p}_2, t)$ does not depend on $\mathbf{R}$:

$$f(\mathbf{r}_1, \mathbf{r}_2, \mathbf{p}_1, \mathbf{p}_2, t) = f(\mathbf{r}, \mathbf{p}_1, \mathbf{p}_2, t). \quad (3.73)$$

Then

$$\left(\frac{\partial}{\partial t} f\right)_{\text{coll}} = \frac{1}{m} \int_{\|\mathbf{r}\| \leqslant r_0} (\mathbf{p}_2 - \mathbf{p}_1) \cdot \frac{\partial}{\partial \mathbf{r}} f(\mathbf{r}, \mathbf{p}_1, \mathbf{p}_2) d\mathbf{r} d\mathbf{p}_2$$

$$= \int d\mathbf{p}_2 \|\mathbf{v}_2 - \mathbf{v}_1\| \int d\phi b db \int_{z_1}^{z_2} \frac{\partial}{\partial z} f(\mathbf{r}, \mathbf{p}_1, \mathbf{p}_2) dz, \quad (3.74)$$

where z_1 is the location where particle $\mathbf{r}_2$ enters into the sphere $\|\mathbf{r}\| = r_0$ before the collision in a cylindrical coordinate with the z-direction defined by the entering momentum $\mathbf{p}$ and z_2 is the location of the same particle after collision exiting the sphere, ϕ is the azimuthal angle with respect to the z-axis [5, 7]. The momenta of the two particles after collision are denoted by $\mathbf{p}_1', \mathbf{p}_2'$ at z_2. We approximate

$$f(\mathbf{r}|_{z=z_1}, \mathbf{p}_1, \mathbf{p}_2) = f(\mathbf{r}|_{z=z_1}, \mathbf{p}_1) f(\mathbf{r}|_{z=z_1}, \mathbf{p}_2),$$

$$f(\mathbf{r}|_{z=z_2}, \mathbf{p}_1', \mathbf{p}_2') = f(\mathbf{r}|_{z=z_2}, \mathbf{p}_1') f(\mathbf{r}|_{z=z_2}, \mathbf{p}_2'). \quad (3.75)$$

Finally,

$$\left(\frac{\partial}{\partial t} f\right)_{\text{coll}} = \int d\mathbf{p}_2 \|\mathbf{v}_2 - \mathbf{v}_1\| \int d\phi b db [f(\mathbf{r}|_{z=z_2}, \mathbf{p}_1') f(\mathbf{r}|_{z=z_2}, \mathbf{p}_2')$$

$$- f(\mathbf{r}|_{z=z_1}, \mathbf{p}_1) f(\mathbf{r}|_{z=z_1}, \mathbf{p}_2)], \quad (3.76)$$

where $\mathbf{v}_i = \frac{\mathbf{p}_i}{m}, i = 1, 2$.

4 Equilibrium distribution of the end-to-end vector in simple polymer models

Polymers are macromolecules comprised of a number of subunits called monomers. The conformation of a polymer can be a linear chain, a side chain, a branched chain, a network of chains, etc. Therefore the starting point for studying polymer physics is the construction of polymer models using simple geometric objects like beads, spheres, rods, springs, etc. For a linear polymer chain, we can coarse-grain it into a bead rod or bead spring chains. The dynamics of the polymer is then mimicked by the dynamics of the bead rod or bead spring chains [2].

We first consider the bead rod chain with $N + 1$ beads located at $\mathbf{r}_i, i = 0, \cdots, N$. Each rod is of the same length $b = \|\mathbf{r}_{i+1} - \mathbf{r}_i\|$. The simplest model is the freely joint model where the distributions of the rods are independent. Let $\phi(\mathbf{q})$ be the pdf for rod or bond $\mathbf{q}$,

$$\phi(\mathbf{q}) = \frac{\delta(\|\mathbf{q}\| - b)}{4\pi b^2}. \tag{4.1}$$

It is a uniform distribution function defined on a sphere of radius b. The joint pdf for the freely joint model of N segments is given by

$$f(\mathbf{q}_1, \cdots, \mathbf{q}_N) = \Pi_1^N \phi(\mathbf{q}_i), \tag{4.2}$$

where we denote $\mathbf{q}_i = \mathbf{r}_{i+1} - \mathbf{r}_i$ [4]. It can be easily shown that

$$\langle \|\mathbf{q}_i\|^2 \rangle = b^2, \tag{4.3}$$

where $\langle \ \rangle$ denotes the ensemble average with respect to the pdf f,

$$\langle\langle \bullet \rangle\rangle = \int (\bullet) f(\mathbf{q}_1, \cdots, \mathbf{q}_N) d\mathbf{q}_1 \cdots d\mathbf{q}_N. \tag{4.4}$$

We denote the end-to-end vector as

$$\mathbf{R} = \mathbf{r}_n - \mathbf{r}_0 = \sum_{i=1}^{N} \mathbf{q}_i. \tag{4.5}$$

Then

$$\langle \mathbf{R} \rangle = 0, \quad \langle \|\mathbf{R}\|^2 \rangle = \sum_{i=1}^{N} \langle \|\mathbf{q}_i\|^2 \rangle = Nb^2. \tag{4.6}$$

So, the average separation between the end-to-end position of the bead-rod chain is $\sqrt{N}b$. This can be viewed as the averaged radius of gyration for the chain.

Next we examine the freely rotating model where the $(i+1)$-th bond $\mathbf{q}_{i+1}$ is connected to the i-th bond $\mathbf{q}_i$ with a rotational degree of freedom, i.e., it can rotate freely with respect to $\mathbf{q}_i$ subject to

$$\mathbf{q}_{i+1} \cdot \mathbf{q}_i = \cos\theta. \tag{4.7}$$

While $\mathbf{q}_{i-1}, \cdots, \mathbf{q}_1$ are held fixed, the conditional probability distribution is uniform with respect to the rotational angle θ. So,

$$\langle \mathbf{q}_{i+1} \rangle = \cos\theta \mathbf{q}_i, \tag{4.8}$$

where $\langle\ \rangle$ is the ensemble average with respect to the conditional pdf while $\mathbf{q}_{i-1}, \cdots, \mathbf{q}_1$ are held fixed. Multiplying both sides of the above equation by $\mathbf{q}_j$, where $j < i$, while taking the average over $\mathbf{q}_j, \cdots, \mathbf{q}_i$, we have

$$\langle \mathbf{q}_{i+1} \cdot \mathbf{q}_j \rangle = \cos\theta \langle \mathbf{q}_i \cdot \mathbf{q}_j \rangle. \tag{4.9}$$

Recursively, we can show that

$$\langle \mathbf{q}_i \cdot \mathbf{q}_j \rangle = \cos^{|i-j|} \langle \mathbf{q}_j \cdot \mathbf{q}_j \rangle = \cos^{|i-j|}\theta b^2. \tag{4.10}$$

It then follows that

$$\langle \|\mathbf{R}\|^2 \rangle = \sum_{ij=1}^{N} \langle \mathbf{q}_i \mathbf{q}_j \rangle = \sum_{i=1}^{N} \sum_{j=1-i}^{N-i} \langle \mathbf{q}_i \cdot \mathbf{q}_{i+j} \rangle$$

$$\approx \sum_{i=1}^{N} \sum_{j=-\infty}^{\infty} \langle \mathbf{q}_i \mathbf{q}_j \rangle = Nb^2 \left(1 + 2\sum_{i=1}^{\infty} \cos^i\theta \right) = Nb^2 \frac{1 + \cos\theta}{1 - \cos\theta}. \tag{4.11}$$

The averaged separation between the ends of the chain is again proportional to $\sqrt{N}b$.

For the freely joint chain, we denote $\phi(\mathbf{R}, N)$ the pdf that the end-to-end vector of the chain consisting of N links is $\mathbf{R}$. Then

$$\phi(\mathbf{R}, N) = \int \delta\left(\mathbf{R} - \sum_{n=1}^{N} \mathbf{q}_n \right) \prod_{n=1}^{N} \phi(\mathbf{q}_n) d\mathbf{q}_1 \cdots d\mathbf{q}_N. \tag{4.12}$$

We use the identity

$$\delta(\mathbf{q}) = \frac{1}{(2\pi)^3} \int e^{i\mathbf{k}\cdot\mathbf{q}} d\mathbf{k}. \tag{4.13}$$

Then

$$\phi(\mathbf{R}, N) = \frac{1}{(2\pi)^3} \int e^{i\mathbf{k}\cdot\mathbf{R}} d\mathbf{k} \left[\int e^{-\mathbf{k}\cdot\mathbf{r}} d\mathbf{r} \right]^N$$

$$= \frac{1}{(2\pi)^3} \int e^{i\mathbf{k}\cdot\mathbf{R}} d\mathbf{k} \Big[\frac{1}{4\pi b^2} \int_0^\infty r^2 dr \int_0^{2\pi} d\phi$$

$$\int_0^\pi \sin\theta e^{-ikr\cos\theta} \delta(r - b) d\theta \Big]^N$$

.

$$= \frac{1}{(2\pi)^3} \int e^{i\mathbf{k}\cdot\mathbf{R}} d\mathbf{k} \left[\frac{\sin kb}{kb} \right]^N$$

$$\approx \frac{1}{(2\pi)^3} \int e^{i\mathbf{k}\cdot\mathbf{R}} d\mathbf{k} e^{-\frac{Nk^2b^2}{6}}$$

$$= \left(\frac{3}{2\pi Nb^2} \right)^{3/2} e^{-\frac{3\|\mathbf{R}\|^2}{2Nb^2}}. \tag{4.14}$$

This is a Gaussian distribution. The pdf for the end-to-end vector in the freely rotational chain model is also Gaussian. In fact, the end-to-end distribution is Gaussian provided

$$\psi(\mathbf{q}_n) = \Pi_n \psi(\mathbf{q}_n, \mathbf{q}_{n+1}, \cdots, \mathbf{q}_{n+k}), \tag{4.15}$$

where k is a fixed integer [4].

5   Kinetic theory for polymers

In this section, we give a crash course on the development of kinetic theories for polymeric liquids. We derive the kinetic theory using a phenomenological approach, which can be justified from the Liouville theorem. We begin with the conservation of polymer number density, an analog of the Liouville theorem. This is the most fundamental conservation law in the development of kinetic theories. Let ψ be the number density of some polymer and $\mathbf{F}$ the flux of the polymer flow in the generalized coordinate or phase space Γ, where $x \in \Gamma$ is a vector. Each point x represents a coarse-grained configuration of the polymer. The conservation law for the number density of polymers in any "material volume" in the phase space yields

$$\frac{\partial \psi}{\partial t} + \frac{\partial \mathbf{F}}{\partial x} = 0. \tag{5.1}$$

Using the instantaneous velocity $\mathbf{v}$ at x, we can rewrite the flux as

$$\mathbf{F} = \mathbf{v}\psi. \tag{5.2}$$

Assume that the motion of polymers is due to a force generated by an external field U and the Brownian force. The inertialess force balance equation reads

$$L^{-1} \cdot \mathbf{v} + \frac{\partial}{\partial x}\mu = 0, \quad \mu = kT \ln \psi + U, \tag{5.3}$$

where μ is the chemical potential, U is the "external" potential, and L^{-1} the friction coefficient matrix, which is assumed invertible. Then

$$\mathbf{v} = -L \cdot \frac{\partial \mu}{\partial x}, \tag{5.4}$$

where L is the phase space mobility matrix. (5.1) becomes

$$\frac{\partial \psi}{\partial t} - \frac{\partial}{\partial x}\left(L \cdot \frac{\partial \mu}{\partial x}\psi\right) = 0. \qquad (5.5)$$

This equation is called the Smoluchowski equation or the kinetic equation.

For a polymer system in which the molecular configuration is described by $x = \{\mathbf{x}_i\}_1^N$, where $\mathbf{x}_i \in \mathbb{R}^3$, the Smoluchowski equation is usually written as

$$\frac{\partial \psi}{\partial t} + \sum_{n=1}^N \frac{\partial}{\partial \mathbf{x}_n}(\mathbf{v}_n \psi) = 0, \quad \mathbf{v}_n = -\sum_m L_{nm}\frac{\partial \mu}{\partial \mathbf{x}_m}, \qquad (5.6)$$

where $L = (L_{nm})$ is the mobility matrix and $L_{nm} = L_{mn}, (L_{mn}) > 0$.

When the polymer or the particle system x is immersed in a viscous solvent, each particle on the polymer is going to be subject to a drag exerted by the solvent and additional forces on each particle caused by the perturbation of the flow field due to the motion of the particles. This is called the hydrodynamic effect. When hydrodynamic effect is included, the total instantaneous velocity consists of two parts:

$$\mathbf{v}_n = \mathbf{v}_n^e + \mathbf{v}_n^v, \quad \mathbf{v}_n^e = -\sum_m L_{nm} \cdot \frac{\partial \mu}{\partial \mathbf{x}_m}, \qquad (5.7)$$

where the mobility matrix depends on the location of the particles and the second part $\mathbf{v}_n^v$ is due to the existence of the macroscopic flow field and often it is well approximated by

$$\mathbf{v}_n^v = \nabla \mathbf{v} \cdot \mathbf{x}_n \qquad (5.8)$$

when the perturbation of the flow due to the motion of other particles is weak (dilute limit), where $\mathbf{v}$ is the mass-averaged velocity field. We assume $\nabla \mathbf{v}$ is a slowly varying function in space in the length scale of the polymer system x.

In the following, we assume that the particle is approximated by a sphere. Due to the presence and motion of the spheres, the flow field around each particle is perturbed, which in turn affect the motion of the other spheres. For very dilute solution where the distance between spheres are sufficiently far so that the hydrodynamic interaction can be neglected, the velocity of the sphere is determined by the external force acting on it alone and the mobility matrix is given by

$$L_{mn} = \frac{\delta_{mn}}{\zeta}\mathbf{I}, \qquad (5.9)$$

where $\zeta = 6\pi\eta_s a$ for spheres of radius a. In the general case, we have to solve the fluid velocity $\mathbf{v}(\mathbf{x})$ with the given external force $\mathbf{F}_n$ exerted on the sphere at $\mathbf{x}_n$, which can be expressed as

$$\mathbf{g}(\mathbf{x}) = \sum_n \mathbf{F}_n \delta(\mathbf{x} - \mathbf{x}_n). \tag{5.10}$$

We simplify the problem by treating the sphere as a point mass.

We assume the solvent is incompressible $\nabla \cdot \mathbf{v} = 0$ and governed by the Stokes equation

$$\nabla \cdot \tau + \mathbf{g} = 0, \tag{5.11}$$

where the stress tensor is given by

$$\tau = -p + 2\eta_s \mathbf{D}. \tag{5.12}$$

A solution of the equation is given by

$$\mathbf{v} = K \cdot \mathbf{x} + \sum_m \mathbf{H}(\mathbf{x} - \mathbf{x}_m) \cdot \mathbf{F}_m, \tag{5.13}$$

where $K = \nabla \mathbf{v}$ is treated as a spatially homogeneous tensor and

$$\mathbf{H}(\mathbf{x}) = \frac{1}{8\eta_s \pi \|\mathbf{x}\|} \left(\mathbf{I} + \frac{\mathbf{x}\mathbf{x}}{\|\mathbf{x}\|^2} \right) \tag{5.14}$$

is called the Oseen tensor. This tensor is singular at $\mathbf{x} = 0$, which is due to the assumption that the particle is a point. For finite size sphere, this singularity would be removed; but, the exact solution of the Stokes equation is not feasible. A compromise is to approximate the Oseen tensor for point mass by

$$\tilde{\mathbf{H}}(\mathbf{x}) = \begin{cases} \dfrac{\mathbf{I}}{\zeta}, & \mathbf{x} = 0, \\[2mm] \mathbf{H}(\mathbf{x}), & \mathbf{x} \neq 0. \end{cases} \tag{5.15}$$

The mobility matrix is then calculated by

$$L_{mn} = \tilde{\mathbf{H}}(\mathbf{x}_n - \mathbf{x}_m). \tag{5.16}$$

The Smoluchowski equation then becomes

$$\frac{\partial \psi}{\partial t} = \sum_{n=1}^{N} \frac{\partial}{\partial \mathbf{x_n}} \cdot \left[\sum_m L_{nm} \frac{\partial \mu}{\partial \mathbf{x_m}} - \nabla \mathbf{v} \cdot \mathbf{x}_n \right]. \tag{5.17}$$

The elastic stress in the polymer system can be derived from a virtual work principle. Let

$$\mathcal{A} = \int_V \int \mu\psi \, dx dv \qquad (5.18)$$

be the free energy over the material volume V. (We remark that this free energy formulation is based on that the potential U is independent of ψ; if U depends on ψ as well, $\mathcal{A}$ has to be modified such that $\frac{\delta \mathcal{A}}{\delta \psi} = \mu$.) We consider an infinitesimal deformation given by

$$\delta\psi = \frac{d\psi}{dt}\delta t = -\sum_n \frac{\partial}{\partial \mathbf{x}_n}[\nabla \mathbf{v} \cdot \mathbf{x}_n]\delta t. \qquad (5.19)$$

Take the variation of the free energy and assume $\mathrm{vol}\,(V)$ finite, we have

$$\delta\mathcal{A} = \mathrm{vol}\,(V)\tau^e : \delta t \nabla \mathbf{v} = -\delta t \nabla \mathbf{v}_{\alpha\beta} \sum_n \langle \mathbf{F}_{n\alpha}\mathbf{x}_{n\beta}\rangle, \qquad (5.20)$$

where

$$\mathbf{F}_n = -\frac{\partial \mu}{\partial \mathbf{x}_n} \qquad (5.21)$$

is the force exerted on the part of the polymer at $\mathbf{x}_n$. The elastic part of stress is then given by

$$\tau^e = -\frac{1}{\mathrm{vol}\,(V)}\sum_n \langle \mathbf{F}_n \mathbf{x}_n\rangle, \qquad (5.22)$$

where

$$\langle(\bullet)\rangle = \int_V \int (\bullet) d\mathbf{x} dv. \qquad (5.23)$$

5.1   Langevin equation

An alternative description of the molecular motion is the Langevin equation. For each particle $\mathbf{x}_n$, there is a stochastic differential equation

$$\dot{\mathbf{x}}_n = \sum_m L_{nm}\left(-\frac{\partial U}{\partial \mathbf{x}_n} + f_m(t)\right) + \frac{1}{2}kT\frac{\partial}{\partial \mathbf{x}_m}L_{nm}, \qquad (5.24)$$

where $f_m(t)$ is a random force subject to the Gaussian distribution and

$$\langle f_m\rangle = 0, \quad \langle f_n f_m\rangle = 2(L^{-1})_{nm}kT\delta(t-t'). \qquad (5.25)$$

This Langevin equation implies the Smoluchowski equation.

Lemma. The Langevin equation

$$\frac{dx}{dt} = -\frac{1}{\xi}\frac{\partial}{\partial x}U(x) + \sqrt{\frac{kT}{\xi}}g(t) + \frac{1}{2}\frac{d}{dx}\left(\frac{kT}{\xi}\right) \qquad (5.26)$$

implies the distribution density function of x satisfies

$$\frac{\partial}{\partial t}\psi = \frac{\partial}{\partial x}\frac{1}{\xi}\left(kT\frac{\partial}{\partial x}\psi + \frac{\partial U}{\partial x}\psi\right) = \frac{\partial}{\partial x}\frac{1}{\xi}\left(\frac{\partial}{\partial x}\mu\psi\right), \qquad (5.27)$$

where

$$\langle g(t)\rangle = 0, \quad \langle g(t)g(t')\rangle = 2\delta(t - t'). \qquad (5.28)$$

5.2   System of constraints

Assume that the motion of the particles is subject to a set of constraints

$$C_p(x) = 0, \quad p = 1, \cdots, P. \qquad (5.29)$$

Then the forces exerted on each particle must include the constraining forces

$$\mathbf{F}_n^{(c)} = \lambda_p\frac{\partial}{\partial \mathbf{x}_n}C_p, \qquad (5.30)$$

where λ_p is the Lagrange multiplier. The velocity of the particle is calculated from

$$\mathbf{v}_m = K \cdot \mathbf{x}_m + H_{mn} \cdot (\mathbf{F}_n + \mathbf{F}_n^{(c)}), \qquad (5.31)$$

where $\mathbf{F}_n = -\frac{\partial}{\partial \mathbf{x}_n}\mu$. Taking the time derivative on the constraints, we have

$$\frac{\partial C_p}{\partial \mathbf{x}_n} \cdot \mathbf{v}_n = 0. \qquad (5.32)$$

Combining the above equations together, we can solve the Lagrange multiplier

$$\lambda_p = (h^{-1})_{pq}\left[\frac{\partial C_q}{\partial \mathbf{x}_m} \cdot \mathbf{H}_{mn} \cdot \frac{\partial}{\partial \mathbf{x}_n}\mu - \frac{\partial C_q}{\partial \mathbf{x}_n} \cdot K \cdot \mathbf{x}_n\right]. \qquad (5.33)$$

The Smoluchowski equation with constraints is then given by

$$\frac{\partial}{\partial t}\psi + \frac{\partial}{\partial \mathbf{x}_n}(\mathbf{v}_n\psi) = 0. \qquad (5.34)$$

Kirkwood showed that the stress formula is quite general and it applies to any forces acting on the particle [4]. Therefore, it can also be used to calculate the viscous stress due to the particle constraints. We assume that each particle is subject to a drag or constraint force linear to the velocity gradient and the velocity gradient is slowly varying,

$$F_{mn}^{(c)} = \mathcal{C}_{mnkl} K_{kl}, \tag{5.35}$$

where m is the index for the particle and n is the index for the component of the force $\mathbf{F}_m$. The viscous stress is given by

$$\tau_{ij}^{(v)} = -\frac{1}{\mathrm{vol}\,(V)} \sum_{m=1}^{N} \langle \mathcal{C}_{mikl} \mathbf{x}_{mj} \rangle K_{kl}. \tag{5.36}$$

In addition, we need to add the viscous stress from the contribution of the solvent to the total stress tensor

$$\tau_s^{(v)} = 2\eta_s \mathbf{D}. \tag{5.37}$$

Using this method, we can derive the kinetic theory for rodlike liquid crystal polymers [1, 4, 8]. Next, we give a specific model for polymers modeled as beads connected by linear elastic springs called the Rouse chain model [2, 4].

5.3   Bead-spring (Rouse) chain model

We assume the bead-spring system is described by the phase space coordinate $\{\mathbf{x}_i\}, i = 1, \cdots, N$. The system can also be uniquely described by the connecting vector and the center of mass:

$$\mathbf{x}_c = \frac{1}{N} \sum_{i=1}^{N} \mathbf{x}_i, \quad \mathbf{q}_i = \mathbf{x}_{i+1} - \mathbf{x}_i, \quad i = 1, \cdots, N-1. \tag{5.38}$$

The elastic potential of the system is

$$U = \frac{\xi k_B T}{2} \sum_{i=1}^{N-1} \|\mathbf{q}_i\|^2, \tag{5.39}$$

where $\xi k_B T$ is an elastic constant. For the bead-spring or Rouse model, we assume that the number density function for the chain configuration is separable:

$$\psi = \nu \phi(\{\mathbf{q}\}_1^{N-1}, t) h(\mathbf{x}_c, t), \tag{5.40}$$

where ν is the constant number density of polymers. The Smoluchowski equation reduces to

$$h\frac{\partial\phi}{\partial t} + \phi\frac{\partial h}{\partial t} + \sum_{n=1}^{N-1}\frac{\partial}{\partial\mathbf{q}_n}(\dot{\mathbf{q}}_n\phi)h + \frac{\partial}{\partial\mathbf{x}_c}(\dot{\mathbf{x}}_c h)\phi = 0. \qquad (5.41)$$

Assuming

$$\frac{\partial h}{\partial t} + \frac{\partial h}{\partial\mathbf{x}_c}(\dot{\mathbf{x}}_c h) = 0, \qquad (5.42)$$

we end up with a decoupled Smoluchowski equation for pdf ϕ:

$$\frac{\partial\phi}{\partial t} + \sum_{n=1}^{N-1}\frac{\partial}{\partial\mathbf{q}_n}(\dot{\mathbf{q}}_n\phi) = 0, \qquad (5.43)$$

in which

$$\dot{\mathbf{q}}_n = \nabla\mathbf{v}\cdot\mathbf{q}_n - A_{nm}\left(k_B T\frac{\partial}{\partial\mathbf{q}_n}\ln\phi + \xi k_B T\mathbf{q}_n\right),$$

$$A_{nm} = \frac{1}{\zeta}\begin{cases} 2, & n=m, \\ -1, & n=m\pm 1, \\ 0, & \text{otherwise,} \end{cases} \qquad (5.44)$$

where ζ is the friction coefficient. Since A is symmetric, there exists an eigenvalue-eigenvector decomposition

$$A = \Omega\Lambda\Omega^T, \qquad (5.45)$$

where $\Lambda = \text{Diag}(\Lambda_{ii})$ with $\Lambda_{ii} = 4\sin^2(\frac{i\pi}{2N})$, $i = 1, \cdots, N-1$, and Ω is the orthogonal matrix whose column vectors are the orthonormal eigenvectors of A. We introduce a new coordinate $\mathbf{q}'_n$, $n = 1, \cdots, N-1$, such that

$$\frac{\partial}{\partial\mathbf{q}_n}\Omega = \frac{\partial}{\partial\mathbf{q}'}. \qquad (5.46)$$

Namely,

$$\mathbf{q}'_n = \sum_m \Omega_{nm}\cdot\mathbf{q}_m. \qquad (5.47)$$

The Smoluchowski equation is then transformed into

$$\frac{\partial\phi}{\partial t} + \sum_{n=1}^{N-1}\frac{\partial}{\partial\mathbf{q}'_n}\left[\nabla\mathbf{v}\cdot\mathbf{q}'_n - \frac{1}{\zeta}\Lambda_{nn}\frac{\partial\mu}{\partial\mathbf{q}'_n}\phi\right] = 0. \qquad (5.48)$$

Let $\phi = \Pi_n^{N-1}\phi_n(\mathbf{q}_n, t)$, where ϕ_n is the solution of

$$\frac{\partial \phi_n}{\partial t} = -\frac{\partial}{\partial \mathbf{q}_n'}\left[\nabla\mathbf{v}\cdot\mathbf{q}_n'\phi_n - \frac{\Lambda_{nn}}{\zeta}\frac{\partial \mu_n}{\partial \mathbf{q}_n'}\phi_n\right],$$

$$\mu_n = kT\ln\phi_n + \frac{\xi k_B T}{2}\|\mathbf{q}_n'\|^2. \tag{5.49}$$

Take the second moment of ϕ_n with respect to $\mathbf{q}_n'$, we obtain

$$\overline{\langle\mathbf{q}_n'\mathbf{q}_n'\rangle} = \frac{\partial}{\partial t}\langle\mathbf{q}_n'\mathbf{q}_n'\rangle + \mathbf{v}\cdot\nabla\langle\mathbf{q}_n'\mathbf{q}_n'\rangle - \mathbf{W}\cdot\langle\mathbf{q}_n'\mathbf{q}_n'\rangle + \langle\mathbf{q}_n'\mathbf{q}_n'\rangle$$
$$\cdot\mathbf{W} - [\mathbf{D}\cdot\langle\mathbf{q}_n'\mathbf{q}_n'\rangle + \langle\mathbf{q}_n'\mathbf{q}_n'\rangle\cdot\mathbf{D}]$$
$$= \frac{2\Lambda_{nn}}{\zeta}[k_B T\mathbf{I} - \xi kT\langle\mathbf{q}_n'\mathbf{q}_n'\rangle], \tag{5.50}$$

where $\mathbf{W}$ and $\mathbf{D}$ are the vorticity and rate of strain tensor respectively [2]. We note that the left-hand side of the equation is the upper convected derivative of the second moment tensor $\langle\mathbf{q}_n'\mathbf{q}_n'\rangle$. The total elastic stress tensor is

$$\tau^e = \nu\xi kT\sum_{n=1}^{N-1}\langle\mathbf{q}_n'\mathbf{q}_n'\rangle. \tag{5.51}$$

When $N = 2$, this is the Oldroyd-B model which is derived from the linear viscoelastic theory by using the convected derivative [2,3].

If the spring length absent of tension is q_0, the elastic potential is given by

$$U = \frac{\xi k_B T}{2}\sum_{i=1}^{N-1}(\|\mathbf{q}_i\| - q_0)^2. \tag{5.52}$$

When the spring is nonlinear and has a finite extensibility up to q_0, the elastic potential is given by

$$U = -\frac{\xi k_B T q_0^2}{2}\sum_{i=1}^{N-1}\ln\left[1 - \frac{\|\mathbf{q}_i\|^2}{q_0^2}\right], \quad \|\mathbf{q}_i\| < q_0. \tag{5.53}$$

This is the well-known FENE chain model.

References

[1] A. N. Beris and B. J. Edwards. *Thermodynamics of Flowing System with Internal Microstructure*, Oxford University Press, New York, 1994.

[2] R. B. Bird, R. C. Armstrong and O. Hassager. *Dynamics of Polymeric Liquids*, v. 1, John Wiley & Sons, New York, 1987.

[3] R. B. Bird, C. F. Curtiss, R. C. Armstrong and O. Hassager. *Dynamics of Polymeric Liquids*, v. 2, John Wiley & Sons, New York, 1987.

[4] M. Doi and S.F. Edwards. *Theory of Polymer Dynamics*, Oxford University Press (Clarendon), 1986.

[5] S. Harris. *An Introduction to the Theory of the Boltzmann Equation*, Dover, Mineola, NY, 1971.

[6] J. O. Hirchfelder, C. F. Curtiss, and R. B. Bird. *The Molecular Theory of Gases and Liquids*, John Wiley and Sons, New York, 1964.

[7] K. Huang. *Statistical Mechanics*, 2nd Edition, John Wiley & Sons, New York, 1987.

[8] R. G. Larson. *The structure and Rheology of Complex Fluids*, Oxford University Press, 1999.

[9] E. Merzbacher. *Quantum Mechanics*, John Wiley & Sons, New York, 1999.